Process Intensification

Also of interest

Process Synthesis and Process Intensification.
Methodological Approaches
Rong (Ed.),
ISBN 978-3-11-046505-1, e-ISBN 978-3-11-046506-8

Chemical Process Synthesis.
Connecting Chemical with Systems Engineering Procedures
Bezerra, 2020
ISBN 978-3-11-046825-0, e-ISBN 978-3-11-046826-7

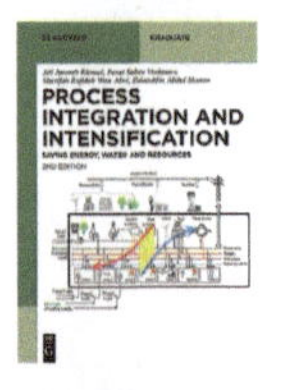

Sustainable Process Integration and Intensification.
Saving Energy, Water and Resources
Klemeš, Varbanov, Wan Alwi, Manan, 2018
ISBN 978-3-11-053535-8, e-ISBN 978-3-11-053536-5

Product and Process Design.
Driving Innovation
Harmsen, de Haan, Swinkels, 2018
ISBN 978-3-11-046772-7, e-ISBN 978-3-11-046774-1

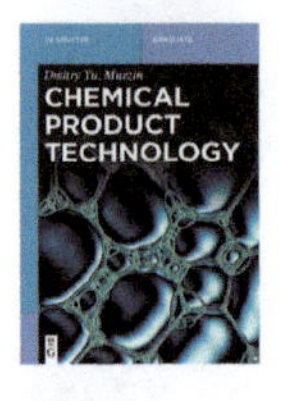

Chemical Product Technology.
Murzin, 2018
ISBN 978-3-11-047531-9, e-ISBN 978-3-11-047552-4

Process Intensification

Design Methodologies

Edited by
Fernando Israel Gómez-Castro,
Juan Gabriel Segovia-Hernández

DE GRUYTER

Editors
Dr. Juan Gabriel Segovia-Hernández
Departamento de Ingeniería Química
División de Ciencias Naturales y Exactas
Universidad de Guanajuato
Noria Alta S/N
36050 Guanajuato,
Mexico
E-Mail: gsegovia@ugto.mx

Dr. Fernando Israel Gómez-Castro
Departamento de Ingeniería Química
División de Ciencias Naturales y Exactas
Universidad de Guanajuato
Noria Alta S/N
36050 Guanajuato,
Mexico
E-Mail: fgomez@ugto.mx

ISBN 978-3-11-059607-6
e-ISBN (PDF) 978-3-11-059612-0
e-ISBN (EPUB) 978-3-11-059279-5

Library of Congress Control Number: 2019947084

Bibliographic information published by the Deutsche Nationalbibliothek
The Deutsche Nationalbibliothek lists this publication in the Deutsche Nationalbibliografie; detailed bibliographic data are available on the Internet at http://dnb.dnb.de.

Cover image: samxmeg/ iStock / Getty Images Plus
Typesetting: Integra Software Services Pvt. Ltd.
Printing and binding: CPI books GmbH, Leck

www.degruyter.com

Contents

List of contributors

Jens Abildskov
Process and Systems Engineering Center
Department of Chemical and Biochemical Engineering
Technical University of Denmark
DK-2800 Kgs. Lyngby
Denmark

J. Rafael Alcántara-Avila
Department of Chemical Engineering
Kyoto University
Katsura Campus Nishikyo-ku
Kyoto 615-8510
Japan
jrafael@cheme.kyoto-u.ac.jp

Roberto Baratti
Dipartimento di Ingegneria Meccanica
Chimica e dei Materiali
Universitá degli Studi di Cagliari
Via Marengo 2
09123 Cagliari
Italy

Abel Briones-Ramirez
Departamento de Investigación y Desarrollo
Exxerpro Solutions
Av. Del Sol 1B
Local 4B El Sol
76113 Querétaro
Querétaro
México

Massimiliano Errico
Department of Chemical Engineering,
Biotechnology and Environmental Technology
University of Southern Denmark
Campusvej 55
5230 Odense M
Denmark
maer@kmb.sdu.dk

Javier Fontalvo
Department of Chemical Engineering
National University of Colombia
Campus La Nubia, Bloque L
170003 Manizales
Caldas
Colombia

Jorge Luis García-Castillo
Department of Chemical Engineering
University of Guanajuato
Noria Alta S/N
Noria Alta
36050 Guanajuato
Guanajuato
Mexico

Claudia Gutiérrez-Antonio
Facultad de Química
Universidad Autónoma de Querétaro
Av. Cerro de las Campanas s/n
Las Campanas
76010 Querétaro
Querétaro
México
claudia.gutierrez@uaq.mx

Jakob Kjøbsted Huusom
Process and Systems Engineering Center
Department of Chemical and Biochemical Engineering
Technical University of Denmark
DK-2800 Kgs. Lyngby
Denmark

Paola Ibarra-Gonzalez
Department of Chemical Engineering
Biotechnology and Environmental Technology
University of Southern Denmark
5000 Odense
Denmark

Juan Álvaro León
Department of Chemical Engineering
National University of Colombia
Campus La Nubia, Bloque L
170003 Manizales
Caldas
Colombia

https://doi.org/10.1515/9783110596120-201

Julio Armando de Lira-Flores
Facultad de Química
Universidad Autónoma de Querétaro
Av. Cerro de las Campanas s/n
Las Campanas
76010 Querétaro
Querétaro
México

Claudio Madeddu
Department of Chemical Engineering
Biotechnology and Environmental Technology
University of Southern Denmark
Campusvej 55
5230 Odense M
Denmark

Jorge C. Melo-González
Department of Chemical Engineering
University of Guanajuato
Noria Alta S/N
Noria Alta
36050 Guanajuato
Guanajuato
Mexico

Kristian Meyer
Process and Systems Engineering Center
Department of Chemical and Biochemical Engineering
Technical University of Denmark
DK-2800 Kgs. Lyngby
Denmark

Nima Nazemzadeh
Department of Chemistry
Materials and Chemical Engineering "Giulio Natta"
Politecnico di Milano
Piazza Leonardo da Vinci
32-20133 Milano
Italy

Rasmus Fjordbak Nielsen
Process and Systems Engineering Center
Department of Chemical and Biochemical Engineering
Technical University of Denmark
DK-2800 Kgs. Lyngby
Denmark

Adriana Palacios Rosas
Department of Chemical and Food Engineering
Universidad de las Americas Puebla
Ex. Hda. Sta. Catarina Mártir
San Andres Cholula
72810 Puebla
México

Eduardo S. Perez-Cisneros
Department of Chemical Engineering
University of Guanajuato
Noria Alta S/N
Noria Alta
36050 Guanajuato
Guanajuato
Mexico

Martin Picón-Núñez
Department of Chemical Engineering
University of Guanajuato
Noria Alta S/N
Noria Alta
36050 Guanajuato
Guanajuato
Mexico
picon@ugto.mx

Oscar Andrés Prado-Rubio
Department of Chemical Engineering
National University of Colombia
Campus La Nubia, Bloque L
170003 Manizales
Caldas
Colombia

Efraín Quiroz-Pérez
Facultad de Química
Universidad Autónoma de Querétaro
Av. Cerro de las Campanas s/n
Las Campanas
76010 Querétaro
Querétaro
México

Nelly Ramírez Corona
Department of Chemical and Food Engineering
Universidad de las Americas Puebla
Ex. Hda. Sta. Catarina Mártir

San Andres Cholula
72810 Puebla
México
nelly.ramirez@udlap.mx

Ben-Guang Rong
bgr@kbm.sdu.dk

Mauricio Sales-Cruz
Departamento de Procesos y Tecnología
Universidad Autónoma Metropolitana -
Cuajimalpa
Avenida Vasco de Quiroga No. 4871
Col. Santa Fe Cuajimalpa
Del. Cuajimalpa de Morelos
C.P. 05300 Ciudad de México
Mexico

Seyed Soheil Mansouri
Process and Systems Engineering Center
Department of Chemical and Biochemical
Engineering
Technical University of Denmark
DK-2800 Kgs. Lyngby
Denmark
seso@kt.dtu.dk

Isuru A. Udugama
Process and Systems Engineering Center
Department of Chemical and Biochemical
Engineering
Technical University of Denmark
DK-2800 Kgs. Lyngby
Denmark

John M. Woodley
Department of Chemical and Biochemical
Engineering
Technical University of Denmark
DK-2800 Kgs. Lyngby
Denmark

Nelly Ramírez Corona and Adriana Palacios Rosas

1 Generalities about process intensification

Abstract: Process intensification (PI) can bring significant reductions not only in the equipment requirement but also in the associated energy demands, the inventory of hazardous materials, and waste streams handling. Although most of these goals are not new to chemical engineering, PI goes beyond the traditional approach, since it focuses on designing innovative process and equipment, as miniaturized units, multifunctional systems, hybrid separations, as well as in the integration of alternative energy sources. This chapter introduces the general principles of PI, comparing the different types of intensified processes. It discusses the potential exploitation of these new technologies and seeks to identify the major barriers for its industrial application.

Keywords: process intensification, miniaturized units, process design, energy savings, sustainable processes

1.1 Introduction

During the past few decades, process intensification (PI) has been gaining interest in the academic and industrial fields, since the implementation of the related methodologies allows the development of more sustainable processes, by the reduction of several unit operations to intensified units in which the same process can be carried out. There is a great variety of definitions for PI and the concept has evolved over the years, from unit miniaturization to integrated approaches for process enhancement. PI can be defined in terms of process shrinking, as a *strategy of reducing the size of a chemical plant needed to achieve a given production objective* [1] or in terms of innovation, as *an integrated approach for process and product innovation in chemical research and development* [2].

PI can be considered as any activity that integrates one or more of the following: (i) smaller equipment for given throughput, (ii) higher throughput for given equipment size or a given process, (iii) less holdup for equipment or less inventory for the processing of certain materials for the same throughput, (iv) less usage of utility materials and feedstock for given throughput or equipment size, and (v) higher performance for given unit size [3].

Research efforts on PI can be categorized into two areas: (i) design of intensified equipment, which includes novel and miniaturized devices, and (ii) the

Nelly Ramírez Corona, Adriana Palacios Rosas, Department of Chemical and Food Engineering, Universidad de las Americas Puebla, Puebla. México

https://doi.org/10.1515/9783110596120-001

development of process-intensifying methods, focused on the synthesis of integrated and/or multifunctional units, within whole processes. The design of new devices includes microreactors, supersonic reactors, rotating devices (reactors and mixers), and microchannel heat exchangers, among others. The development of new intensified processes has been focused on multifunctional units (reactive-separation or reaction-heat transfer) and hybrid separations. Moreover, the potential benefits of using alternative sources of energy are also an opportunity area in PI, for instance in those systems that make use of ultrasound, microwaves, or electric fields [4].

1.2 PI principles

The differences among PI, Process System Engineering (PSE), and Process Optimization (PO) have been described in literature [5, 6]. As stated in such reviews, even if there is a significant similarity among these areas, each of them involves particular aims. PO looks for performance improvement by implementing existing concepts, while PI and PSE involve the selection of design and processing methods. PI is based on the development of new concepts, including the synthesis of innovative process stages and the design of novel equipment. In order to clearly identify the PI fundamentals, four general principles of PI can be stated [6]:
1. Maximize the effectiveness of intra and intermolecular events.
2. Give each molecule the same processing experience.
3. Optimize the driving forces at every scale and maximize the specific surface area to which these forces apply.
4. Maximize the synergistic effects from partial processes.

Furthermore, during the PI design, the relevance of the scale level (molecular scale, mesoscale, and macroscale) has been emphasized by describing four approaches, associated with the following domains: spatial, thermodynamic, functional, and temporal.
1. Spatial domain refers to the importance of well-defined structures or environments that allow maximization of process synergy (i.e., activity and selectivity of a certain catalyst are highly dependent on its geometrical structure).
2. The thermodynamic domain is focused on the optimum use of energy, considering alternative forms of energy and/or ways for its transfer.
3. In the functional domain, PI looks to maximize the synergy from different units, processes, or forms of energy, bringing multiple functions within a single component (i.e., reactive distillation units).
4. Temporal domain considers the manipulation of time scales of different processes in order to significantly reduce the processing times or process periodicity. For

example, by using multitasking units or by switching from batch to continuous processing.

It is important to emphasize that during the synthesis of intensified process, the challenge is not only centered on the design of compacted units but also on how to guarantee their functionality. Taking into account process objectives, chemical processing can be described by the elementary process function of each unit operation or by the related process phenomena. These different phenomena can be classified into eight categories: (i) mixing, (ii) phase contact, (iii) phase transition, (iv) phase change, (v) phase separation, (vi) reaction, (vii) energy transfer, and (viii) stream dividing [7].

As a complementary perspective, PI might be categorized in two main groups: (i) local intensification and (ii) global intensification. Local intensification aims to use PI techniques to improve the efficiency of a single unit in isolation from the whole process. On the other hand, global intensification considers the performance of the whole process, taking into account the interactions among all units within the system [8].

1.3 Devices and equipment in PI

From its basic definition, PI is characterized by the development of novel equipment, wherein the miniaturization and multifunctionality are fundamental targets. PI equipment can be divided into two categories: (i) equipment involving chemical reactions and (ii) equipment for operation not involving reactions.

Tables 1.1 and 1.2 summarize the emblematic equipment in PI reported in literature [4, 5], their application, advantages, and disadvantages, aiming to illustrate their potential applications.

In addition, there is a great variety of multifunctional systems that, compared to those traditionally used, are expected to substantially improve the process performance in terms of processing time, energy requirements, or process selectivity. For instance, the implementation of internally heat integrated distillation columns and thermally coupled distillation systems allows high energy savings compared to conventional distillation columns [9]. Multifunctional and multiphase reactors that can incorporate not only reaction and heat transfer but also phase transition or separation, offer advantageous conditions for complex reactive systems [10]. Reactive distillation, a technology that incorporates both phase separation and chemical reaction in a single column, is, so far, the most important industrial application of PI [11]. Other hybrid systems such as reactive absorption, combining absorption of gases with chemical reactions in the liquid phase; reactive extraction, wherein the reaction enhances the capacity of the solvent; reactive crystallization, combining

Table 1.1: Equipment for PI for reacting systems [4, 5].

Equipment	Applications	Advantages	Disadvantages
Static mixer reactor	Useful in the polymer industry for melt viscosity adjustments, among other applications.	Good mixing and radial heat transfer characteristics.	Low specific geometrical area and high sensitivity to obstruction by solids.
Monolithic reactor	For partial oxidations of hydrocarbons, catalytic combustion, and removal of soot from diesel engines.	Negligible pressure drop, high geometrical area, improves selectivity as reduces mass transfer resistances.	Difficult heat removal due to the absence of radial dispersion.
Microreactor	Different reaction systems as fluorination, chlorination, nitration, sulfonation, biodiesel production, enzymatic reactions, and vitamin precursor synthesis.	High heat transfer rates (recommended for highly exothermic processes). Very low volume/surface area ratios (suggested for process involving toxic or explosive components). Allow precise control of operating conditions.	In some operating conditions, catalyst deactivation may occur relatively fast. Formation of solids increases pressure drop and may block reactor channels.

nucleation and crystal growth with chemical reactions, among others, are few examples of intensified technologies that may drastically transform the challenges in process engineering in the coming years [5].

1.4 PI methodologies

PI is focused on designing novel equipment or developing new conceptual processes, as well as on its implementation for retrofitting existing chemical plants. Designing intensified processes implies the definition of a number of degrees of freedom, which is not a straightforward task because of the large number of alternatives [12]. Given the complexity of the problem, several methodologies for designing intensified process have been developed during the past decade, including heuristic-based approaches for conceptual design, process screening via simulation, and sophisticated mathematical problem formulations.

Literature on the development of intensified process is wide-ranging, and different approaches aim to solve different problems. For instance, thermodynamic

Table 1.2: Equipment for PI for nonreactive systems [4, 5].

Equipment	Applications	Advantages	Disadvantages
Static mixer	In plastic industry for blending of plasticizers, stabilizers, colorants, flame retardants, etc. In food factories can be used to mix food formulations. In chemical industry for gas mixing prior to processing.	Divide and redistribute streamlines using only the pumping energy of the flowing fluid. Produce a complex vortex system in which concomitant phenomena simultaneously enhance mass and heat transfer.	There is lack of information on the spatial mixing quality, which can result in underestimation or overestimation of mixing times.
Compact heat exchanger	As peripheral equipment in microreactors.	Exhibit high heat fluxes and convective heat transfer coefficients.	Limited choice, particularly for high pressure. Susceptibility to fouling.
Rotating packed bed	Deaeration of flooding water, separation processes, and reaction systems.	High centrifugal forces intensify momentum and mass transfer.	Compared to conventional columns requires additional power and maintenance for the rotating system within the range of power required by the feed and reflux pumps.
Centrifugal absorber	Ion exchange and absorption processes.	Centrifugal field enables use of very small absorbent particles with short contact times.	Significant PI is achieved only in the liquid side resistance, controlled by mass transfer.

characteristics of the system may be utilized to generate feasible flowsheets to fulfill the design task. Holtbruegge et al. [13] presented a systematic procedure for the conceptual design of intensified processes, wherein the boundaries for the design task and the process indicators must be defined in terms of thermophysical properties. Their method uses thermodynamic insights to create a group of promising flowsheet options, and then use mathematical algorithms for their optimization. They divided process design into three steps: (i) system identification and analysis, (ii) generation of flowsheet options, and (iii) evaluation of flowsheet options. At the same time, generation of flowsheets considers three levels: (i) identification of reaction/separation techniques, (ii) detailed investigations of separation steps/techniques, and (iii) generation of flowsheets.

This approach allows the generation of a variety of promising flowsheet options, based on thermodynamic fundamentals, and provides a tool not only for the process synthesis but also for a deeper understanding of intensified techniques and processes.

Regarding the development of PI for process retrofitting, some synthesis techniques and design concepts have been evaluated, with the aim of guaranteeing the sustainability of existing plants. In this field, Niu and Rangaiah [14] presented a heuristic-based approach for retrofitting, via PI. It consists of four steps: (i) the first is to create a base case analysis, where the retrofit target is defined by using selected metrics, which can include utility costs, component losses, conversion of reactants, and selectivity of products. In this first stage, the objective is to identify the main contributors for each metric; (ii) the second step focuses on generating improved solutions, by optimizing the operating conditions, without capital investment (if no improved solution is attained, the original process remains as a basis); (iii) the third step considers the generation of integrated solutions; and (iv) the fourth step involves the comparison of the retrofit solutions for decision-making.

Currently, process retrofitting through PI usually involves the optimization of operating conditions or process integration, rather than replacing existing equipment with novel intensified units.

Besides, in order to quantitatively analyze the feasible alternatives, new modeling and computer-aided tools to perform multiobjective optimization have been developed by several authors. Ponce-Ortega et al. [3] proposed a general mathematical programing formulation for PI, considering the use of existing units and the implementation of additional units. These authors take into account two different PI cases: (i) unit intensification and (ii) plant intensification. In the first case the goal is to minimize the unit size for a given throughput or to maximize the performance for given size. For plant intensification, the objective is to simultaneously intensify the whole process, minimizing the process inventory, utilities, and feedstock, or maximizing the process throughput.

The methodologies previously discussed are a few examples among a number of contributions on PI development. It is evident that the field presents a high grade of development and that will certainly continue to grow in the coming years.

1.5 PI safety and control properties

As has been noted, the benefits of PI lie in cost reductions and compactness. In particular, it makes it possible to increase production capacity per unit of processing area, unlocking greater opportunities through sustainable growth and innovation. Furthermore, as PI may substantially reduce the process size, by implementing smaller equipment and inventories, process safety, and control properties can also been improved.

Several processes, mainly those in the chemicals and petrochemicals industries, involve the production, handling, and use of large quantities of hazardous materials. For hazardous processes, PI encourages the inventory of dangerous materials, and the consequences of processes failures, to be significantly reduced; thus, increasing intrinsic plant safety. From a simple point of view, the best way to deal with the problem is not to have it in the first place [15]. This has been also been pointed out by Kletz [16], through the phrase "what you don't have can't leak!". Significant approaches to the design of inherently safer processes and plants could be grouped into four major strategies [16, 17]:

1. Minimize, use small quantities of hazardous substances.
2. Substitute, replace material with a less hazardous substance.
3. Moderate, use less hazardous conditions, a less hazardous form of a material, or facilities, which minimize the impact of a release of hazardous material or energy.
4. Simplify, design facilities that eliminate unnecessary complexity and make operating errors less likely, and which are forgiving of errors which are made.

According to Etchells [18], some of the safety benefits that may arise from PI are:

- In some cases, the number of process operations can be reduced, leading to fewer transfer operations, and less pipework (which can be a source of leaks).
- It may be easier to design a smaller vessel to contain the maximum pressure of any credible explosion, so that further protective devices, such as emergency relief systems, are not needed (or the duties placed upon them are less onerous).
- Many incidents are associated with process transients such as startup and shutdown. These are reduced during continuous (and intensified) processes.
- For exothermic reactions, the heat evolution should be much less variable than in batch reactions, and should be easier to control. Furthermore, the enhanced specific surface area of intensified plant makes heat transfer easier. Certainly, very few runaway reactions occur in continuous processes (although there have been some notable exceptions [19]).

Thus, the applications and methods of PI are various, and therefore it is dangerous to assume that safety is always improved through intensification, even though this is true in many cases [20].

1.6 Sustainability

Sustainability has become a key concept for new processes and product development that requires a multidisciplinary approach, in which the sustainable development of energy systems accounts for six goals [21]:

1. Prosperity (economy) presenting a strong, inclusive, and transformative economy.

2. Justice, promoting safe and peaceful societies and strong institutions.
3. Partnership, to catalyze global solidarity for sustainable development.
4. Planet (environment), to protect ecosystems for societies and future generations.
5. People (social), to ensure healthy lives, knowledge, and the inclusion of women and children.
6. Dignity, to end poverty and fight inequality, which should definitely not be the last one.

Yong et al. [22] have summarized the future trends of the sustainable development of energy systems in three main areas: (i) higher efficiency and waste reduction of biofuel production, (ii) CO_2 removal and conversion, and (iii) process integration.

As discussed elsewhere [21, 22], the scope of PI is becoming much wider, considering the integration of not only heat and power but also water, safety, and other aspects of processes.

Environmentally, Reay et al. [15] have noticed that the most telling impact of PI is likely to be in the development of reactor designs; since the reactor dictates both the product quality and the extent of the downstream separation and treatment equipment, during any chemical process. Thus, the development of reactor designs enables PI to be beneficial for the energy use, impacting on the environment, through the delivery of a high-quality product, without extensive downstream purification sequences.

1.7 Main drawbacks for PI implementation

Stankiewicz and Moulijn [23] recognize some obstacles for PI implementation. Currently, the chemical industry is primarily focused on trade strategies, looking for improving their market visibility by merging complementary product portfolios of different industries, more than by implementing more fundamental changes in processing via R&D. Furthermore, most R&D efforts are focused on the development of new products (new materials and specialty products) rather than new processing methods. This last drawback is directly associated to the lack of familiarity of process engineers with the latest developments in PI and emerging technologies implementation. In addition, much novel equipment and many devices have not yet been assessed on an industrial scale, mainly because of the nature of emerging technologies and processing methods, with drastic differences between them, such that standard methodologies for their scaling up are still under development.

Furthermore, from a safety point of view, in some cases, hazards may remain, or new ones may be created, in the development of PI processes. According to Etchells [18], some potential problems may include:

- Some PI technologies require high-energy inputs, e.g., from microwaves, high voltages or electromagnetic radiation, or require to be operated at higher temperatures and pressures. Although expertise associated with the handling of high-energy sources is present in some industries, the new technology may also bring less familiar groups into contact with this hazard.
- The processes may be more complex or call for more complex control systems, and safety may suffer.
- As the residence time for many intensified processes will be of the order of seconds rather than hours, the subject of control and monitoring has to be addressed. It has been suggested that process control may become easier but more importantly, as the system is smaller, it may become more responsive to changes in process conditions. In some cases, it may mean that the process is unsuitable, or that new and novel techniques are needed to control them and these may not yet be available.
- In some cases, process pipework may be more complex, with a higher potential for equipment failure or operator error.
- Intensified reactors have the potential to significantly enhance reaction rates as a result of the improved mixing. This could lead to a much greater rate of energy release than in traditional reactors and, in some cases, may result in a change in the reaction chemistry.
- Rotating equipment may not be suitable for friction-sensitive substances (i.e., substances that can either deflagrate or detonate due to friction). Certainly the hazard of ignition needs to be addressed.
- Where fouling can occur on complex heated surfaces, then thermally unstable materials can overheat, possibly leading to high pressures being generated.

As previously discussed, PI has been considered as a strategy for reducing the inventory of hazardous chemicals contained inside the process, so that the safety and environmental consequences are reduced. However, as process robustness is defined in terms of process ability to tolerate significant disturbance from its surroundings, designing compacted plants may represent potential operational problems [17, 24].

1.8 PI assessment examples

It is clear that in order to replace conventional units with more efficient ones within the existing processes, it is necessary to verify their implementation feasibility, cost-effectiveness, controllability, and safety properties. Nowadays, several approaches are being developed to provide decision-making tools, considering both quantitative and qualitative elements. A couple of examples of “fast assessment” tools reported in literature are presented in this section [12, 20].

Table 1.3: Comparison of design variables for two intensified processes for obtaining biodiesel. Adapted from Rivas et al. [12].

Design parameters	Reactive Distillation (RD)	Reactive Absorption (RA)
Column		
Number of stages	15	15
HETP (m)	0.5	0.6
Column diameter (m)	0.4	0.4
Column volume (m^3)	0.94	1.13
Heat exchangers duty (kW)		
Fatty acid heaters	95	81 and 27
Methanol heaters	8	65
Biodiesel cooler	38	14
Reboiler duty	133	0
Condenser/decanter duty	72	77
Energy consumption as steam (E)		
Steam consumption (kg steam/ ton FAME)	168	34

Example 1.1 A simple method based on a weighted factor namely Intensification Factor (IF) can be used as decision-making element to compare different intensified processes [12]. Consider the Biodiesel production by integrated reactive technologies. In this case, biodiesel is produced by the reversible reaction of esterification of waste oils considering Reactive Absorption (RA) and Reactive Distillation (RD) processes. Table 1.3 summarizes the design variables of each intensified process [12].

It is clear that investment cost and energy requirements vary with the implemented technique. In order to compare both technologies, an intensification factor can be calculated as in Eq. 1.1.

$$IF = \prod_{i=1}^{n} \left(\frac{F_{bi}}{F_{ai}} \right)^{di} \tag{1.1}$$

Where F is the evaluation criteria or factor, b represents the new technology and a the previous (or existing) technology. For instance, the IF associated to the column cost is proportional to the ratio of columns volumes (V) with a scale exponent (Rivas et al. [12] reported a d value of 0.85). Considering the sizing values reported in Table 3, IF_{col} can be estimated as:

$$IF_{col} = \left(\frac{V_{RD}}{V_{RA}}\right)^{0.85} = \left(\frac{0.94}{1.13}\right)^{0.85} = 0.85 \quad (1.2)$$

The IF_{hx} associated to the total exchangers cost is determined by the ratio of weighted heat load at power of 0.6

$$IF_{hx} = \frac{\sum_i^{RD} Q_i^{0.6}}{\sum_i^{RA} Q_i^{0.6}} = \frac{59.8}{51.8} = 1.15 \quad (1.3)$$

Where Q_i is the heat load of each heat exchanger for the corresponding process (RD or RA). Finally, the *IF* for operating cost are computed as the ratio of steam consumption:

$$IF = \frac{E_{RD}}{E_{RA}} = \frac{168}{34} = 4.94 \quad (1.4)$$

Then, IF_{total} can be calculated as

$$IF_{total} = \prod_{(i=1)}^{p} (IF_i)^{c_i} = (0.85)(1.15)(4.94) = 4.83 \quad (1.5)$$

Where p is the number of potential intensification strategies and the c_i is a weigthening factor, if available info is limited or inexistent then c_i takes a value of 1. For this case, as investment costs for columns and exchangers are relative similar for both processes and the operating cost are dominant, reactive absorption seems to be more cost-effective. Although this factor provides a simple method to evaluate different intensified processes, it must be weighted by experts in the field [12].

Example 1.2 Regarding safety on intensified processes, a simple way to evaluate how PI affects safety consists in developing a checklist using the concept of layers of protection [20]. Consider the case in which the oxidation reactor (bubble column) of the anthraquinone process is replaced by a tubular reactor with static mixers that enhance mass transfer.

This unit replacement reduces the reaction volume to about one tenth of the conventional reactor. The main reason for this size reduction is the use of oxygen instead air. However, due to the fast oxidation, oxygen is rapidly consumed and it needs to be continuously fed to the reactor through several inlets. Furthermore, the reactor is very sensitive to certain process conditions. In order to compare both processes from a safety point of view, the criteria of the checklist are used and the obtained review is presented in Table 1.4 [20].

As can be observed from the check list, the intensified process are safer in three criteria, but become worse for five criteria. It is important to highlight that the final decision should consider that these criteria have different degrees of importance, such that, some additional quantitative measurements must be also considered [20].

Replace selected units in existing plants aiming intensifying the process, requires an in-depth analysis about the implications of its implementation. Different

Table 1.4: Effects on safety, if the bubble column is replaced by a tubular reactor. Adapted from Ebrahimi et al. [20].

Safety criteria	Bubble column	Tubular reactor	Comparison between the systems	Effect
1. Inherent safety layer				
Reaction route				
Chemicals	Air	Oxygen	The flammable solvent has a lower ignition energy with oxygen than with air.	Negative
Process novelty	Traditional	Novel	Novelty reduces safety because of less experience.	Negative
2. Passive layer				
Reactor type and geometry:				
Complexity	Simple column	More complex	The intensified reactor has several gas feed points which makes it more complex.	Negative
Fluid hold-up	Liquid hold-up	Liquid hold-up about 10 times less.	Lower inventory reduces the risk.	Positive
Internal patterns	Channelling, axial dispersion	More regular	Flow pattern in the tubular reactor with static mixers is more regular with less axial dispersion. In the bubble column, channelling, axial dispersion, and irregular flow appear in a larger extent.	Positive
Surface to volume ratio		Higher surface to volume ratio.	Intensified process has more capability to heat exchange.	Positive
Layout	Simple column	Tube elbows, several gas inlets.	Layout of the new reactor is more complex. The reactor tube has several elbows where liquid might be trapped at least in unusual conditions.	Negative
3. Active layer				
Process control		Several inlets to control	Control of several oxygen feed points has a risk of malfunction.	Negative

criteria assessment for the process development decisions should be taken into account to guarantee that these new arrangements provide at least a similar performance than that the original.

1.9 Final remarks

Many advances have been made in terms of PI, engineering tools for innovative process design, and novel PI technology. Nevertheless, their industrial implementation is still limited, mainly due to the lack of standard methodologies for their scaling. As discussed previously, another important issue is the lack of familiarity of process engineers with the latest developments in PI and their implementation. To accomplish the skills necessary to design and implement PI for existing and new chemical plants, it is imperative that new chemical engineers embrace innovative approaches during their academic training; since, there is no doubt that PI will be an important key factor in supporting a sustainable future in the chemical engineering sector. In the following chapters, different PI strategies will be discussed in depth.

References

[1] Cross, W. T., Ramshaw, C. Process intensification: laminar flow heat transfer. *Chemical Engineering Research and Design*, 1986, 64(4), 293–301.

[2] Becht, S., Franke, R., Geißelmann, A., Hahn, H. An industrial view of process intensification. *Chemical Engineering and Processing: Process Intensification*, 2009, 48(1), 329–332.

[3] Ponce-Ortega, J. M., Al-Thubaiti, M.M., El-Halwagi, M. Process intensification: New understanding and systematic approach. *Chemical Engineering and Processing: Process Intensification*, 2012, 53 63–75.

[4] Stankiewicz, A. I., Moulijn, J. A. Process intensification: Transforming chemical engineering. *Chemical Engineering Progress*, 2000, 96(1), 22–34.

[5] Keil, F., Process Intensification, *Reviews in Chemical Engeneering*, 2018, 34(2), 135–200.

[6] Van Gerven, T., Stankiewicz, A. Structure, Energy, Synergy, Time-The Fundamentals of Process Intensification. *Industrial & Engineering Chemistry Research*, 2009, 48(5), 2465–2474.

[7] Lutze, P., Babi, D. K., Woodley, J. M., Gani, R. Phenomena based methodology for process synthesis incorporating process intensification. *Industrial & Engineering Chemistry Research*, 2013, 52(22), 7127–7144.

[8] Portha, J. F., Falk, L., Commenge, J. M. Local and global process intensification. *Chemical Engineering and Processing: Process Intensification*, 2014, 84, 1–13.

[9] Vazquez–Castillo, J. A., Venegas–Sánchez, J. A., Segovia–Hernández, J. G., Hernández-Escoto, H., Hernandez, S., Gutiérrez–Antonio, C., Briones–Ramírez, A. Design and optimization, using genetic algorithms, of intensified distillation systems for a class of quaternary mixtures. *Computers & Chemical Engineering*, 2009, 33(11), 1841–1850.

[10] Utikar, R. P., Ranade, V. V. Intensifying multiphase reactions and reactors: strategies and examples. ACS *Sustainable Chemistry & Engineering*, 2017, 5(5), 3607–3622.

[11] Segovia-Hernández, J. G., Hernández, S., Petriciolet, A. B. Reactive distillation: A review of optimal design using deterministic and stochastic techniques. *Chemical Engineering and Processing: Process Intensification*, 2015, 97, 134–143.
[12] Rivas, D. F., Castro-Hernández, E., Perales, A. L. V., Van der Meer, W. Evaluation method for process intensification alternatives. *Chemical Engineering and Processing: Process Intensification*, 2018, 123, 221–232.
[13] Holtbruegge, J., Kuhlmann, H., Lutze, P. Conceptual Design of Flowsheet Options Based on Thermodynamic Insights for (Reaction–) Separation Processes Applying Process Intensification. *Industrial & Engineering Chemistry Research*, 2014, 53(34), 13412–13429.
[14] Niu, M. W., Rangaiah, G. P. Process retrofitting via intensification: a heuristic methodology and its application to isopropyl alcohol process. *Industrial & Engineering Chemistry Research*, 2016, 55(12), 3614–3629.
[15] Reay, D., Ramshaw, C., Harvey A., Process Intensificatio: Engineering for Efficiency, Sustainability and Flexibility. In: Process Intensification – An Overview. Butterworth-Heinemann Elsevier, Oxford UK, 2013, 27–55.
[16] Kletz, T. A., *Plant Design for Safety: A User-friendly Approach*. Taylor & Francis, NY, USA, 1991.
[17] Hendershot, D.C., (1997). Inherently safer chemical process design. *J. Loss Prev. Process. Ind.* 1997, 10, 51–157.
[18] Etchells, J.C., Process Intensification: Safety Pros and Cons. *Trans IChemE*, Part B, Process Safety and Environmental Protection, 2005, 83(B2): 85–89.
[19] Etchells, J.C., Why reactions run away. IChemE Symp. Series No 147, Hazards XIII Process safety-sharing best practice. IChemE, Rugby, UK, 1997, 361–366.
[20] Ebrahimi, F., Virkki-Hatakka, T., Turunen, I. Safety analysis of intensified processes, *Chemical Engineering and Processing: Process Intensification*, 2012, 52 28–33.
[21] Nemet, A., Varbanov, P.S., Kleme, J.J. Cleaner production, Process Integration and intensification, *Clean Techn. Environ. Policy*, 2016, 18: 2029–2035.
[22] Yong, J.Y., Kleme, J.J., Varbanov, P.V., Huisingh, D. Cleaner energy for cleaner production: modelling, simulation, optimization and waste management. J. *Clean Prod.* 2016, 111: 1–16.
[23] Stankiewicz, A. I., Moulijn, J. A. Process intensification. *Industrial & Engineering Chemistry Research*, 2002, 41 (8), 1920–1924.
[24] Luyben, W. L., Hendershot, D. C. Dynamic disadvantages of intensification in inherently safer process design. *Industrial & Engineering Chemistry Research*, 2004, 43(2), 384–396.

Efraín Quiroz-Pérez, Julio Armando de Lira-Flores
and Claudia Gutiérrez-Antonio

2 Microreactors: Design methodologies, technology evolution, and applications to biofuels production

Abstract: In the past few years, the application of process intensification strategies has attracted attention, since it leads to the development of production processes with reduced energy consumption and increased efficiency. In the production processes, reactors constitute a key element for the conversion of raw materials to products; therefore, the efforts have been focused on the proposal of novel and improved equipment. In particular, microreactors have small size, minor residence times, an efficient mass and heat transfer, and high yields. This technology has been applied for the generation of several products, such as organic compounds, nanoparticles, polymers, among others. In particular, the application of microreactors for the production of biofuels is a powerful technology to obtain competitive carburant, both technically and economically. Thus, in this chapter, a revision of the design methodologies for microreactors is presented, along with the applications and technological advances of microreactors for biofuels production.

Keywords: microreactor, design methodology, biofuels production, microreactor technology

2.1 Introduction

Climate change and decline in oil production have driven the development of new processes to generate energy, fuels, and chemicals with renewable raw materials. In particular, the proposal of biofuels production processes has received a lot of attention, due to its importance for transportation and power and heat sectors. An important aspect to remark is that the carbon dioxide emissions generated during the combustion of the fuel are the ones captured by the crops during its growth. Therefore, a considerable decreasing in carbon dioxide emissions is obtained. Considering this fact, the development of conversion processes with reduced energy consumption and capital costs is necessary to minimize the carbon footprint of biofuels. In this context, process intensification is a potential mean for process improvement, to meet the increasing demands for sustainable production [1].

Efraín Quiroz-Pérez, Julio Armando de Lira-Flores, Claudia Gutiérrez-Antonio, Facultad de Química, Universidad Autónoma de Querétaro, Querétaro, México

https://doi.org/10.1515/9783110596120-002

According to Stankiewicz and Moulijn, process intensification consists of any chemical engineering development that leads to a substantially smaller, cleaner, and more energy-efficient technology [2], and it is one of the five principles to get an inherently safer design [3]. Thus, there is novel and multifunctional equipment, where mass and heat transfer are significantly improved. Respect to novel equipment, the development of reactors has attracted much attention.

Reactors are key equipment in any production processes because they perform the conversion of raw materials to products [4], especially for biofuels production. In conventional reactors, mass and heat transfer are limited, especially in heterogeneous reactions; due to this, reactants in excess and devices to create turbulence are needed to guarantee the contact between reactants along with a better heat transfer.

In the development of new and more efficient reactors, from process intensification point of view, there are important advances. According to Tian et al. [5], there are several intensified reactor technologies, which include structured catalytic reactors, oscillatory flow reactors, reverse flow reactors, catalyst that supply heat in endothermic reactions, and microreactors. In particular, microreactors, as the name suggests, are reactors with channel sizes of the order of micrometers, at which diffusion is the dominant mixing mechanism [6], as can be observed in Figure 2.1.

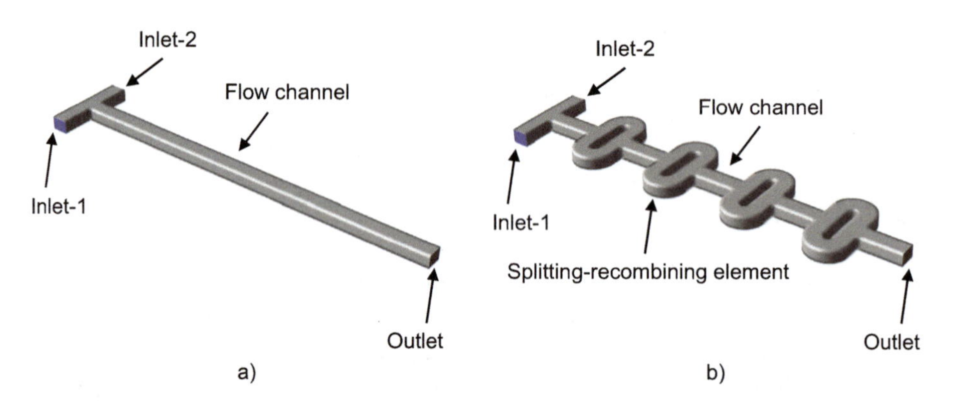

Figure 2.1: Main components in a microreactor with (a) a straight flow channel and (b) a flow channel with splitting-recombining elements.

As a consequence, little or no excess of reactants is required, and the safety increases due to the small residence times. An important advantage is that the desired total productivity is obtained simply by a linear increase in the productivity of individual microreactors, circumventing conventional scaling-up problems [7]. Microreactors have been used on the synthesis of nanoparticles, organics, polymers, and biosubstances [8], such as erythrulose [9], hydroxymethyl furan [10], nanocrystals of barium sulfate and boehmite [11], losartan potassium-loaded nanoparticles [12], palladium nanoparticles [13], azo dyes [14], phenylethanol [15], hydrogenation of ethylpyruvate [16], oxidation of dibenzothiophene with hydrogen peroxide [17], cell lysis and DNA extraction [18], combustion [19], direct H_2O_2 synthesis [20], synthesis in liquid-liquid systems [20], and polymerization reactors [21].

Regarding microreactors, several reviews have been published focused on their applications on the synthesis of nanoparticles, organics, polymers, and biosubstances [8], methods and applications of heterogeneous catalysis [22], and microstructured catalytic reactors for gas-phase reactions [23]. Moreover, there are two reviews related to bioenergy area. One of them presented the use of microreactors for the production of biofuels (methanol and biodiesel considering Fischer-Tropsch technology) and electrical energy (hydrogen storage for fuel cells) [24]. That review includes a compilation of scientific developments reported until 2012. The other work was published in 2019, and it presents an analysis of the technical challenges for the development of microfluidic bioreactors for the production of biofuels (biodiesel and bioethanol) [25]. Both reviews are very interesting; in spite of none of these reviews include the production of gaseous biofuels. These reviews discuss the technical challenges to the development of microreactors for the production of biofuels; however, design methodologies are necessary to increase the implementation of intensified technologies in the industry, since they allow addressing the complexity and offering systematic solution procedures [26]. To the authors' knowledge, there is no available a compilation of design methodologies and applications to biofuels including a patent revision.

Therefore, this chapter focuses on the review of design methodologies for microreactors, mainly those where computational fluid dynamics (CFD) is used. Moreover, a revision of the applications of microreactors for production of biofuels along with technological advances of this technology are included.

The chapter is organized as follows. Section 2.2 includes the reported methodologies based on CFD for the design of microreactors. Later, the applications of microreactors for biofuels production are presented in Section 2.3. Finally, the technological advances for the production of biofuels are discussed in Section 2.4, while a recapitulation of these topics is addressed in Section 2.5.

2.2 Design methodologies

The design of microreactors requires extensive knowledge of the related phenomena such as fluid flow, chemical kinetics, and transport of heat and mass [27]. In this sense, CFD has been proposed as an attractive method for the design and optimization of these units [28].

A CFD study consists of obtaining the numerical solution of the equations that describe momentum, heat, and mass transfer in a system using computational methods. One of the most interesting characteristics of techniques based on CFD is the possibility to model and represent the aforementioned phenomena in systems with both conventional and complex geometries. Therefore, the performance of microreactors with different geometric designs can be evaluated in terms of the described mechanisms.

Based on the previous discussion, the most common strategies based on CFD for the design of microreactors are presented in the following subsections.

2.2.1 Design based on the analysis of parameters of microreactors performance

Microreactors performance can be determined using the evaluation of mixing degree, mass and heat transfer, and reaction yield in these units [27]. Thus, the most commonly used strategy for the design of microreactors is described in the following lines:

a) First of all, different designs for the microreactors are proposed. These designs can be obtained by modifying some of the following geometric parameters: the shape of the flow channel, the shape of the channel cross-section, the shape and arrangement of the mixing elements located through the flow channel, the shape of the premixing elements at the inlet of the flow channel, the height of the flow channel, and the width of the flow channel. It has been demonstrated that the aforementioned geometric parameters have a significant effect on the performance of microreactors. Then, simulations are carried out for each one of these designs considering the operating conditions for the process to be represented (pressure, temperature, inlet velocities, concentrations of chemical species, etc.), the properties of fluids and materials involved (density, viscosity, heat capacity, etc.), and the boundary and initial conditions.
b) From the analysis of the results obtained, in terms of profiles for the mean values of velocity, temperature, concentration of chemical species, etc., the performance of each design is determined. In this case, the best microreactor design is the one that exhibits flow uniformity and the highest values for heat and mass transfer rate.
c) Additionally, some specific parameters such as mixing degree, figure of merit (FoM), and nonuniformity of flow can be estimated to support the findings obtained in the previous step. These parameters are defined in the following subsections. In this case, the best microreactor design is the one that offers the best compromise between reaction yield and energy consumption.
d) Finally, experiments with the best design can be carried out to validate the results obtained from the simulations through the comparison of the behavior of different variables (velocity, temperature, species concentration, etc.). Another usual way to validate the results obtained is the comparison with data reported in experimental works developed in similar conditions.

2.2.1.1 Mixing index

Given the importance of an appropriate mixing between chemical species to improve heat and mass transfer in microreactors, the degree of mixing is a parameter used to evaluate the performance of these devices. The degree of mixing can be evaluated with the mixing index (M_i), which can be obtained from the standard deviation of species concentration for each cross-section in the direction of flow:

$$M_i = 1 - \sqrt{\frac{\tau^2}{\tau_{\max}}} \tag{2.1}$$

$$\tau^2 = \frac{1}{n}\sum_{i=1}^{n}(\omega_i - \omega_\infty)^2 \tag{2.2}$$

where τ indicates the variation of concentration for each cross-section, $\tau_{\max}$ is the maximum variance over the range of data, n is the number of sampling points inside the cross-section, ω_i is the mass fraction at the sampling point i, and ω_∞ is the mass fraction at infinity. The mixing index has a value of 1 for complete mixing and 0 where there is no mixing.

2.2.1.2 Mean variables

The mixed mean temperature is given by:

$$T_{\text{mean}} = \frac{1}{VA_c}\int_{A_c}^{0} Tu dA_c \tag{2.3}$$

while the mean mass fraction is written as:

$$\omega_{i,\text{mean}} = \frac{1}{VA_c}\int_{A_c}^{0} \omega_i u dA_c \tag{2.4}$$

where A_c is the cross-sectional area of the channel, u is the flow velocity, and V is the mean velocity, which is given by:

$$V = \frac{1}{A_c}\int_{A_c}^{0} u dA_c \tag{2.5}$$

2.2.1.3 Figure of merit

As mentioned before, the FoM is a parameter used to evaluate the effect of Reynolds number and the reactor geometry based on the pressure drop and reaction rate. This parameter is defined as the ratio of the mass consumption rate per unit pumping power required, given by:

$$\mathrm{FoM} = \frac{m_{r,\mathrm{in}} - m_{r,\mathrm{out}}}{P_{\mathrm{pump}}} \tag{2.6}$$

where $m_{r,\mathrm{in}}$ is the mass flow of the reactant at the reactor inlet, whereas $m_{r,\mathrm{in}}$ is the mass flow of the reactant at the reactor outlet, P_{pump} is the pumping power required to drive the fluid through the microreactor, calculated by:

$$P_{\mathrm{pump}} = \frac{1}{\eta_{\mathrm{pump}}} Q \Delta p \tag{2.7}$$

In the above equation, η_{pump} is the pump efficiency, Q is the volume flow rate of the fluid, and Δp is the pressure drop.

Some authors have proposed an alternate form for the calculation of FoM. Solehati et al. [29] have presented a definition of this parameter in terms of the mixing index (M_i):

$$\mathrm{FoM} = \frac{M_i}{\Delta p} \tag{2.8}$$

2.2.1.4 Nonuniformity of flow

Nonuniformity of flow is another parameter used to determine the performance of microreactors. According to Mies et al. [30], for microreactors with compartments it can be characterized by introducing a special parameter, s, defined as the mean square deviation from the average flow value:

$$s = \sqrt{\frac{\sum_{i=1}^{4} (u_i - u_{\mathrm{ave}})^2}{3}} \tag{2.9}$$

where u_i is the area averaged velocity in the compartment i, and u_{ave} is the average flow value in the compartments considered for the simulations. For four compartments, for instance, it is defined as:

$$u_{\mathrm{ave}} = \frac{1}{4} \sum_{i=1}^{4} u_i \tag{2.10}$$

Thus, the degree of nonuniformity of flow in percent, δ, from the total flow is evaluated using Eq. (2.11) in order to compare the results obtained at different flow velocities.

$$\delta(\%) = \frac{s \times 100}{u_{\text{ave}}} \tag{2.11}$$

2.2.1.5 Studies based on the analysis of parameters related to microreactor performance

Different microreactor designs for a wide variety of applications have been proposed from the evaluation of the aforementioned parameters (Table 2.1).

Sasmito et al. [31] studied the effect of different channel geometries in a T-junction microreactor on mixing and reactant conversion for the heterogeneous oxidation of methane. The geometries considered for the microreactor channel were straight, conical spiral, in-plane spiral, and helical spiral. The authors found that spiral channels provide a better performance, compared to the straight channels, in terms of mixing uniformity, reactants conversion, and heat transfer. These channel designs also exhibited a pressure drop higher than that estimated for the straight channel. However, from the comparison of the FoM values for the different channel designs, it was found that the spiral channels exhibited lower values of this parameter compared to that observed for the straight channel. The authors concluded that if the performance of the microreactor is of greater interest, designs with a spiral channel can be considered based on their enhancement of heat and mass transport.

On the other hand, if pumping power is the major constraint, designs with a straight channel are recommended. Similar findings have been reported by An et al. [32] in their study focused on the analysis of microreactors with different channel designs of square cross-section. The effect of different shapes for the cross-section of the channels on the microreactors performance has also been studied. Kurnia et al. [33] considered microreactors with six different designs for this geometric parameter in straight and spiral channels: circular, half circular, rectangular, square, trapezoidal, and triangular. In this case, microreactors with spiral channels showed a better performance than those with straight channels, which is in agreement with the findings reported in similar studies. From the comparison of the FoM values, it is established that spiral reactors with a circular cross-section offer the best balance between reaction rate and pumping power requirement.

In order to improve the mixing degree, with the subsequent enhancement on heat and mass transfer in microreactors, some authors have proposed the placing of static mixing elements through the flow channel. In 2012, Shaker et al. [34] presented a study where different designs of circular and square elements in both centric and eccentric configurations were analyzed: sudden expansions, elements that add recirculation zones and redirect the flow, and elements that split and recombine the flow. In this case, it was found that these last were the mixing elements that exhibited a better

Table 2.1: Studies based on the analysis of parameters used to evaluate microreactors performance.

Authors	Case	Conditions	Geometric parameters studied	Parameters used to evaluate microreactor performance
Sasmito et al. [31]	Methane oxidation at the microreactor surface coated with a platinum catalyst.	Inlet velocity for air and methane: 1, 5, 10 m/s Inlet temperature for air and methane: 300 K Inlet mass fractions: Methane – 0.9; oxygen – 0.21 Outlet pressure: 101,325 Pa Temperature at the wall: 1,290 K	Shape of the channel: straight and spiral	Figure of merit (FoM) Mean velocity Mixed mean temperature Mixed mean mass fraction
An et al. [32]	Methane oxidation at the microreactor surface coated with a platinum catalyst.	Inlet velocity for the air-methane mixture: 1, 5, 10, 15 m/s Inlet temperature for the air-methane mixture: 300 K Inlet mass fractions: Methane – 0.9; oxygen – 0.21 Outlet pressure: 101,325 Pa Temperature at the wall: 1,290 K	Shape of the channel: straight and spiral	FoM
Kurnia et al. [33]	Methane oxidation at the microreactor surface coated with a platinum catalyst.	Inlet velocity: Values equivalent to different Reynolds number – 100, 300, 500, 1,000.	Shape of the channel cross-section: circular, half-circular, rectangular, square, trapezoidal, and triangular.	FoM Mean velocity Mixed mean temperature Mixed mean mass fraction

Shaker et al. [34]	Methane oxidation at the microreactor surface coated with a platinum catalyst.	Inlet velocity for air and methane: 0.01, 0.05, 0.1, 0.5, 1, 2, 3 m/s Inlet temperature for air and methane: 300 K Inlet mass fractions: Methane – 0.9; oxygen – 0.21; hydrogen – 0.1 Outlet pressure: 101,325 Pa Temperature at the wall: 1,290 K	Shape of square and circular mixing elements in the flow channel: sudden expansions, elements that add recirculation zones and redirect the flow, and elements that split and recombine the flow.	FoM
Bawornruttanaboonya et al. [35]	Methane oxidation at the microreactor surface coated with a platinum catalyst.	Inlet velocity for the air-methane mixture: 1, 5, 10, 15 m/s Inlet temperature for the air-methane mixture: 300 K Inlet mass fractions: Methane – 0.9; oxygen – 0.21 Outlet pressure: 101,325 Pa Temperature at the wall: 1,290 K	Different values of height and width of splitting-recombining elements, and separation distance between elements through the flow channel.	FoM
Fang and Yang [36]	Mixing of fluids with properties similar to water.	Inlet velocity: Values equivalent to different Reynolds number – 0.01 to 100.	Different designs for the flow channel: T-microreactor and slanted groove microreactor.	Mixing index
Alam and Kim [37]	Mixing of water and ethanol.	Inlet velocity: Values equivalent to different Reynolds number – 0.1, 1, 15, 30, 45, 90.	Curved microreactor with grooves and without grooves.	Mixing index
Wang et al. [38]	Alkylation of isobutane/1-butene.	Inlet flow rates: Sulfuric acid – 11.4 ml/min; hydrocarbon – 7.6 ml/min Outlet pressure: 0.5 MPa	Different values for height in a microreactor with grooves.	Mixing index

performance compared to the other designs, despite their high values for the pressure drop. According to the authors, when the values of Reynolds number in these designs increase, flow mixing leads to a maximum reactants conversion and a higher pressure drop. In the same conditions, simpler designs have lower requirements of pumping power, but also a lower reactants conversion. Therefore, in the described operating regime the values of FoM for both designs will be very similar. In a later study, Bawornruttanaboonya et al. [35] analyzed the effect of different geometric parameters of splitting-recombining elements in microreactors used for the catalytic methane oxidation. According to the authors, larger values for the splitting-channel height and width, and shorter distances between mixing elements through the flow channel promote a higher methane conversion, but also a higher pressure drop. However, at high values of Reynolds number (about 100), the authors observed similar FoM values between the microreactor with splitting-recombining elements and the one without those elements, which is the same behavior reported previously by Shaker et al. [34]. Other authors have also presented Splitting-recombining configurations for microreactors designs. In 2009, Fang and Yang [36] presented a study where the effectiveness of their proposed design was evaluated with the mixing index. The authors reported that based on the estimated values of this parameter for their design, a conventional T-microreactor, and a slanted groove microreactor, the former exhibited superior performance. These findings were also observed in a series of experiments developed by the authors considering the aforementioned designs.

The performance of microreactor designs with grooves in the flow channel has also been evaluated with the mixing index by other authors. Alam and Kim [37] have proposed a microreactor design with a curved channel and rectangular grooves for the mixing of water and ethanol. The authors compared the results for this design with those obtained for a curved microreactor without grooves. In this case, it is observed that the grooves promote a better mixing degree, which is corroborated by the experimental observations corresponding to the distribution of ethanol mass fraction. Recently, Wang et al. [38] presented a study where different values of the flow channel height in a microreactor with grooves were considered. The authors reported that microchannels with smaller cross-section exhibit higher values of mixing index compared to channels with a larger cross-section.

From the revision of these studies, it can be observed that some authors use a combination of parameters related to the performance of microreactors to characterize their designs. On the other hand, some of them only use one of these parameters as the only criteria to propose a better design. In this case, it is important to mention that, while mixing index and mean variables provide a representation of the interaction between chemical species as well as the changes in temperature, concentration, and velocity through the microreactor, that information may not be enough to choose the best design among different proposed configurations, especially for industrial-scale processes, where pumping power requirements can be high. In this sense, given that FoM establishes a relationship between reaction rate and power consumption, it has been proven to be a parameter that offers a decision criteria based on the priorities for each particular case (i.e., there may be processes where pumping power is a major constraint and processes where product yield is more important). Moreover, when FoM is analyzed along with other parameters (mixing index, nonuniformity of flow), it is possible to obtain a better perspective of the performance of different microreactors designs.

2.2.2 Design based on optimization techniques

As discussed in the previous section, it has been demonstrated that the aforementioned parameters, which depend on the microreactor design, are directly related to the performance of these units. Taking this into account, some authors have proposed the optimization of these parameters in order to obtain reactor designs that guarantee the best performance of such units (Table 2.2).

In 2017, Grundtvig et al. [39] presented a study where they performed the shape optimization of a biocatalytic square microreactor. The strategy proposed by the authors consisted of performing CFD simulations in ANSYS CFX coupled with an optimization algorithm in MATLAB. In this case, script files containing the mesh files, the simulation setup file, and the post-processing file from the CFD simulations were exported to MATLAB. Then, with an user-programmed code the script files were converted into arrays, the points that define the microreactor geometry were modified, and the start of CFD simulations was commanded. The optimization procedure was based on an evolutionary procedure, where the objective function was defined as the concentration of product at the outlet of the microreactor, which had to be maximized. On the other hand, two restrictions were established for the optimization problem: either the amount of product formed was 10 times higher than the amount of product formed for the initial microreactor configuration, or the number of iterations between two local maxima exceeded 1,500; according to the authors, this value was established in order to limit the optimization time. In the study, the optimization routine stopped when either one of these constraints was satisfied, or when the system reached convergence. From the described optimization procedure, a configuration characterized by a series of expansions and shrinkages of the surface of the microreactor was obtained. This design exhibited a better performance regarding to mixing and reaction yield, compared to a straight square microreactor. However, it was also concluded that the optimized design cannot be easily fabricated due to the complexity of the resulting structure.

Design of microreactors based in the optimization of some geometric parameters has also been proposed in the literature. Solehati et al. [29] presented a study where some geometric parameters in a wavy microreactor such as height, width, wavy frequency, and wavy amplitude were varied according to a Taguchi method. Taguchi method is an engineering tool for experimental optimization used to find

Table 2.2: Studies based on optimization techniques.

Authors	Case	Conditions	Geometric parameters studied	Parameters used to evaluate microreactor performance
Grundtvig et al. [39]	Synthesis of the chiral product (*S*)-1-phenylethylamine and acetone, from acetophenone and isopropylamine catalyzed by amine transaminase.	Concentrations at inlet: Acetophenone – 20 mM; isopropylamine – 1 M; transaminase – 0.15 M.	Shape of the walls of a microreactor.	Concentration profiles
Solehati et al. [29]	Mixing of fluids.	Inlet velocity: 0.04 m/s	Dimensions of a wavy microreactor: height, width, wavy frequency, and wavy amplitude.	FoM Mixing index

the sensitivity of each parameter and determine the combination of the design factors [40]. According to the authors, the optimal designs for the microreactors obtained with the Taguchi method were considered in CFD simulations. From this comparison, it was found that the maximum error between Taguchi predictions and CFD simulations was less than 6%.

From the revision of these studies, it can be concluded that, nowadays, thanks to the possibility to link software of different type (i.e., CFD and optimization/programming), it is possible to develop strategies focused on obtaining better microreactors designs based on the conjoint analysis of fluid flow, heat and mass transfer, and the optimization of the aforementioned geometric parameters or some operating variables (maximizing product yield or reactants conversion, for instance). However, the main disadvantage of the study presented in this section is that the optimized design obtained through this strategy has a very irregular wavy shape, which is difficult to build. Regarding the optimization strategies based on the Taguchi method, it is important to mention that currently is one of the most popular methodologies used for the design of chemical equipment, when it is used along with CFD. Given that this method focuses on finding the sensitivity of each parameter to determine the combination of the design factors, some authors usually tend to choose a limited number of geometric parameters, ignoring the rest of them. This could be a disadvantage of the method since the optimized design could not take into account the effect of important geometric parameters that were previously ignored for the analysis.

2.2.3 CFD-based studies applied to the design of microreactors for biofuels production

The aforementioned approaches for microreactors design have also been applied to biofuels production, specifically biodiesel obtained from the transesterification of vegetable oil with alcohols (Table 2.3).

Given the complex nature of the fluid system formed by these two reactants, an efficient mixing degree in the microreactors used is of relevant importance [41]. As discussed before, optimal mixing can be achieved not only by increasing the flow velocities but also by modifying the reactor design. Considering this, Santana et al. [42] developed a computational model to compare different microreactor designs and analyze their effect on biodiesel production. To evaluate the mixing degree in the different reactor designs, the authors used a mixing index. In this case, the authors found that a flow channel with a cross-shape promoted the highest mixing index and also the highest oil conversion. Hence, a proportional relationship was established between these two variables. However, the highest calculated value of oil conversion was lower than the average value reported in experimental studies. According to the authors, this can be attributed to the short residence time achieved with the proposed design. Taking into account that internal obstructions can split and recombine the

Table 2.3: Studies focused on the design of microreactors for biodiesel production.

Authors	Case	Conditions	Geometric parameters studied	Parameters used to evaluate microreactor performance
Santana et al. [42]	Transesterification of *Jatropha curcas* oil with ethanol.	Inlet velocity – Given by different Reynolds numbers from 10 to 100; static pressure at outlet – 0.	Different designs for the mixing section: a T-shape, a cross-shape, and a double-T-shape	Mixing index
Santana et al. [43]	Transesterification of *Jatropha curcas* oil with ethanol.	Inlet velocity – Given by different Reynolds numbers from 1 to 160; static pressure at outlet – 0.	Different arrangements of circular obstructions in the flow channel: a linear pattern and an alternating pattern.	Mixing index
Santana et al. [44]	Transesterification of *Jatropha curcas* oil with ethanol.	Conditions: 25 to 75 °C; inlet velocity for oil – 6.66, 7.40, 8.88 m/s; inlet velocity for ethanol – 5.18, 4.44 and 2.96; alcohol-to-oil molar ratio – 6:1, 9:1 and 12:1; static pressure at outlet – 0.	Presence of flow deflectors, perpendicular to the reactor walls.	Mixing index

flow stream through microreactors, which enhance the mixing between reactants, the authors modified their previous computational model to analyze the effect these elements have on microreactors performance [43]. In this case, it was found that a design with obstacles in an alternating pattern performed better than designs with obstacles in a linear pattern. According to the authors, this could be related to the differences in the level of flow perturbation observed between these two designs. It is also reported that obstacles in an alternate pattern promote an increase in the contact surface area between the oil and alcohol. This behavior was also observed in simulations where the width of a microreactor without obstructions was decreased. This confirms the importance of the contact area in alcohol-oil systems. A different configuration for the internal obstructions in microreactors has been tested by the same authors in a recent study [44]. In this work, the modeling of a reactor with alternating flow deflectors located in the walls was performed. In this case, the design with flow deflectors exhibited a superior performance regarding mixing index and oil conversion compared with a reactor without internal elements. From these observations, the authors confirmed that flow deflectors induce a series of fluid flow phenomena, which results in better mixing and enhances mass transfer. In addition, with the proposed modeling strategy, an acceptable agreement between the calculated and experimental values for oil conversion was observed.

2.3 Applications of microreactors to biofuels production

In this section, we present the revision of the scientific advances related to the use of microreactors for biofuels production. According to the revision of the literature, we found the use of microreactors for the production of liquids and gaseous biofuels, which are presented in Sections 2.3.1 and 2.3.2, respectively.

2.3.1 Liquid biofuels

In the literature, there are applications of microreactors for the production of biomethanol, Section 2.3.1.1, biodiesel, Section 2.3.1.2, and other processes, Section 2.3.1.3. This information is presented next.

2.3.1.1 Biomethanol

Biomethanol is an alcohol with the same chemical composition of its counterpart from fossil origin. Although biomethanol is not used as biofuel or additive, such as biobutanol and bioethanol, respectively, it has attracted interest recently since it is one of the products of the photocatalysis of carbon dioxide. In particular, the application of microreactors for the production of biomethanol, through the photocatalysis of carbon dioxide, was reported by Cheng et al. [45]. As result, they found that the use of a membrane microreactor for this reaction is feasible and the maximum yield reported was 111 μmol/g•cat•h, which is competitive with other results already reported in the literature. The catalyst used was TiO_2 and the membrane was made from carbon paper. The source light was a 365 nm UV light.

This reference was the only reported for the production of this biofuel; however, it is an interesting work where the most promissory conversion route is intensified through the use of microreactors.

2.3.1.2 Biodiesel

Biodiesel is composed of monoalkyl esters of long-chain fatty acids derived from vegetable oils or animal fats (through transesterification reactions). Biodiesel is used in internal combustion engines mixed with fossil diesel. The production process of biodiesel is simple and it is performed at low conditions of pressure and temperature. Therefore, this biofuel has been widely studied considering different topics. Next, we present the applications to biodiesel production where microreactors are employed.

Sun et al. [46] is one of the first reported applications for the production of biodiesel using a capillary microreactor. They analyzed the residence time, reaction temperature, catalyst concentration (KOH), and ratio methanol to oil on the yield of biodiesel. The maximum yield of biodiesel was 95%, with a residence time of 10 min. From the report of this study, several works have been published, some of them are summarized by chronological order in Table 2.4

From Table 2.4, we observe an interesting trend in the increase in the biodiesel yield along with a reduction in residence time and catalyst oil ratio as the time progresses. Moreover, the reported works are focused on homogeneous catalyst mainly with ethanol and methanol and edible oils. An important aspect is that just one work includes experimentation and CFD modeling [52], where a minimum residence time and maximum biodiesel yield is observed.

There are also modeling works focused on the analysis of mass transfer in the transesterification process [53, 54] and on the determination of kinetic coefficients [55]. As it can be observed, modeling and simulation in microfluidics are still new compared with conventional scale, the literature being more scarce for biodiesel synthesis [56].

An interesting recent contribution is related to the production of biodiesel in a semi-industrial pilot microreactor [57]. In this plant, 50 microtubes are included for a total volume of 200 ml, waste cooking oil is used as raw material, and KOH is

Table 2.4: Experimental studies for the production of biodiesel in microreactors.

Type of microreactor	Oil	Alcohol	Catalyst	Residence time	Catalyst/ oil ratio	Biodiesel yield	Ref.
Capillary microreactor	Rapeseed Cottonseed	Methanol	KOH	600 s	6	95%	[46]
Circular PerFluoroAlkoxy tubes	Sunflower	Ethanol	NaOH	240 s	45.4	96%	[47]
Microreactors at constant volume and constant reaction time	Soybean	Ethanol	KOH	360 s	1	99%	[48]
Circular tubes	Soybean	Methanol	KOH	180 s	9	98%	[49]
Capillary Millichannel	Palm oil	Methanol	KOH	180 s	23	91%	[50]
Microtube and milichannel reactors	Palm oil	Methanol	KOH	180 s	21	95%	[51]
Microchannel	Sunflower	Ethanol	NaOH	12 s	9	99.99%	[52]

used as catalyst. The maximum purity of biodiesel was 98.26% at 120 s of residence time and a ratio methanol/oil of 9.4. Moreover, the biodiesel obtained satisfy the ASTM D6751 standard.

The production of biodiesel with microreactors is the most advanced application reported until now. However, more efforts are needed in order to analyze another type of processes, such as heterogeneous catalysis with special emphasis on enzymatic processes.

2.3.1.3 Other processes

We also found an article where the use of microreactors was proposed for a hydrodeoxygenation process [58]. The raw material was acetic acid, which is used as model compound for the pyrolysis oil. As product of the reaction, acetaldehyde and ethyl acetate were generated. A maximum conversion of 70% is reported at 2.07 MPa and 450 °C with sulfided $NiMo/Al_2O_3$ catalyst. Although this study does not generate a biofuel, it is important since it involves a conversion route that can be used to produce green diesel and biojet fuel.

Moreover, the conversion of cellulose through fast pyrolysis was studied in a microreactor using ZSM-5 catalyst [59]. As products of the reaction, aromatics and coke were obtained. The maximum yield for aromatics was 38.2%, while for coke was 27%; gaseous products also were obtained, mainly CO. This work is very interesting, since aromatic compounds have a high demand worldwide. Moreover, renewable aviation fuel may or may not contain aromatic compounds depending on the conversion route [60]; when this biofuel does not contain aromatic compounds, these must be added. In this context, the development of intensified processes is valuable.

2.3.2 Gaseous biofuels

In the literature, there are applications of microreactors for the production of renewable hydrogen, Section 2.3.2.1, synthetic gas natural, Section 2.3.2.2, and synthesis gas, Section 2.3.2.3. This information is presented next.

2.3.2.1 Renewable hydrogen

Renewable hydrogen is considered as a renewable fuel for the transport sector and as energy storage. Due to this and its outstanding properties, after numerous considerations, hydrogen emerged as a hopeful ideal sustainable future energy carrier [61]. There are several processes for the production of hydrogen from different sources. For instance, hydrogen is generated through dark fermentation [62], sonochemistry [63], thermal decomposition [64], photocatalysis [64], biosynthesis with microorganisms [65], PEM water electrolysis [66], gasification [67], and photoelectrochemistry [68]. As it can be observed, this biofuel has been widely studied. Next, we present the applications to renewable hydrogen production where microreactors are employed.

In 2012, renewable hydrogen was produced from biogas using steam reforming process [69]. The microreactor system operates at 1,073 K and atmospheric pressure, and the catalysts used were Rh–Ni/Ce–Al_2O_3 and Ni/Ce–Zr–Al_2O_3. The hydrogen yield was higher than 80% for all catalysts.

In 2013, Sanz et al. [70] presented a comprehensive revision of the recent advances for the production of hydrogen through microstructured catalytic reactors. In their work, they describe the use of microreactors to produce renewable hydrogen from dimethyl ether, biomethanol, biodiesel, bioglycerol, biogas, producer gas, bioethanol, and bio-oil.

Later, the production of biohydrogen from the autothermal reforming of methane using spiral multicylinder microreactor was reported [71]. The authors simulated this process using CHEMKIN and CFD software to analyze catalytic wall temperature, oxygen/carbon ratio, water-carbon ratio, and catalyst active site density on the yields hydrogen and methane. They reported yields of 93%–99% and 65%–80% for methane and hydrogen, respectively.

Entezary and Kazemeini [72] performed the conversion of ethylene glycol and glycerol in a structured catalyst microreactor to produce hydrogen. The catalysts used were Pt/Al_2O_3 and $Pt/CeO_2Al_2O_3$. As results, they obtained 75.3% conversion of glycerol with a yield of 92.4% of hydrogen; these results are higher than those obtained in a fixed-bed reactor.

Moreover, in 2018, the steam reforming of glycerol in a heat exchanger integrated to a microchannel reactor was reported [73]. As main result, they found that the integration of the heat exchanger increases the conversion of glycerol in 10%, this value being 73.3% with a selectivity of 65% to hydrogen.

Recently, the steam reforming of methanol through a microreactor was reported for the production of hydrogen [74]. The authors reported a maximum conversion of 90% for methanol at 300 °C and the constant generation of 0.58 mol/h of hydrogen flow rate.

The production of renewable hydrogen with microreactors has had an important advance until now. However, more efforts are needed in order to analyze another type of processes, such as photocatalysis of carbon dioxide.

2.3.2.2 Synthetic gas natural

Synthetic natural gas (SNG) is another biofuel that has attracted interest recently. SNG contains mainly methane and it can be generated from the methanation of syngas. The use of microreactor for this production pathway was presented by Liu et al. [75]. The catalyst used was Ni. They found that at 500 °C the conversion reached was above 98%.

2.3.2.3 Synthesis gas

The synthesis gas is generated as product of the gasification process. Syngas can be used as raw material for the production of other biofuels. Recently, the use of microreactors to perform the autothermal reforming of biogas to produce gas synthesis was reported [76]. This simulation considered the use of $NieRe/Al_2O_3$ catalyst in a coated wall microreactor, and also in a fixed-bed microreactor for comparisons purposes. They found that the performance of the coated wall microreactor was superior than the one of the fixed-bed microreactor.

2.4 Technological advances for biofuels production

There are several sources available to identify the number of patent publications. In this section, AcclaimIP patent research software [77] was used in order to present and analyze some statistic information about microreactor patents. There are more than 30,663 patents related to microreactors, and this number is increasing every day. However, patent systems are national or regional, i.e., the authors apply to patent the same invention in several countries; even each country has different types of documents arise according to the requirements of each legislation. For this reason, the number of patents increases significantly with each new invention. Thus, a "family" of patents is a set of documents related to the same invention. The number of family patents for microreactors are 13,664. Figure 2.2 shows the number of patents since 1949 by published year until May 29, 2019.

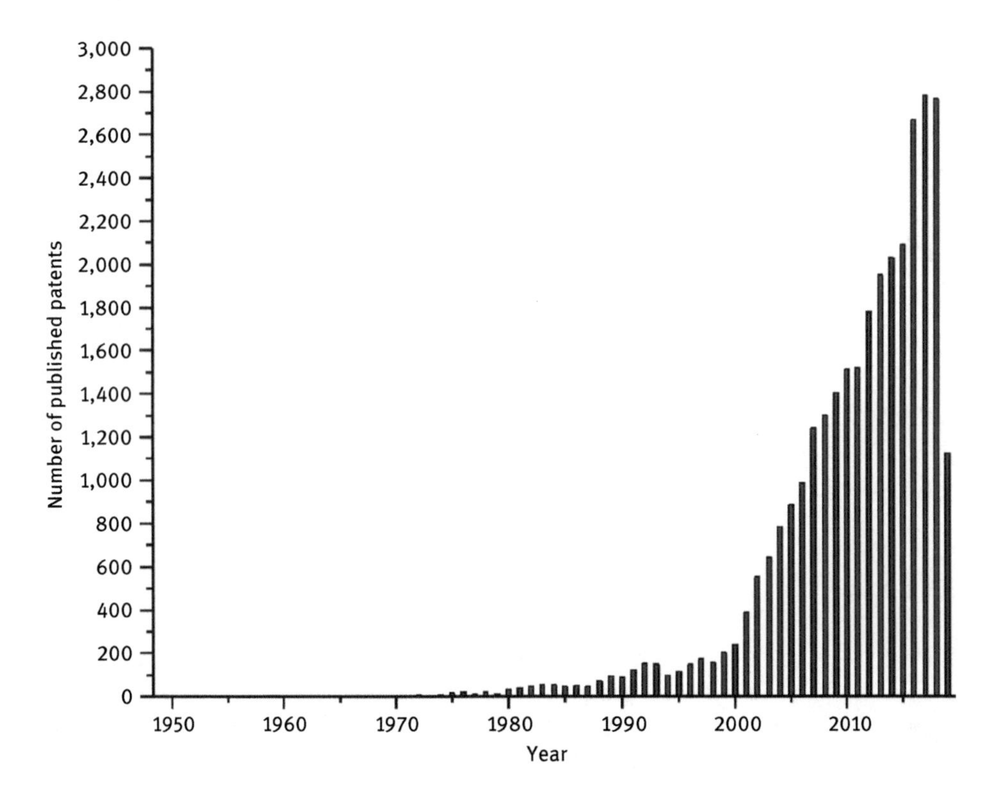

Figure 2.2: Number of patents related to microreactors by published year.

In the first published patent, a microreactor was used to carrying out X-ray diffraction studies to identify active elements presented in catalytic materials [78]. This apparatus was handled with powdered substances under several conditions of pressure and temperature, and in the presence of various fluids materials. A second microreactor patent was published in 1970 (21 years later), and after that the number of patents increased exponentially. Actually, the patents published from 2016 to 2018 summarized 8,215. The total number of patents are distributed by country as showed in Figure 2.3.

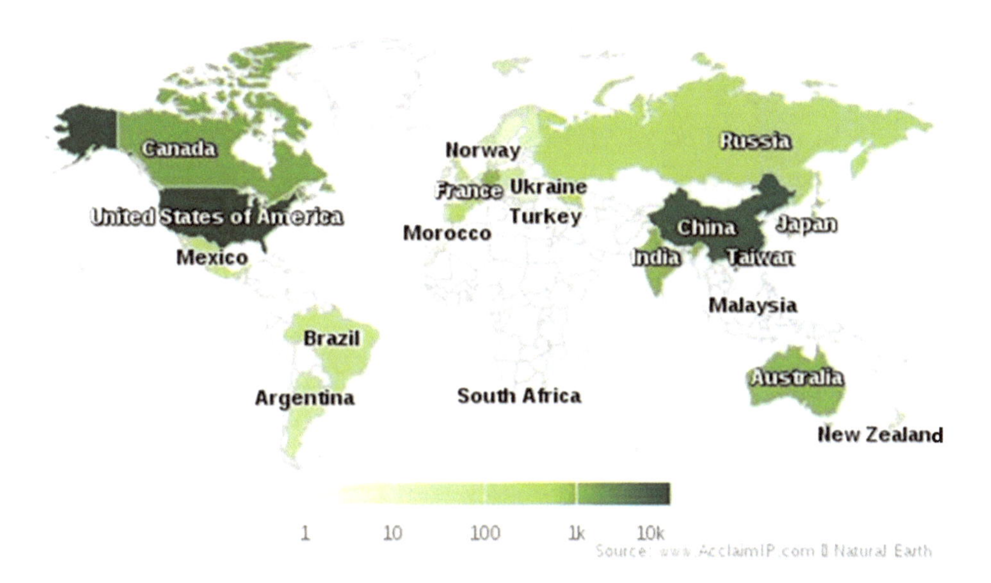

Figure 2.3: Top microreactors patenting countries of issuance.

Four patent offices have more than 70% of the total microreactors published patents. These offices are the United States Patent and Trademark Office (USPTO) with 30% (9,204); the National Intellectual Property Administration (CNIPA), also known as the Chinese Patent Office, with 21.3% (6,529); the World Intellectual Property Organization, or WIPO, with 10.8% (3,309); and the European Patent Office (EPO) with 8.4% (2,564). The top 10 owners are presented in Figure 2.4.

This data shows the dominance of Chinese companies regarding published patents. Also, the main inventors are from Asia. Table 2.5 lists the top inventors and their number of patents involving microreactors. Moreover, Figure 2.5 shows the top 20 companies or institutions with most cited and citing patents.

It is important to remark that Shell Oil Company and China Petroleum Chemical Company have more citing documents; this could be attributable to the relevance and number of patents.

An analysis of the microreactor patents, using the Cooperative Patent Classification (CPC), shows the main topics related to microreactors from 2016 to 2018, they are:

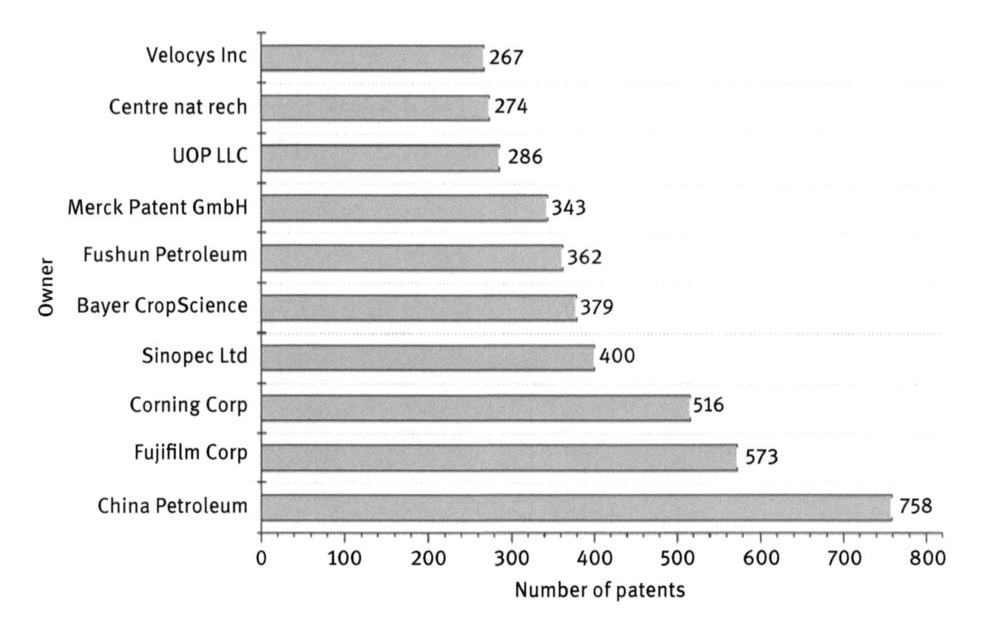

Figure 2.4: Top 10 owners of patents.

Table 2.5: Top 10 microreactor inventors.

Inventor name	Number of patents
Kai Guo	78
Huiping Tian	72
Rosinger, Christopher Hugh	68
Wei Lin	61
Gatzweiler, Elmar	60
Jinbing Li	58
Rongsheng Li	54
Tian Huiping	49
Tonkovich, Anna Lee	49
Zheng Fang	49

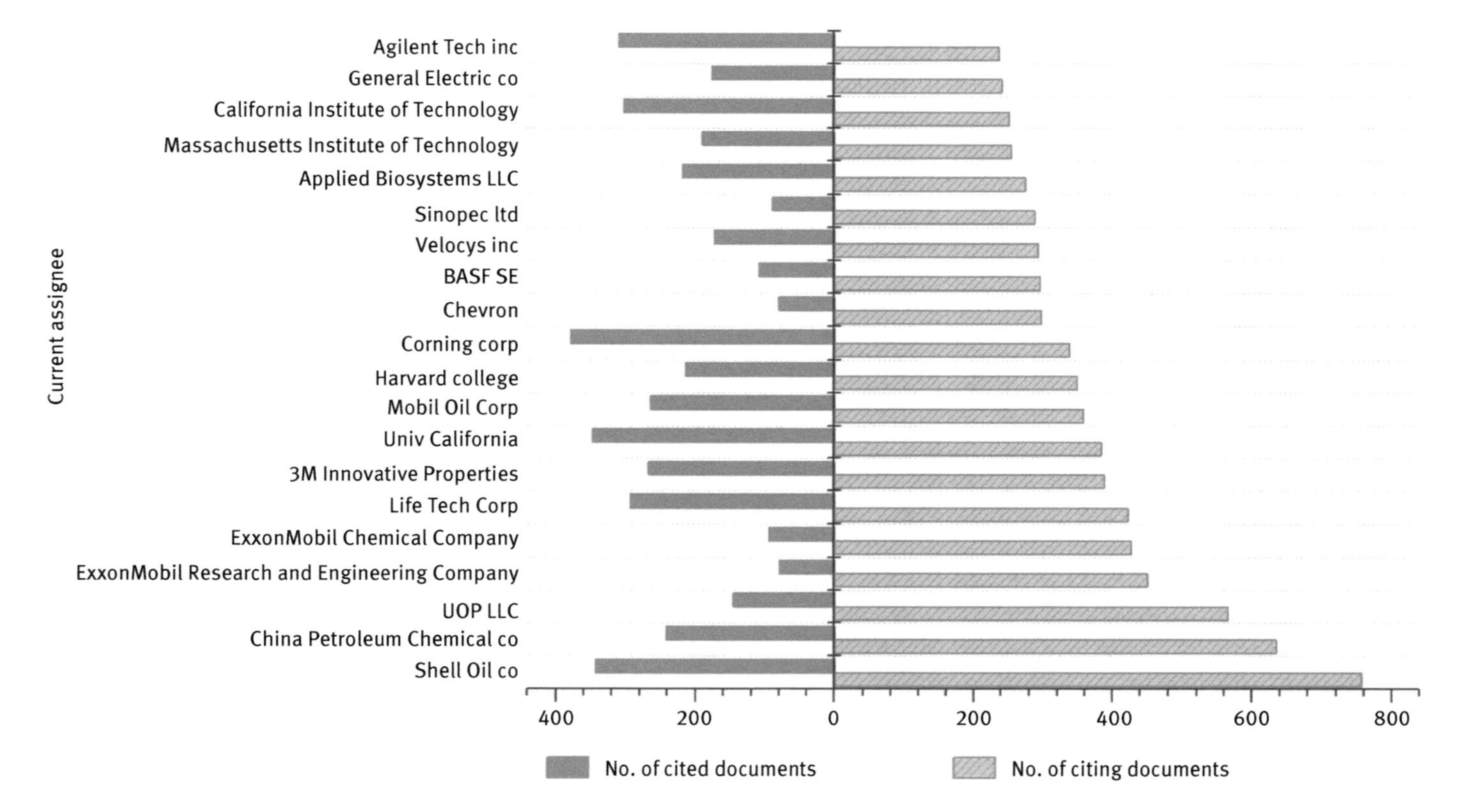

Figure 2.5: Top 20 companies or institutions with most cited and citing patents.

- apparatus for physical or chemical processes (B01J19 and B01J2219) and additional constructional details of the equipment (B01L2300);
- apparatus for physical or chemical laboratory (B01L3) and the solutions for specific problems of the equipment (B01L2200);
- measuring or testing processes involving enzymes, nucleic acids, or microorganisms (C12Q1);
- technologies relating to the chemical industry (Y02P20)
- moving or stopping fluids (B01J400)
- catalysts comprising metals or metal oxides or hydroxides (B01J23); and
- investigating or analyzing materials by specific methods (G01N33).

Figure 2.6 shows the number of patents related to these topics. This data is evidence that the number of patents related to the design of microreactors was the main topic but this number increases and decreases during this period. However, the patents number of measuring or testing processes involving enzymes, nucleic acids, or microorganisms was the only one that had an increasing monotonic trend.

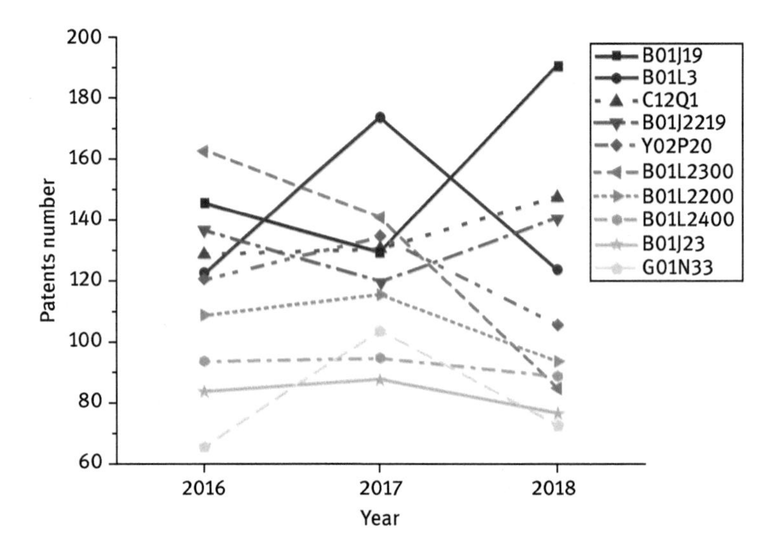

Figure 2.6: Number of patents for top 10 CPC class related to microreactors from the year 2016 to 2018.

The microreactors are fabricated using several materials and methods. The materials could be, e.g., glass, quartz, diamond, silicon, stainless steel, nickel ceramics, and polymers such as polymethylmethacrylate, polydimethylsiloxane (PDMS), and polyimide. Table 2.6 shows the number of microreactor patents related to these materials.

The primary materials are glass, silicon, and polymers. An analysis of the advantages and disadvantages of these typical materials used for microreactor fabrication is presented

Table 2.6: Material for microreactor fabrication.

Material	Number of patents
Glass	15,580
Quartz	5,814
Diamond	2,097
Silicon	15,348
Stainless steel	6,912
Nickel	8,697
Ceramics	8,082
Polymethylmethacrylate (PMMA)	741
Polydimethylsiloxane (PDMS)	2,422
Polyimide	2,668
Polyvinyl chloride	3,130
Polystyrene	5,062
Polycarbonate	3,905

in [79]. Also, the microreactors can be fabricated as microcapillaries and chips for complex structures. There are 696 patents of microreactors fabricated as microcapillaries and 9,115 patents as chips. For the fabrication, the methods most commonly used are:

- Micromachining method: 2,980 patents.
- Lithography, electroplating and molding (Lithographie, Galvanoformung, und Abformung LIGA) technique: 3,293 patents
- Etching technique: 7,098 patents.

The microreactor design can be mainly classified according to the operation condition, batch or continuous. For batch operation, there are 2,303 of micromixers patents. On the other hand, the continuous reactor design is divided into (a) continuous flow, 20,211 patents and (b) segmented flow, 5,257 patents.

The microreactors are utilized in applications such as pharmaceutical industry (7,103 patents), biological process (11,261), the synthesis of nanoparticles (3,325 patents), and synthesis of organic chemicals and polymers (9,060 patents). Other important applications of the microreactors are the hydrogenation processes (17,445 patents), the oxidation processes (20,982 patents), and catalyst processes (5,720 patents).

The microreactors can also be used to produce biofuel with 516 patents, mainly biodiesel (490 patents). Figure 2.7 present the number of published patents since 1994.

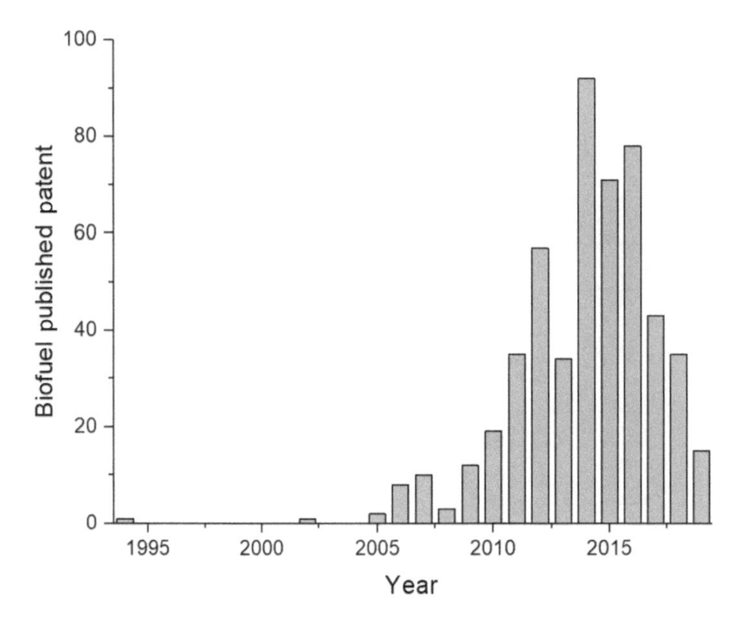

Figure 2.7: Number of patents related to microreactors to produce biofuels by published year.

The first patent to produce biofuels, from 1994, details a fixed-bed continuous flow microreactor, which is operating at atmospheric pressure [80]. In this patent, the catalyst is supported in a particular form of zeolite ZSM-5, and the reaction products are aliphatic hydrocarbons. In 25 years, the highest number of published patents to produce biofuels using microreactors was reported in 2014, with 92 documents, followed by 2016 with 78 documents. Table 2.7 shows this number of documents presented by each patent office in the past 10 years, where the USPTO was the main contributor.

According to Table 2.7, Shell companies are the principal developing technologies to produce biofuels using microreactors (Table 2.8). For academic institution, the University of Alberta has the highest number of patents, i.e., 16. The published patents related with the production of biofuels and microreactors also include biokerosene (3 patents), biomethanol (2 patents), bioethanol (2 patents), and biobutanol (there are not specific patents to produce it).

There are 208 family patents related to biodiesel production involving microreactors. For example, Jovanovic et al. [81] developed a method to produce biodiesel mixing alcohol and oil inside a microreactor. This biodiesel is produced under supercritical conditions using a catalyst, such as metal oxides, metal hydroxides,

Table 2.7: Number of published documents by patent office in the past 10 years.

Publication year	OTHER GLOBAL	US	EP	JP	WO	DE
2009	6	3	0	0	3	0
2010	5	6	1	0	6	1
2011	8	19	0	0	8	0
2012	21	17	1	2	16	0
2013	15	12	1	0	6	0
2014	38	35	1	1	17	0
2015	36	27	4	2	2	0
2016	33	40	4	0	1	0
2017	17	23	2	0	1	0
2018	11	16	6	0	1	1
2019	5	10	0	0	0	0

US: United States Patent and Trademark Office (USPTO)
EP: European Patent Office
JP: Japan Patent Office
WO: World Intellectual Property Organization, or WIPO
DE: German Patent and Trade Mark Office

Table 2.8: Top ten of owners patents to produce biofuels.

Owners	Number of patents
SHELL OIL CO	90
SHELL INT RES C	63
ELEVANCE RENEWA	20
GENOMATICA INC	19
DANISCO USA INC	16
UNIV ALBERTA	16
GIGAGEN INC	15
RES TRIANGLE IN	13
GOODYEAR TIRE &	11
POWERDYNE INC	10

metal carbonates, alcoholic metal carbonates, alkoxides, mineral acids, and enzymes. The conversion to biodiesel increases when residence time is elevated. Also, there is an invention related to the production of biodiesel in marine conditions from living aquatic resources. A microreactor single channel is used for preliminary purification of fish oil [82]. Jianying et al. [83] disclose a method for continuous synthesis of biodiesel through a microreactor. The method consists of adding an alkaline catalyst with short-chain alcohol and grease into a continuously heated microreactor. This equipment includes a function of mixing and operates under normal pressure. Dennis et al. [84] invented a catalytic transesterification process to produce biodiesel. The microreactor operates under a pressure and a temperature less than 5 psi and 70 °C, respectively. Several microreactors are coupled in parallel to improve the biodiesel yield. Also, there are patents focused in provide a method of biodiesel adsorption deacidification for environment protection [85–87]. Li et al. [88] developed a catalyst wall loadization microreactor for biodiesel production. The reaction is carried out at 0.3–1.0 MPa and 80–150 °C, and a higher conversion ratio can be observed in the feedstock oil in a relatively short period of time.

For biokerosene, Rubens et al. [89–91] invented a process for the production of two fuels compatible with kerosene of fossil origin, and a process for the production of biogasoline. These authors mention the possibility to produce these biofuels in microreactors, which can be used to carried out the reactions to generate a mixture of esters under different conditions of temperature and pressure. These conditions are specified depending on the type of feedstock.

There are two patents for biomethanol diesel production. Adler [92] developed a continuous method for the production of biomethanol diesel, which requires minimal investment in equipment. Microreactors are used for both the washing process and the transesterification. The method operates at temperatures of 65 °C. The unit can also be used for washing oils. Also, Dunkel and Adler invented a method for continuous production of biomethanol and ethanol diesel in a multistage microreactor cascade of standardized glass components for the preferred application in agriculture [93]. Bioethanol production is also presented in a large number of patents. However, only two developed a method using a microreactor directly. Chen et al. [94] developed an integrated catalytic bed microreactor that operates at uniform temperature; the aim is the conversion of bioethanol to ethylene with a selectivity higher than 98%. Piotr et al. invented a catalytic method for producing fuel biocomponent from bioethanol [95]. Other patents only suggest using microreactors, but the designs are not included.

The trend of family patent number is decreasing for biofuels. For biodiesel, the family patents number was 51 in 2016, 28 in 2017, basically half of 2016, and 22 in 2018, but there is 19 family patent so far (may) 2019. The same situation for bioethanol, 23 in 2016, 9 in 2017, and 3 in 2018. For others alcohols and biojet there not a trend, because the number of patents is a small amount each year. The focuses of the biofuels patents, in this period, are the reaction products, the biomass type used (mainly fatty acid and cellulosic solid), the biomass separation and the catalyst.

2.5 Recapitulation

Microreactors technology is an interesting alternative to increase the conversion and reduce reaction times along with energy consumption for the production of biofuels. This technology is used mainly in pharmaceutical industry, biological process, the synthesis of nanoparticles, and synthesis of organic chemicals and polymers. However, the application of microreactors for the production of biofuels is relatively recent; in 1994 the first process to produce biofuels through microreactors technology was patent. Among all liquid and gaseous biofuels, microreactor technology has been applied for the production of biodiesel and renewable hydrogen mainly. Other biofuels such as biomethanol, bioethanol, synthetic natural gas, and synthesis gas have been less explored. This evidence the great opportunity area to evaluate the feasibility of use microreactors for the production of these biofuels. Moreover, other biofuels have not been studied, such as bioethanol, biobutanol, green diesel, and biojet fuel. In particular, these last biofuels are considered key in the energetic transition for its potential to replace gasoline, diesel, and jet fuel from fossil origin.

Regarding the design methodologies, CFD has represented a powerful tool for the design of microreactors in different applications where reaction yield and transfer of heat and mass must be enhanced. For instance, the complex nature of the fluid system in biodiesel production makes difficult its analysis in experimental studies, while CFD-based studies can provide a more detailed perspective of the involved phenomena in different reactor designs. In addition, given that one of the main characteristics of CFD is the possibility to model and simulate physical systems with nonstandard geometries, complex configurations could be proposed and tested. According to the revised methodologies of design, the analysis of some important parameters such as mixing index and FoM, which could be obtained from CFD simulations, can be helpful in the evaluation of microreactors performance. In addition to this, some strategies based on the optimization of these parameters and some geometric features have been proposed for the design of these units.

In this sense, the increase in computational capacity as well as the capabilities of commercial codes to interact with different types of software have made possible the coupling between process modeling and optimization. As mentioned before, the main disadvantage of some studies focused on the optimization of the geometric design of microreactors is the very complex resulting configuration. This difficulty, however, could be overcome by imposing a constraint that prevents irregular designs (i.e., establishing limits for curvature angles, and the width and height of the waves, etc.).

Regarding biofuels production, biodiesel synthesis is the process that has received major attention in the analysis and design of microreactors. As mentioned earlier, alcohol and oil tend to form a complex mixture, which in the majority of reported studies is assumed as a single-phase mixture. However, it has been observed in experiments that the flow conditions, as well as the microreactors design, may promote the existence of different two-phase flow patterns (bubbly, Taylor, etc.). Given that many CFD codes have different models to represent the interaction between phases, the study of microreactor designs operating at these flow conditions could be possible. This approach could also be applied for the modeling and analysis of hydrotreating processes for the production of green diesel and biojet fuel, since oil and hydrogen form a two-phase system. In the case of bioethanol, some of the experimental studies reported for the production of this biofuel in microreactors have proposed a configuration where yeast is immobilized in a thin film on the microreactor surface. This could be modeled and simulated through CFD, considering that the existing codes allow the incorporation of customized surface reaction kinetics. What is more, the mesh refinement in zones near the surfaces of microreactors could provide a better representation of relevant phenomena in these devices. Since the microreactor surface is of great importance for the production of bioethanol, methodologies based on the optimization of the geometric shape, as the one previously discussed, could be applied with appropriate constraints to obtain an efficient reactor design. Experimentally, the production of green diesel and biojet fuel has been carried out in fixed-bed microreactors. In this case, it is well known that fluid flow through porous media is one complex phenomena that directly affect the product yield. Taking this into consideration, the existing methodologies of design could consider other parameters to evaluate the performance of different microreactors design. For instance, liquid distribution is a key parameter that could be estimated through CFD-based methods to evaluate the efficiency of those designs.

The microreactors technology is growing; microreactors patents numbers are increasing exponentially; four patent offices have more than 70% of the total patents. These microreactors are fabricated using several materials and methods. The etching technique is used more times than others to fabricated microreactors. The family patent numbers using the CPC class indicate that the design of microreactors for processes and laboratory classes are the most important because the number of it. However, the patents number of measuring or testing processes involving enzymes, nucleic acids, or microorganisms is the only one that has an increasing monotonic trend. Also, there are more patents to fabricated microreactors as chips than microcapillaries. For biofuels production, the microreactors are used to produce mainly biodiesel and bioalcohols. Although there are large number of patents related with biofuels and microreactors, only a few of these patents are used to produce biofuels in microreactors, i.e., many patents only suggest the use of microreactors.

Moreover, the patent number related to biofuels and microreactors is decreasing. The highest number of published patents was in 2014. The family patent number for biodiesel and bioethanol was reduced significantly, decreasing more than 50% for biodiesel and 85% for ethanol in 3 years. Then for the technology advances, the number of patents is still increasing every year, mainly for the microreactors design, but it is decreasing for biofuels production proposes, where the production reaction is the main topic.

Finally, another important aspect for the application of microreactors is the economic one. In this point, the market for microreactors seems promissory. According to Reuters [96], the forecasts indicate that the value of the market of microreactor technology will grow from 32,456 million USD in 2015 to 108,927.1 million USD in 2022. This indicates the viability of this technology for its industrial use.

Acknowledgments: Financial support provided by FOMIX Gobierno del Estado de Querétaro-CONACyT, through grant 279753, for the development of this project is gratefully acknowledged.

References

[1] Lutze P, Gani R, Woodley JM. Process intensification: A perspective on process synthesis. Chemical Engineering and Processing: Process Intensification 2010, 49(6), 547–558.
[2] Stankiewicz AI, Moulijn JA. Process intensification: Transforming Chemical Engineering. Chemical Engineering Progress 2000, January, 22–34.
[3] García-Serna J, Pérez-Barrigón L, Cocero MJ. New trends for design towards sustainability in chemical engineering: Green engineering. Chemical Engineering Journal 2007, 133(1–5), 7–30.
[4] Li J, Huang W, Ge W. Multilevel and multiscale PSE: Challenges and opportunities at mesoscales. Computer Aided Chemical Engineering 2018, 44, 11–19.
[5] Tian Y, Demirel SE, Hasan MMF, Pstikopoulos EN. An overview of process systems engineering approaches for process intensification: state of the art. Chemical Engineering and Processing: Process Intensification 2018, 133, 160–210.
[6] Reay D, Ramshaw C, Harvey A. Chapter 5 – Reactors. Process Intensification – Engineering for Efficiency, Sustainability and Flexibility. Butterworth-Heinemann 2008, 103–186.
[7] Fernandes del Pozo D, Van Daele T, Van Hauwermeiren D, Gernaey KV, Nopens I. Quantifying the importance of flow maldistribution in numbered-up microreactors. Computer Aided Chemical Engineering 2016, 38, 1225–1230.
[8] Yao X, Zhang Y, Du L, Liu J, Yao J. Review of the applications of microreactors. Renewable and Sustainable Energy Reviews 2015, 47, 519–539.
[9] Lawrence J, O'Sullivan B, Lye GJ, Wohlgemuth R, Szita N. Microfluidic multi-input reactor for biocatalytic synthesis using transketolase. Journal of Molecular Catalysis B: Enzymatic 2013, 95, 111–117.
[10] Zhang X, Liu H, Samb A, Wang G. CFD simulation of homogeneous reaction characteristics of dehydration of fructose to HMF in micro-channel reactors. Chinese Journal of Chemical Engineering 2018, 26(6), 1340–1349.
[11] Ying Y, Chen G, Zhao Y, Li S, Yuan Q. A high throughput methodology for continuous preparation of monodispersed nanocrystals in microfluidic reactors. Chemical Engineering Journal 2008, 135(3), 209–215.
[12] Patil P, Khainar G, Naik J. Preparation and statistical optimization of Losartan Potassium loaded nanoparticles using Box Behnken factorial design: Microreactor precipitation. Chemical Engineering Research and Design 2015, 104, 98–109.
[13] Gioria E, Wisniewski F, Gutierrez L. Microreactors for the continuous and green synthesis of palladium nanoparticles: Enhancement of the catalytic properties. Journal of Environmental Chemical Engineering 2019, 7(3), 103136.

[14] Wang F, Huang J, Xu J. Continuous-flow synthesis of azo dyes in a microreactor system. Chemical Engineering and Processing: Process Intensification 2018, 127, 43–49.
[15] Kawakami K, Ueno M, Takei T, Oda Y, Takahashi R. Application of a Burkholderia cepacia lipase-immobilized silica monolith micro-bioreactor to continuous-flow kinetic resolution for transesterification of (R, S)-1-phenylethanol. Process Biochemistry 2012, 47(1), 147–150.
[16] Abdallah R, Fumey B, Meille V, de Bellefon C. Micro-structured reactors as a tool for chiral modifier screening in gas–liquid–solid asymmetric hydrogenations. Catalysis Today 2007, 125(1–2), 34–39.
[17] Cui X, Yao D, Li H, Yang J, Hu D. Nano-magnetic particles as multifunctional microreactor for deep desulfurization. Journal of Hazardous Materials 2012, 205–206, 17–23.
[18] Grobler A, Levanets O, Whitney S, Booth C, Viljoen H. Rapid cell lysis and DNA capture in a lysis microreactor. Chemical Engineering Science 2012, 81, 311–318.
[19] Ju Y, Maruta K. Microscale combustion: Technology development and fundamental research. Progress in Energy and Combustion Science 2011, 37(6), 669–715.
[20] Yue J. Multiphase flow processing in microreactors combined with heterogeneous catalysis for efficient and sustainable chemical synthesis. Catalysis Today 2018, 308, 3–19.
[21] Jin S, Shenglong Z, Kai W, Yundong W. Synthesis of million molecular weight polyacrylamide with droplet flow microreactors. Journal of Taiwan Institute of Chemical Engineers 2019 98, 78–84.
[22] Tanimu A, Jaenicke S, Alhooshani K. Heterogeneous catalysis in continuous flow microreactors: A review of methods and applications. Chemical Engineering Journal 2017, 327, 792–821.
[23] Kolb G, Hessel V, Cominos V, Hofmann C, Löwe H, Nikolaidis G, Zapf R, Ziogas A, Delsman ER, de Croon MHJM, Schouten JC, de la Iglesia O, Mallada R, Santamaria J. Selective oxidations in micro-structured catalytic reactors – For gas-phase reactions and specifically for fuel processing for fuel cells. Catalysis Today 2007, 120(1), 2–20.
[24] Kolb G. Review: Microstructured reactors for distributed and renewable production of fuels and electrical energy. Chemical Engineering and Processing: Process Intensification 2013, 65, 1–44.
[25] Banerhee R, Kumar SPJ, Mehendale N, Sevda S, Garlapati VK. Intervention of microfluidics in biofuel and bioenergy sectors: Technological considerations and future prospects. Renewable and Sustainable Energy Reviews 2019, 101, 548–558.
[26] Lutze P. PSE Tools for Process intensification. Computer Aided Chemical Engineering 2015, 37, 35–40.
[27] Seelam PK, Huuhtanen M, Keiski RL. Microreactors and membrane microreactors: fabrication and applications. In: Basile A., ed. Handbook of membrane reactors: Reactors type and industrial applications. Cambridge, UK, Woodhead Publishing, 2013, 188–235.
[28] Erickson D. Towards numerical prototyping of labs-on-chip: modeling for integrated microfluidic devices. Microfluidics and Nanofluidics 2005, 1, 301–318.
[29] Solehati N, Bae J, Sasmito AP. Geometrical optimization of micro-mixer with wavy structure design for chemical processes using Taguchi method. In: Azevedo A., ed. Advances in Sustainable and Competitive Manufacturing Systems. Springer, 2013, 1173–1184.
[30] Mies MJM, Rebrov EV, de Croon MHJM, Schouten JC. Design of a molybdenum high throughput microreactor for high temperature screening of catalytic coatings. Chemical Engineering Journal 2004, 101, 225–235.
[31] Sasmito AP, Kurnia JC, Mujumdar AS. Numerical evaluation of transport phenomena in a T-junction microreactor with coils of different configurations. Industrial & Engineering Chemistry Research 2012, 51, 1970–1980.
[32] An H, Li A, Sasmito AP, Kurnia JC, Jangam SV, Mujumdar AS. Computational fluid dynamics (CFD) analysis of micro-reactor performance: Effect of various configurations. Chemical Engineering Science 2012, 75, 85–95.

[33] Kurnia JC, Sasmito AP, Birgersson E, Shahim T, Mujumdar AS. Evaluation of mass transport performance in heterogeneous gaseous in-plane spiral reactors with various cross-section geometries at fixed cross-section area. Chemical Engineering and Processing: Process Intnsification 2014, 82, 101–111.
[34] Shaker M, Ghaedamini H, Sasmito AP, Kurnia JC, Jangam SV, Mujumdar AS. Numerical investigation of laminar mass transport enhancement in heterogeneous gaseous microreactors. Chemical Engineering and Processing: Process Intensification 2012, 54, 1–11.
[35] Bawornruttanaboonya K, Devahastin S, Mujumdar AS, Laosiripojana N. A computational fluid dynamic evaluation of a new microreactor design for catalytic partial oxidation of methane. International Journal of Heat and Mass Transfer 2017, 115, 174–185.
[36] Fang WF, Yang JT. A novel microreactor with 3D rotating flow to boost fluid reaction and mixing of viscous fluids. Sensors and Actuators B:Chemical 2009, 140, 629–642.
[37] Alam A, Kim KY. Analysis of mixing in a curved microchannel with rectangular grooves. Chemical Engineering Journal 2012, 181, 708–716.
[38] Wang D, Zhang T, Yang Y, Tang S. Simulation and design microreactor configured with micromixers to intensify the isobutene/1-butene alkylation process. Journal of the Taiwan Institute of Chemical Engineers 2019, 98, 53–62.
[39] Grundtvig IPR, Daugaard AE, Woodley JM, Gernaey KV, Krühne U. Shape optimization as a tool to design biocatalytic microreactors. Chemical Engineering Journal 2017, 322, 215–223.
[40] Fowlkes WY, Creveling CM. Engineering methods for robust product design: Using Taguchi methods in technology and product development, Massachusetts, NE, USA, Addison-Wesley Publishing Company, 1995.
[41] Han W, Charoenwat R, Dennis BH. Numerical investigation of biodiesel production in capillary microreactor. Proceedings of the ASME 2011 International Design Engineering Technical Conferences and Computers and Information in Engineering Conference. Washington, DC, USA, 2011, 28–31.
[42] Santana HS, Silva JL, Taranto OP. Numerical simulation of mixing and reaction of Jatropha curcas oil and ethanol for synthesis of biodiesel in micromixers. Chemical Engineering Science 2015, 132, 159–168.
[43] Santana HS, Silva JL, Taranto OP. Numerical simulations of biodiesel synthesis in microchannels with circular obstructions. Chemical Engineering and Processing: Process Intensification 2015, 98, 137–146.
[44] Santana HS, Tortola DS, Silva JL, Taranto OP. Biodiesel synthesis in micromixer with static elements. Energy Conversion and Management 2017, 141, 28–39.
[45] Cheng X, Chen R, Zhu X, Liai Q, He X, Li S, Li B. Optofluidic membrane microreactor for photocatalytic reduction of CO2. International Journal of Hydrogen Energy 2016, 41(4), 2457–2465.
[46] Sun J, Ju J, Ji L, Zhang L, Zu N. Synthesis of Biodiesel in Capillary Microreactors. Industrial and Engineering Chemistry Research 2008, 47(5), 1398–1403.
[47] Richard R, Thiebaud-Roux S, Prat L. Modelling the kinetics of transesterification reaction of sunflower oil with ethanol in microreactors. Chemical Engineering Science 2018, 87, 258–269.
[48] Schwarz S., Borovinskaya ES, Reschetilowski W. Base catalyzed ethanolysis of soybean oil in microreactors: Experiments and kinetic modeling. Chemical Engineering Science 2013, 104, 610–618.
[49] Rahimi M, Aghel B, Alitabar M, Sepahvand A, Ghasempour HR. Optimization of biodiesel production from soybean oil in a microreactor. Energy Conversion and Management 2014, 79, 599–605.

[50] Rashid WNWA, Uemura Y, Kusakabe K, Osman NB, Abdullah B. Synthesis of Biodiesel from Palm Oil in Capillary Millichannel Reactor: Effect of Temperature, Methanol to Oil Molar Ratio, and KOH Concentration on FAME Yield. Procedia Chemistry 2014, 9, 165–171.
[51] Azam NAM, Uemura Y, Kusakabe K, Bustam MA. Biodiesel Production from Palm Oil Using Micro tube Reactors: Effects of Catalyst Concentration and Residence Time. Procedia Engineering 2016, 148, 354–360.
[52] Santana HS, Silva Jr JL, Tortola DS, Taranto OP. Transesterification of sunflower oil in microchannels with circular obstructions. Chinese Journal of Chemical Engineering 2018, 26(4), 852–863.
[53] Pontes PC, Chen K, Naveira-Cotta CP, Costa Junior JM, Tostado CP, Quaresma JNN. Mass transfer simulation of biodiesel synthesis in microreactors. Computers and Chemical Engineering 2016, 93(4), 36–51.
[54] Pontes PC, Naveir-Cotta CP, Quaresma JNN. Three-dimensional reaction-convection-diffusion analysis with temperature influence for biodiesel synthesis in micro-reactors. International Journal of Thermal Sciences 2017, 118, 104–122.
[55] Costa Jr JM, Naveira-Cotta CP. Estimation of kinetic coefficients in micro-reactors for biodiesel synthesis: Bayesian inference with reduced mass transfer model. Chemical Engineering Research and Design 2019, 141, 550–565.
[56] Santana HS, Silva Jr JL, Taranto OP. Development of microreactors applied on biodiesel synthesis: From experimental investigation to numerical approaches. Journal of Industry and Engineering Chemistry 2019, 69, 1–12.
[57] Mohadesi M, Aghel B, Maleki M, Ansari A. Production of biodiesel from waste cooking oil using a homogeneous catalyst: Study of semi-industrial pilot of microreactor. Renewable Energy 2019, 136, 677–682.
[58] Joshi N, Lawal A. Hydrodeoxygenation of acetic acid in a microreactor. Chemical Engineering Science 2012, 84, 761–771.
[59] Qiao K, Shi X, Zhou F, Chen H, fu J, Ma, Huang H. Catalytic fast pyrolysis of cellulose in a microreactor system using hierarchical zsm-5 zeolites treated with various alkalis. Applied Catalysis A: General 2017, 547, 274–282.
[60] Gutiérrez-Antonio C, Romero-Izquierdo AG, Gómez-Castro FI, Hernández S. Energy integration of a hydrotreatment process for sustainable biojet fuel production. Industrial and Engineering Chemistry Research 2016, 55(29), 8165–8175.
[61] Abe JO, Popoola API, Ajenifuja E, Popoola OM. Hydrogen energy, economy and storage: Review and recommendation. International Journal of Hydrogen Energy 2019, 44, 15072–15086.
[62] Yang G, Wang J. Various additives for improving dark fermentative hydrogen production: A review. Renewable and Sustainable Energy Reviews 2018, 95, 130–146.
[63] Rashwan SS, Dincer I, Mohany A, Pollet BG. The Sono-Hydro-Gen process (Ultrasound induced hydrogen production): Challenges and opportunities. International Journal of Hydrogen Energy 2019, 44(29), 14500–14526.
[64] Reverberi AP, Klemeš JJ, Varbanov PS, Fabiano B. A review on hydrogen production from hydrogen sulphide by chemical and photochemical methods. Journal of Cleaner Production 2016, 136 Part B, 72–80.
[65] Sharma A, Arya SK. Hydrogen from algal biomass: A review of production process. Biotechnology reports 2017, 15, 63–69.
[66] Kumar SS, Himabindu V. Hydrogen production by PEM water electrolysis – A review. Materials Science for Energy Technologies 2019, 2(3), 442–454.
[67] Zhang Y, Xu P, Liang S, Liu B, Shuai Y, Li B. Exergy analysis of hydrogen production from steam gasification of biomass: A review. International Journal of Hydrogen Energy 2019, 44(28), 14290–14302.

[68] Ahmed M, Dincer I. A review on photoelectrochemical hydrogen production systems: Challenges and future directions. International Journal of Hydrogen 2019, 44(5), 2474–2507.
[69] Izquierdo U., Barrio VL, Lago N, Requies J, Cambra JF, Güemez MB, Arias PL. Biogas steam and oxidative reforming processes for synthesis gas and hydrogen production in conventional and microreactor reaction systems. International Journal of Hydrogen 2012, 37 (8), 13829–13842.
[70] Sanz O., Echave FJ, Romero-Sarria F, Odriozola JA, Montes M. Chapter 9 – Advances in Structured and Microstructured Catalytic Reactors for Hydrogen Production. Renewable Hydrogen Technologies – Production, Purification, Storage, Applications and Safety 2013, 201–224.
[71] Yan Y, Zhang Z, Zhang L, Wang X, Liu K, Yang Z. Investigation of autothermal reforming of methane for hydrogen production in a spiral multi-cylinder micro-reactor used for mobile fuel cell International Journal of Hydrogen Energy 2015, 40(45), 1886–1893.
[72] Enterazy B, Kazemeini M. Improved H2 production from the APR of polyols in a microreactor utilizing Pt supported on a CeO2Al2O3 structured catalyst. International Journal of Hydrogen Energy 2018, 43(48), 21777–21790.
[73] Delparish A, Avci AK. Modeling of intensified glycerol steam reforming in a heat-exchange integrated microchannel reactor. Catalysis today 2018, 299, 328–338.
[74] Ke Y, Zhou W, Chu X, Yuan D, Wan S, Yu W, Liu Y. Porous copper fiber sintered felts with surface microchannels for methanol steam reforming microreactor for hydrogen production. International Journal of Hydrogen Energy 2019, 44(12), 5755–5765.
[75] Liu Z, Chu B, Zhai X, Jin Y, Cheng Y. Total methanation of syngas to synthetic natural gas over Ni catalyst in a micro-channel reactor. Fuel 2012, 95, 599–605.
[76] Bawornruttanaboonya K, Devahastin S, Mujumdar AS, Laosiripojana N. Comparative evaluation of autothermal reforming of biogas into synthesis gas over bimetallic NiRe/Al2O3 catalyst in fixed-bed and coated-wall microreactors: A computational study. International Journal of Hydrogen Energy 2018, 43(29), 13237–13255.
[77] ANAQUA, AcclaimIP Patent Search & Analytics Software. (Accessed May 29, 2019, at http://www.acclaimip.com/)
[78] Warren, L.R. Apparatus for obtaining X-ray diffraction patterns. Patent number US2483500 A. Application number US76179147A. U. S. 4 Oct 1949.
[79] Suryawanshi PL, Gumfekar SP, Bhanvase BA, Sonawane SH, Pimplapure MS. A review on microreactors: Reactor fabrication, design, and cutting-edge applications. Chemical Engineering Science 2018, 189, 431–448.
[80] Timmons, R.B., et al. Catalytic hydrodehalogenation of polyhalogenated hydrocarbons. Patent number 5276240. Application number 962997. Jun 4, 1994.
[81] Jovanovic, G.N., et al. Microreactor Process for Making Biodiesel. Oregon State University. Patent number US20090165366. Application number 12/227804. USA. 2009
[82] ГАЛЫНКИН et al. Fishing vessel for the production of biodiesel from living aquatic resources. Russian Ministry of Industry & Trade. Patent number CN109621858A. Application number RU2011115028U. Russia 2012.
[83] Jianying et al. Method for continuous synthesis of biodiesel through microreactor. Dalian University of Technology. Patent number CN103103025A Application number CN20131036534 20130130. China. 2013.
[84] Dennis, B.H., et al. Methods and systems for improved biodiesel production. University of Texas System. Patent number Application number US12/556,857. USA. 2013.
[85] Zhang et al. Biodiesel adsorption deacidification and regeneration method. China Petroleum Chemical CO. Patent number CN105349260 A. Application number CN201410411863A. China 2016.

[86] Zhang et al. A method for adsorbing deacidification of biodiesel. China Petroleum Chemical CO. Patent number CN104560388 B. Application number CN201310503049A. China 2017.
[87] Zhang et al. Biodiesel adsorption deacidification method. China Petroleum Chemical CO. Patent number CN105273838 A. Application number CN201410283294A. China 2019.
[88] Li et al. Method for synthesizing and separating biodiesel by using catalyst wall-loading microreactor and micro-separator with difference in interface properties. Fuzhou University. Patent number CN109621858A. Application number CN201811590844A. China 2016.
[89] Rubens, M.F., et al. Process for the production of biokerosene by means of an integrated route, and biokerosenes thus produced. UNICAMP Patent number WO/2013/13889. Application number BR2013/000087. 2013.
[90] Rubens, M.F., et al. Process biokerosene production in integrated and route bioquerosenes so obtained. UNICAMP Patent number BR132012032606 E2. Application number BR132012032606A. 2014.
[91] Rubens, M.F., et al. Process of obtaining bioquerosene and bioquerosene so obtained. UNICAMP Patent number BRC10803465 F1. Application number BRC10803465-6. 2019.
[92] Adler, B. Method for the continuous production of biomethanol diesel. Dracowo Forschungs & Entwicklungs Gmbh. Patent number EP1576078A2. Application number EP20030714685. Germany. 2007.
[93] Dunkel J, and Adler, B. Production of biomethanol and bioethanol diesel comprises continuous re-esterification, removing glycerin after dosing the catalyst-alkanol mixture with removal of methanol in the biodiesel, and washing with phosphoric acid and water. Dracowo Forschungs & Entwicklungs Gmbh. Patent number DE10043644A1. Application number DE2000143644. Germany. 2002.
[94] Chen et al. Packed bed microreactor for preparing ethylene by dehydration of bioethanol. Dalian Chemical Physics Inst. Patent number CN102068950 A. Application number CN200910220045A. China 2011.
[95] Piotr, D. et al. Catalytic method for producing fuel biocomponent from bioethanol. Politechnika Lodzka. Patent number PL408327 A1. Application number PL40832714A. Poland 2017.
[96] Reuters. Micro Reactor Technology Market Global Segmentation 2019 by Size Estimation, Industry Share, Top Key Players and Top Regions Forecast to 2022. (Accessed Jun 6, 2019, at https://www.reuters.com/brandfeatures/venture-capital/article?id=82704)

Martín Picón-Núñez, Jorge Luis García-Castillo
and Jorge C. Melo-González

3 Heat transfer enhancement technologies for improving heat exchanger performance

Abstract: The enhancement of heat transfer consists of the augmentation of the rate of heat transfer to achieve a fixed heat duty in a smaller surface area within the limitations of pressure drop. The main feature of a suitable application is characterized by the maximization of the heat transfer with a minimum increase of pressure drop. Broadly speaking, heat transfer can be enhanced via the increase of fluid velocity, using new heat transfer surfaces or using turbulence promoters. This chapter describes each of these alternatives. It starts by analyzing the fundamental principles of single-phase heat transfer and the relation between velocity, heat transfer coefficient, and pressure drop. Next, the advent of new and compact heat transfer technologies is discussed, then the use of turbulence promoters and their thermohydraulic performance is covered. Finally, the extension to heat recovery networks and the application to multistream heat exchanger technology is described.

Keywords: Enhanced surfaces, heat transfer augmentation, turbulence promoters, thermohydraulic model, performance comparison

3.1 Introduction

The drive for increasing heat recovery in the process industry, especially in highly thermal energy intensive processes is the key for the development of new heat transfer technologies [1]. The main feature of new technologies is that for the same fluid velocities, higher heat transfer rates are obtained at the expense of pressure drop [2]. So, new heat exchanger technology incorporates the design of new surfaces that create high local turbulence without a substantial increment of the demand for pumping power. New surface design is commonly characterized by a large hydraulic diameter and small flow passages [3].

The area density of a heat exchanger is a term that indicates how much surface area is available for heat transfer within a certain volume [4]. The higher this ratio, the more compact the exchanger is. Therefore, the level of compactness of a heat exchanger is related to how much surface area can be fitted within a unit volume. With this concept in mind, the way surface area can be increased in a heat exchanger

Martín Picón-Núñez, Jorge Luis García-Castillo, Jorge C. Melo-González, Department of Chemical Engineering, University of Guanajuato, Mexico

https://doi.org/10.1515/9783110596120-003

is by means of extended surfaces or fins or by using corrugated surfaces. A slightly different approach consists of the use of turbulence promoters [5].

The use of extended heat transfer surfaces gives rise to what is known as compact heat exchanger. For a certain velocity, these exchangers exhibit high heat transfer coefficients with relatively low pressure drops [4]. In the case of conventional heat exchangers, such as the shell and tube exchanger, recent techniques for heat transfer enhancement are the use of internal inserts or turbulence promoters. Whatever the technique, the purpose of using secondary surfaces or inserts is to create local turbulence maintaining the fluid velocity almost unchanged. Pressure drop is increased mainly due to friction, which becomes a suitable trade-off between pumping power and heat transfer.

Heat transfer enhancement techniques are important since they have been developed to be an aid to achieve one of the following two exchanger design goals [6]:
a) Design a heat exchanger with the smallest surface area for the given heat duty and pressure drop
b) Increase the heat recovery capacity of an existing heat exchanger with fixed dimensions and pressure drop

Heat transfer enhancement techniques can be used at a grassroot design stage or as a retrofit strategy [7]. In both cases, knowledge of the thermohydraulic performance of the extended surfaces is fundamental for the sizing of the unit. Thermohydraulic performance refers to the way the heat transfer coefficient and the pressure drop behave as a function of the fluid velocity. In terms of dimensionless numbers, the thermohydraulic performance is represented by the variation of the Nusselt number and the friction factor as a function of the Reynolds number. When a new extended surface is produced, the thermohydraulic performance can be determined either by experimental or numerical methods. Since the validation of numerical methods depends on the availability of experimental data, most thermohydraulic performance published to date is in the form of empirical or semiempirical correlations.

Heat transfer enhancement can be achieved via active or passive methods [8]. Active methods use external power consumption, whereas passive methods consist of the modification of the heat transfer surface or the use of mechanical inserts. The physics behind heat transfer enhancement is explained using the boundary layer theory. In single-phase heat transfer, apart from fouling, the main resistance to heat transfer is caused by the boundary layer, where heat transfer takes places mainly by conduction. If this boundary layer could be removed, the rate of heat transfer would increase. One of the mechanisms available to disrupt the boundary layer is through the creation of local turbulence. Passive enhancement techniques fulfill this task by disrupting or even removing the boundary layer to bring about a rapid increase on the heat transfer coefficient [9]. New exchanger technologies are designed bearing this concept in mind so that in most cases, they include new surface geometries. In the case of existing conventional technology, such as the shell and tube heat exchangers which are still the main type of equipment in use today, turbulence promoters have been developed to achieve this boundary layer removal. Important to mention is that the local turbulence created by the extended surfaces or the turbulence promoters acts as a continuous force that not only removes the film of stagnant fluid attached to a surface but it also removes particles of sedimented material on the walls. Thus, apart from increasing the heat transfer performance of a surface, heat transfer enhancement also brings benefits in terms of fouling mitigation [10].

This chapter discusses the thermal and hydraulic principles of heat transfer enhancement, looks at the development of compact exchanger technology, and addresses the use of turbulence promoters in shell and tube heat exchangers.

3.2 Thermal and hydraulic principles

A design parameter that strongly determines the thermal and hydraulic performance of a single-phase heat exchanger is fluid velocity. The dependence of heat transfer and friction factor on velocity becomes evident in the expressions for the Nusselt number and the pressure drop, as shown by Eqs. (3.1) and (3.2):

$$Nu = function(velocity) \tag{3.1}$$

$$f = function(velocity) \tag{3.2}$$

In most heat transfer extended surfaces or inserts, the rate of growth of the pressure drop with velocity is larger than the rate of rise of heat transfer. Therefore, pressure drop imposes a limit to the extent to which velocity can be increased to achieve improved heat transfer capacity. The heat transfer coefficient is obtained from the expression of the Nusselt number and expressed as a function of velocity in Eq. (3.3). The pressure drop is a function of the friction factor and the mass flux (mass flow rate per unit sectional area) and can be expressed as a function of velocity as given by Eq. (3.4). Therefore, since the heat transfer coefficient and the pressure drop are both functions of velocity, these two terms can be related by means of the same design variable. This is shown in Eq. (3.5).

$$h = function(v) \tag{3.3}$$

$$\Delta P = function(v) \tag{3.4}$$

$$h = function\ (\Delta p) \tag{3.5}$$

The design of a new heat exchanger or the retrofit of an existing one can be conducted based on the relation described by Eq. (3.5). In the design of a heat exchanger, velocity can be increased in two ways: (a) by reduction of the cross-sectional area and (b) by the increase of the number of passes.

In most heat exchanger geometries, the pressure drop grows faster with velocity than the heat transfer coefficient. If F_t is the rate of change of velocity between the new condition (v^N) and original condition (v^0), then the rate of growth of heat transfer is determined by Eq. (3.7).

$$F_t = \frac{v_t^N}{v_t^0} \tag{3.6}$$

$$\frac{h_t^N}{h_t^0} = (F_t)^n \tag{3.7}$$

Where n is the exponent of the Reynolds number in the Nusselt correlation. Similarly, for the pressure drop it can be shown that:

$$\frac{\Delta P_t^N}{\Delta P_t^0} = (F_t)^m \tag{3.8}$$

Where m is the exponent of the Reynolds number in the expression for pressure drop. For a bare tube, the exponents n and m are 0.8 and 2, respectively. The rate at which the heat transfer coefficient and the pressure drop grow is graphically depicted in Figure 3.1. As can be seen, pressure drop grows more rapidly than heat transfer coefficient. The maximum rate of velocity change that can be accepted depends on the value of the maximum pressure drop that can be allowed. For instance, if the original pressure drop is 10 kPa, and the allowable value is 50 kPa, velocity can be increased 2.23 times for pressure drop to fall within an acceptable

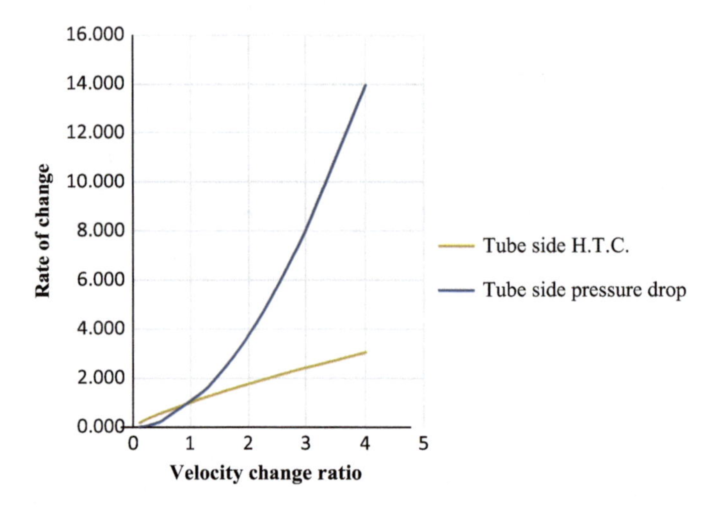

Figure 3.1: Tube side change on heat transfer coefficient and pressure drop as a function of velocity change.

value. This indicates that while pressure drop increases five times, the heat transfer coefficient barely doubles its value. From Figure 3.1 it can also be observed that the opposite effect takes place as velocity reduces; this is, pressure drop falls more rapidly than the heat transfer coefficient. This feature can be taken advantage of in the case of the retrofit of multipass shell and tube exchangers. As mentioned earlier, a way of increasing heat recovery is by means of turbulence promoters. As the tubes of the unit are fitted with such elements, the heat transfer coefficient rises, and so does the pressure drop. In such case, the number of passes could be reduced, thus reducing the velocity. Despite this velocity reduction, the heat transfer is still larger than the original, due to the enhanced effect of the inserts while the pressure drop falls to values lower than the original. Illustrative examples showing this application are presented in Section 3.4.

3.3 Extended surfaces

The basic principle behind the design of a new heat transfer surface is: maximize the local turbulence with the lowest increase of pressure drop. Since pressure drop becomes a determining factor in heat transfer enhancement, it is important to analyze it in more detail. The total pressure loss experienced by a fluid at its pass through a heat exchanger is made up of the following main components: (a) pressure losses due to changes of direction, (b) pressure losses due to changes of velocity, (c) pressure losses due to change in fluid density, and (d) pressure losses due to friction. Out of these four types of pressure loss, the one that is directly related to heat transfer is the pressure loss due to friction. In fact, strictly speaking, the term ΔP in Eq. (3.5) refers to it. The capacity of a plain heat transfer surface to convert friction energy into heat transfer is relatively poor; however, if the surface is modified so that local turbulence is increased, the heat transfer capacity is also increased. In design, care must be taken so that the fluid does experience abrupt changes of direction to avoid pressure losses that do not contribute to heat transfer. Common types of heat transfer surfaces are: flat plates and tubular plates. The modifications to these types of surfaces are described below.

3.3.1 Flat plate surfaces

Flat plate surfaces can be modified by means of corrugations or by the incorporation of secondary surfaces. When corrugations are used, technologies such as the plate and frame and compabloc exchangers have emerged. Figure 3.2 shows a typical plate with a chevron corrugation. The main geometrical features of such plate are: (a) Plate length (L), (b) plate width (W), (c) Chevron angle (β), and (d) port

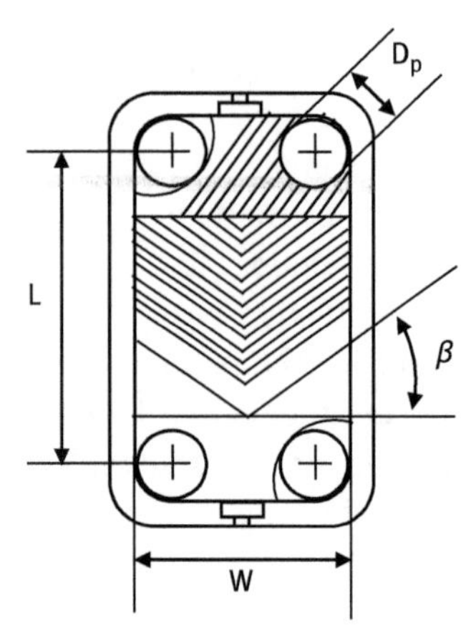

Figure 3.2: Heat transfer surface of a plate and frame heat exchanger. Corrugated plate with a Chevron angle (β).

diameter (D_p). Typical commercial dimensions are given in Table 3.1. It is important to mention that this technology presents some operational situations that reduce its thermal effectiveness. One is flow maldistribution, which is the result of unequal pressure drop distribution between the channels of the heat exchanger. A second situation is the effect created by end plate [11] and by the intermediate plates in the case of units with multiple passes [12].

Table 3.1: Typical plate dimensions.

Dimension	Range
Length, L (m)	0.5–3.0
Width W, (m)	0.2–1.5
Port diameter, Dp (m)	0.254–0.4
Chevron angle, β	25–65
Plate spacing, b (mm)	1.5–5.0
Plate thickness, τ (mm)	0.5–1.2

A technology variation that gives the exchanger improved resistance to operation under high pressures and temperatures is the compabloc heat exchanger. This unit consists of the stack of corrugated plates welded and covered by a highly resistant flat shell. Multipass units are typical flow configurations. In each pass, fluids move in cross-flow fashion with respect to the other. Figure 3.3 shows the internal array of the corrugated plates.

When in a flat surface the flow direction is modified in a soft manner, the constant direction change creates sheer forces that tend to remove the stagnant layer attached to the walls, hence increasing the heat transfer coefficient. An example of such type of geometry is the spiral heat exchanger (SHE) shown in Figure 3.4.

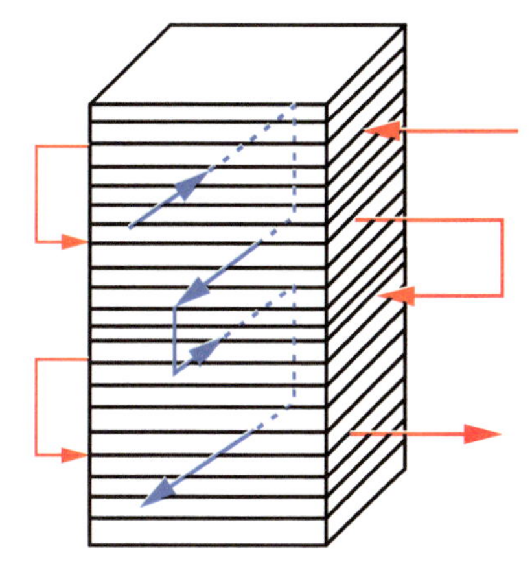

Figure 3.3: The compabloc exchanger consists of the assembly of square corrugated plates.

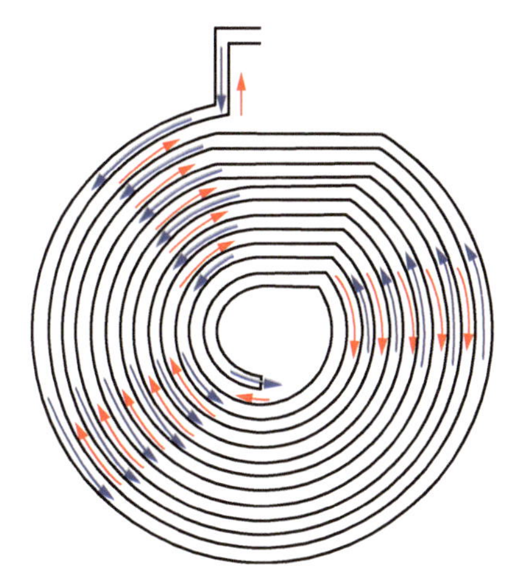

Figure 3.4: Spiral heat exchanger.

Spiral heat exchangers are especially suitable for applications with highly viscous and fouling prone fluids. In this type of geometry, the fluid undergoes a continual change of direction as it flows through the channels of the unit as depicted in Figure 3.4. This flow feature reduces fouling since solid particles are kept in suspension and removed from the surface [13].

From the geometrical point of view, the main variables of a spiral heat exchanger are: plate spacing of the two streams (δ_1 and δ_2), plate width (W), inner diameter (D_i), outer diameter (D_s), and plate length (L), as shown in Figure 3.5.

The thermohydraulic performance of a SHE depends strongly on the average curvature of the unit. The curvature is represented by the Dean number (K) [14, 15]. The heat transfer coefficient increases with the curvature; this effect is more significant in laminar flow regimes than for turbulent regimes [16].

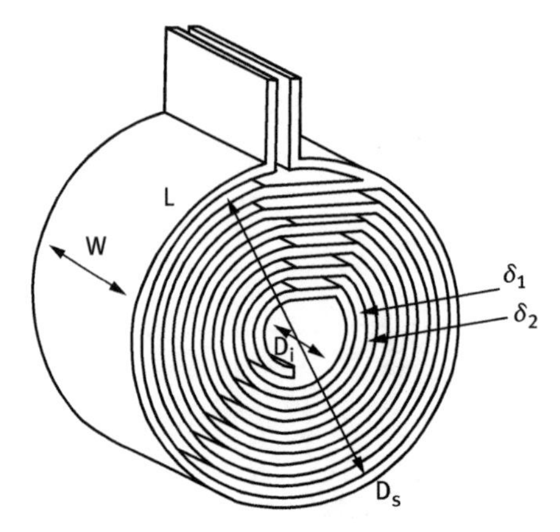

Figure 3.5: Geometry of a spiral heat exchanger.

3.3.2 Secondary surfaces

Another way of intensifying the heat transfer performance of flat surfaces consists in the incorporation of extended or secondary surfaces. This type of technology is well represented by the plate and fin exchanger shown in Figure 3.6.

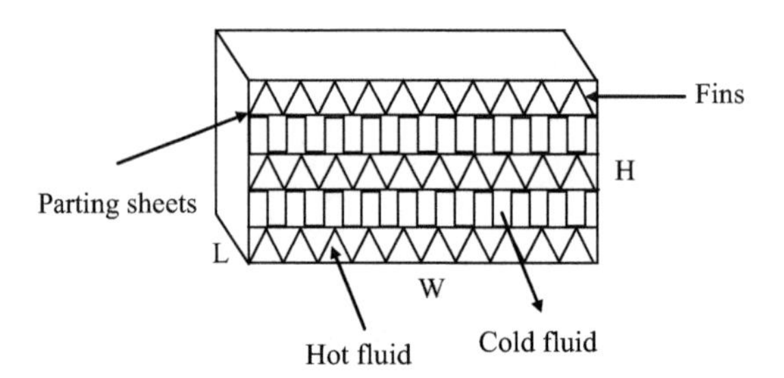

Figure 3.6: Plate and fin heat exchanger technology.

There are many types of secondary surfaces available for this type of exchanger [17]. The most common types geometries are (Figure 3.7): Louvered fin, triangular plain-fin, rectangular-plain fin, off set strip fin, and wavy fin.

Given the wide range of surfaces available for design, a selection tools becomes necessary when sizing a unit. This is of paramount importance if the exchanger geometry is limited to a confined space. When highly performing secondary surfaces are used, they result in exchangers with large frontal area and short flow length. On the other hand, lower performing surfaces deliver long flow lengths and smaller frontal areas. One way of comparing the thermohydraulic performance of secondary surfaces is by means of the Volume Performance Index [18].

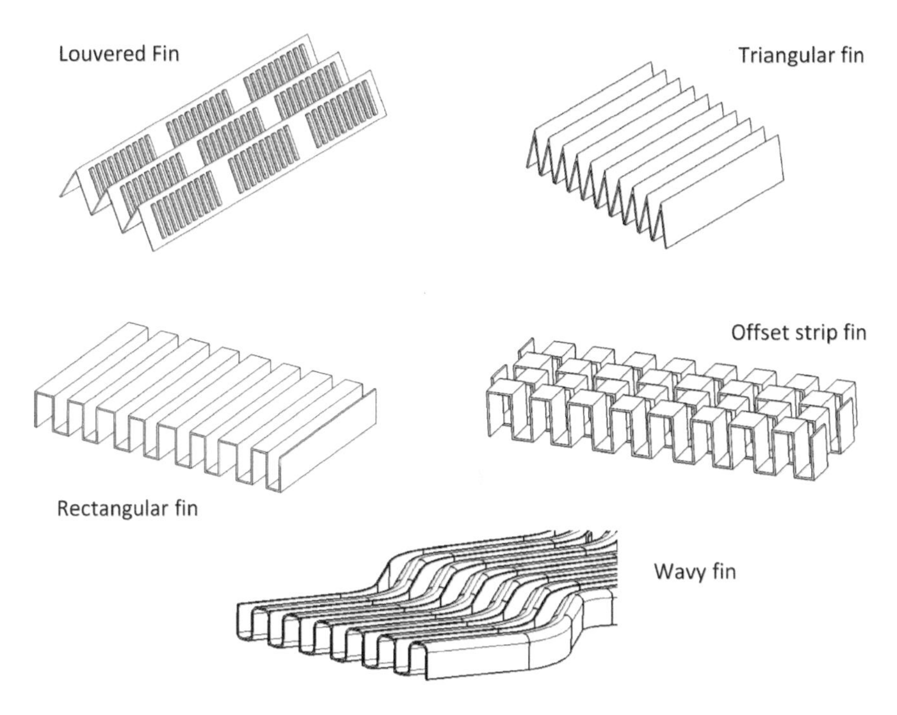

Figure 3.7: Most common types of secondary surfaces used in plate and fin heat exchangers.

$$VPI = \frac{\left(\frac{St^3}{f}\right)^{1/2}}{d_h} \tag{3.9}$$

Where f is the friction factor, d_h the hydraulic diameter, and St the Stanton number given by:

$$St = \frac{hA_c}{mC_p} \tag{3.10}$$

The term h is the film heat transfer coefficient, A_c is the free flow area, m is the mass flow rate, and C_p is the fluid heat capacity.

3.3.3 Extended surfaces for tubes

Depending on what side of a tube the fluid with the lowest heat transfer characteristics flows on, the heat transfer performance can be enhanced on the inside, on the outside, or simultaneously on both sides. Heat transfer can be augmented via the attachment of fins. Typical fin geometries are shown in Figure 3.8.

The twisted tube is another effective way of intensifying the thermohydraulic performance of tubes (Figure 3.9). In this geometry, the inner and outer fluids are forced

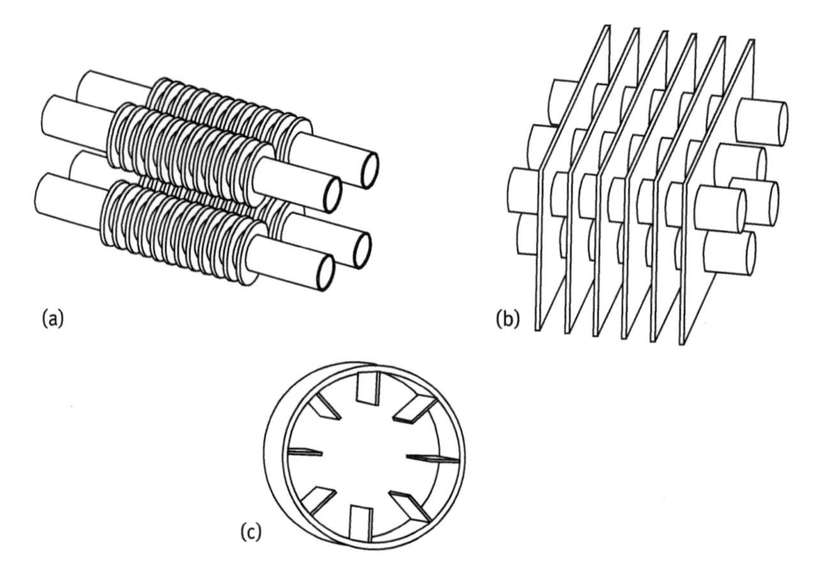

Figure 3.8: Extended surfaces in tubular heat exchanger applications: (a) external individual fins; (b) external common fins; (c) internal fins.

Figure 3.9: The twisted tube technology increases the heat transfer coefficient as the fluid moves in a swirl flow motion.

to flow in a swirling flow motion that results in the removal of the boundary layer with the subsequent increase of the heat transfer coefficient with minimal increase of the pressure drop. This geometry involves both, the internal and the external fluid.

In conventional shell and tube exchanger technology, the baffle design also offers an important area of opportunity for improvement. Baffles with surfaces that force the fluid to change direction and velocity abruptly should be avoided. A novel modification to the typical segmental baffle is the helical one. This geometry produces a more uniform flow distribution for the same pressure drop [19]. Although the manufacturing costs of this type of baffle are higher, the lower maintenance costs in the long term makes the investment worth [20]. As shown in Figure 3.10, the principal geometrical dimensions are: helical pitch, P_{hel} (distance between two consecutive baffles); the helical angle, β (the angle formed between the helix and the vertical), and the shell diameter, D_i [21].

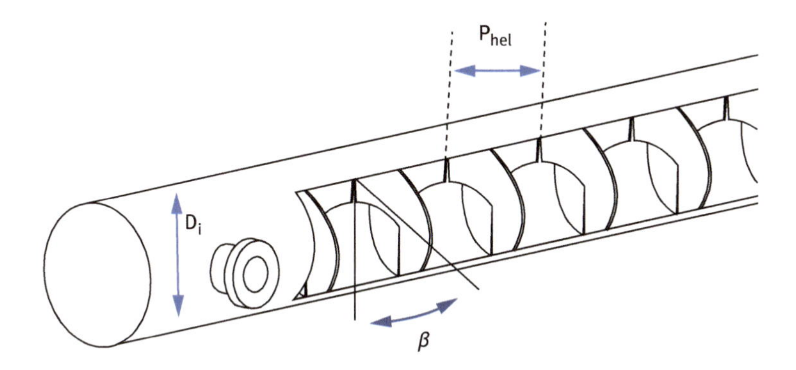

Figure 3.10: Features of helical baffles in tubular heat exchangers.

The helix angle that gives the highest thermohydraulic performance is 40° according to Zhang et al. [21]. With respect to segmental baffles, it has been experimentally and numerically demonstrated that the heat transfer rate of a helical baffle is superior in a range 9%–23% [19, 21]. Variations of the helical baffle include the use of multiple shell passes [22]. The thermohydraulic performance of helical baffles as a function of the helical angle in the form of correlations obtained from experimental data has been reported by Gao et al. [23]. These correlations are reproduced in Table 3.2.

Table 3.2: Correlations for the thermohydraulic performance as a function of the helical angle.

Correlation	Error (%)	Angle (β)
$Nu_s = 0.1259Re^{0.6063}$	1.23	8°
$Nu_s = 0.1692Re^{0.5945}$	1.03	12°
$Nu_s = 0.3406Re^{0.5032}$	0.68	20°
$Nu_s = 0.3708Re^{0.5122}$	0.81	30°
$Nu_s = 0.3063Re^{0.5636}$	0.83	40°
$f_s = 0.1893Re^{-0.4684}$	2.18	8°
$f_s = 0.1779Re^{-0.4686}$	3.64	12°
$f_s = 0.1392Re^{-0.4432}$	3.02	20°
$f_s = 0.1821Re^{-0.5192}$	1.65	30°
$f_s = 0.0928Re^{-0.4331}$	2.07	40°

Based on the concept of maximizing the allowable pressure drop, a shortcut design approach for helical baffles shell and tube exchangers has been derived. To illustrate the benefits of the use of these baffles with respect to conventional segmental baffles, the design results of a case study are reproduced here from [24]. Table 3.3 shows the operating conditions and geometrical data and the final design results are shown in Table 3.4. It can be appreciated that for the same heat duty, the helical baffle exchanger consumes 28.5% less pressure drop and requires 27.1% less surface area compared with a segmental heat exchanger.

Table 3.3: Operating conditions.

	Kerosene	Crude oil
Mass flow rate (kg/h)	19,867.24	67584.91
Inlet temperature (°c)	198.9	37.78
Outlet temperature (°c)	93.3	76.67
Allowable pressure drop (pa)	68,947.6	68,947.6
Volumetric flow rate (m3/h)	24.4	79.1
Fouling factor (m2°c/w)	0.00015	0.00015
Geometrical data for segmental baffle exchanger		
Inner diameter (m)	0.54	0.0205
Outer diameter (m)	0.57	0.0254
Pitch (m)		0.0318
Tube thickness		13 bwg
Tube arrangement		square 45°

Table 3.4: Results of the design of a helical heat exchanger.

	Segmental baffles	Helical Baffles
Heat transfer area (m^2)	61.5	44.8
Heat duty (W)	1,494,662.67	1,494,662.67
Tube length (m)	4.877	4.6
Overall heat transfer coefficient (W/m^2K)	317.98	524.48
Shell side pressure drop (Pa)	68,947.5	29,350.3
Tube side pressure drop (Pa)	68,947.5	49,264.5
Baffle angle	–	15°
Number of baffles	36	36
Shell side heat transfer coefficient (W/m^2K)	919.88	878.79
Tube side heat transfer coefficient (W/m^2K)	687.07	3,015.3
No. shell passes	1	2
No. of tube passes	4	2
No. of tubes	158	61

3.4 Turbulence promoters for tubular geometries

Since a large majority of existing heat exchangers in the processing industry are of the shell and tube type, turbulence promoters for increased heat transfer performance have been extensively developed over the past years. One of the most successful turbulence promoter in use in industry is the wire matrix type. Among this type of geometry, HiTRAN is the technology that stands out [25]. Figure 3.11 shows the mechanical construction of this type of insert.

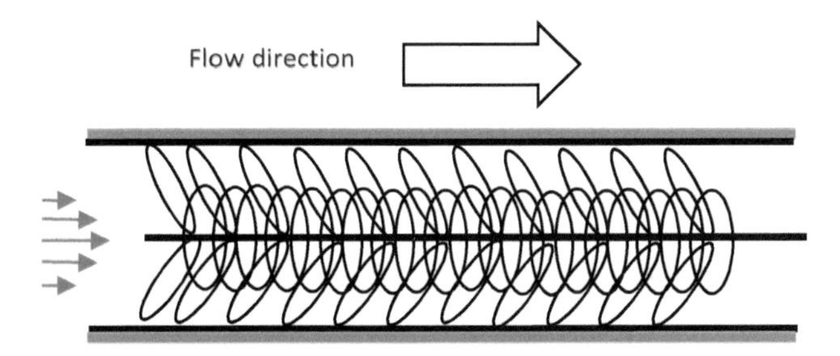

Figure 3.11: Wire matrix turbulence promoter (HiTRAN).

A wire matrix turbulence promoter takes advantage of the velocity profile inside the tube (Figure 3.11). The operation is as follows: as the fluid in the central part of the tube hits the central part of the insert, where the matrix is denser than the sections near the walls, the fluid is redirected in the radial direction to the walls where the fluid in the stagnant laminar sublayer is replaced by high velocity fluid. In so doing, the heat transfer coefficient increases substantially in a range that goes from 30% to 500%.

The thermohydraulic performance of turbulence promoters is generally presented in the form of potential expressions as the ones shown below.

$$f = aRe^{-b} \tag{3.11}$$

$$j = xRe^{-y} \tag{3.12}$$

Where j is the Colburn factor defined as (St $Pr^{2/3}$) where St is the Stanton number and Pr is the Prandtl number. The terms a, b, x, and y are the coefficients and exponents. They are obtained from curve fitting to experimental data. Table 3.5 shows the values of exponents y coefficients reported in the literature for various types of inserts [9].

The application of these passive intensification methods has become a very effective tool to achieve two purposes: (a) the reduction of the size of new shell and tube heat exchangers and (b) the increase of the heat transfer capacity of existing units.

Table 3.5: Exponents and coefficients for some turbulence promoters.

Insert type	a	b	x	y
Twisted Tape S&L H/D 8.5	0.2192	0.3166	0.1078	0.3203
Twisted Tape S&L H/D 3.62	0.5821	0.3787	0.3423	0.4206
Coil Wire H/D 5.5	0.2276	0.2953	0.0964	0.3000
Coil Wire H/D 2.8	0.2514	0.2345	0.1221	0.3000
Coil wire H/D 1.12	0.2615	0.1568	0.1682	0.3000

3.4.1 Benefits of the use of turbulence promoters. Illustrative example 1

A four-tube-pass heat exchanger of the shell and tube type is designed for the process data shown in Table 3.6. For the specified heat load, the exchanger required area is 17.14 m^2. A new exchanger for the same application is designed using Twisted Tapes S&L H/D 3.62 (see Table 3.5). The maximum permitted pressure drop on the cold side is 12 kPa. When inserts are installed, the surface area is 14.64 m^2 and pressure drop increases to 14.93 kPa. Since the pressure drop is larger than the permitted values, the design to consider is to reduce the number of tube-passes from 4 to 2. Table 3.7 shows the comparative results. After reducing the number of passes, the resulting surface area (16.22 m^2) is still smaller than the design without inserts and the pressure

Table 3.6: Operating conditions and physical properties for the illustrative example.

	Cold stream (Shell-side)		Hot stream (Tube-side)
Inlet temperature, °C	15		35
Outlet temperature, °C	20		25
Mass flow rate, kg/s	10		5.6
Pressure drop, kPa	18		12
Density, kg/m^3	1000		980
Viscosity, kg s/m^2	0.0001019		0.0001019
Thermal conductivity, W/m °C	0.620		0.630
Heat capacity, J/kg °C	4,186		3,558.38
Thermal conductivity of material, W/m °C		86	
Shell diameter, m		0.3084	
Baffle spacing, m		0.20	
Tube inner diameter, m		0.0175	
Number of tubes		109	
Tube thickness, m		0.001524	
Fouling factor, °C m^2/W		0.00052	

Table 3.7: Comparison of design results for the design of a four-tube pass exchanger with and without inserts and with reduced number of passes.

	Design bare tube	Design with inserts	Reducing the number of passes with inserts
Surface td32#area,td46#m^2	17.14	14.64	16.22
No. of passes	4	4	2
Tube length, m	2.6	2.2	2.5
Heat transfer coefficient (tube-side), W/m^2 °C	3,317.1	5,910.4	3,955.5
Heat transfer coefficient (shell-side) W/m^2 °C	7,634.0	7,634.0	7,634.0
Pressure drop (tube-side), kPa	10.25	14.93	2.79
Pressure drop (shell side), kPa	17.83	16.96	17.83

drop is now even smaller than the allowable value (2.79 kPa). Consequently, the design using inserts and with reduced number of passes is still a better option.

3.4.2 Benefits of the use of turbulence promoters. Illustrative example 2

A different type of situation arises when it is desired to increase the heat recovery capacity of an existing exchanger. The initial design with bare tubes of Table 3.8 is considered for this study. If the geometry of the exchanger is kept without changes and turbulence promoters are installed, the thermo-hydraulic performance of the exchanger is modified. The five different types of inserts of Table 3.5 are installed and the resulting exchanger performance is displayed in Table 3.8. As the performance of the shell-side remains unchanged, only the changes in the tube-side are presented.

Table 3.8: Thermohydraulic performance of an existing heat exchanger with different type of inserts.

Type of turbulence promoter	Heat load (kW)	Tube-side heat transfer coefficient (W/m^2°C)	Tube-side pressure drop (kPa)
Bare tube	198.89	3,317.1	10.25
Twisted Tape S&L H/D 8.5	213.59	4,887.2	12.57
Twisted tape S&L H/D 3.62	218.37	5,910.4	16.96
Coil Wire H/D 5.5	215.78	5,313.4	15.18
Coil Wire H/D 2.8	221.22	6,729.9	27.12
Coil wire H/D 1.12	227.03	9,270.8	55.96

A turbulence promoter increases the heat transfer capacity with a corresponding increase in pressure drop. The type of design determines the rate at which these parameters grow for a given application. It is desirable that the heat transfer be enhanced more than the pressure drop. Table 3.8 indicates that all the five turbulence promoters tested result in increased heat duty, however, Twisted Tape S&L H/D 8.5 results in a pressure drop that is slightly over the permitted valued. All other inserts result in higher pressure drops. For practical purposes, the use of Twisted Tape S&L H/D 8.5 is acceptable.

In recent years, the number of different turbulence promoters that have emerged is so large that it is necessary to develop performance comparison methods so that in design, the best choice can be made at a glance. One of the most widely used performance comparison methods is presented in the section below.

3.4.3 Thermo-hydraulic performance comparison of turbulence promoters

In the search of improved thermo-hydraulic performance, a wide variety of turbulence promoters have been developed. Examples are: twisted tapes [26–30], winglet tapes [31, 32], circular rings [28, 33], horseshoe baffles [34], helical inserts [35], and coil wires [36]. The focus of research on these types of devices is centered around geometry variations that result in higher heat transfer rates and lower pressure drops. Bhuiya et al. [35] modified the design of the twisted tape by including different size perforations along the tape. Murugusen et al. [29] incorporated pins to the twisted tapes. Other design variations consist of inserts formed by circular rings combined with twisted tapes [28]. Additionally, the twisted tape insert has been modified including v-cuts along its length [29]. Other variations of the twisted tape have been studied by Wongcharee and Eiamsa-ard [30].

The thermo-hydraulic performance of turbulence promoters can be compared using the Thermal Enhancement Factor (η) that is defined as the ratio between the heat transfer coefficient (h_p) of the tube with the insert and that of the bare tube (h_s). This parameter allows to identify the level of improvement in heat transfer and pressure drop and provides a specific means for comparison and selection purposes. It is determined under the condition that the pumping power between the bare tube and the tube with inserts are the same. The term is expressed as:

$$\eta = \left.\frac{h_p}{h_s}\right|_{pp} = \frac{\left(\frac{Nu_P}{Nu_S}\right)}{\left(\frac{f_P}{f_S}\right)^{\frac{1}{3}}} \tag{3.13}$$

It has been observed that values of the coefficient of thermal improvement equal to 1 or above indicate a good performance. Table 3.9 shows a sample of the many geometries reported in the open literature and the calculated values of η.

Table 3.9: Geometrical features of some turbulence promoters reported in the literature.

Turbulence promoter configuration	Name	Reynolds range	Nu_p/Nu_s	f_p/f_s	TEF (η)
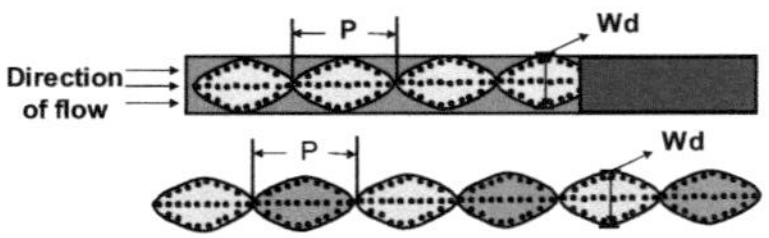 Direction of flow; P; Wd; P; Wd	Perforated twisted tape[26]	7,200–49,800	2.07–4-4	2.1–4.6	1.28–1.59
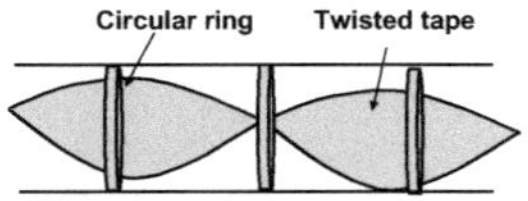 Circular ring; Twisted tape	Circular-rings and twisted tapes[19]	6,000–20,000	CR alone 2.36–2.80 CR and TT 2.44–4.70	CR alone 7.4–13.75 CR and TT 11.94–35.83	CR alone 1.09–1.25 CR and TT 1.04–1.42
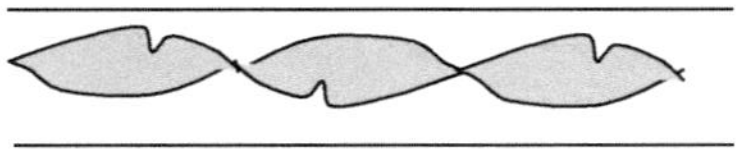	V-Cut twisted tape [20]	2,000–20,000	1.36–2.46	2.49–5.82	1.07–1.27
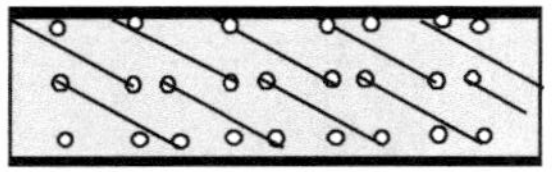	Staggered-winglet perforated-tapes (WPT) and Staggered-winglet tape (WTT) [23]	4,180–26,000	4,180–26,000	WPT 2.39–4.78 WTT 4.63–4.90	WPT 4.87–42.69 WTT 33.46–49.80
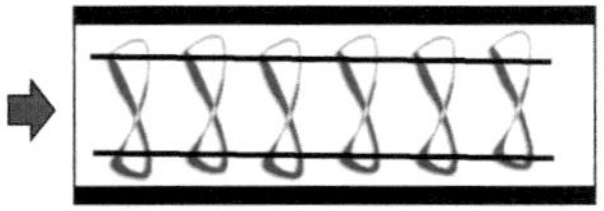	V-Shapped rings [24]	5,000–25,000	2.47–5.77	6.57–82.01	1.15–1.63

3.5 Heat transfer intensification in heat recovery networks

The heat recovery duties in an entire processing plant is carried out through a set of heat exchangers. Consider the flow diagram of the process shown in Figure 3.12. The heating and cooling needs of the process are carried out by a network of heat exchangers. A heat exchanger network is integrated by hot and cold streams, process-to-process heat exchangers, coolers, and heaters [37]. Methodologies for the design of heat recovery networks include those that are based on basic thermodynamic principles to the more elaborated mathematical optimization approaches [38]. The basic principle behind the design of a heat recovery network is the construction of the Composite Curves from the process stream data (mass flow rates, supply and target temperatures, and heat capacities). A Composite Curve shows that for a given amount of heat recovery, determined by the minimum temperature approach (ΔT_{min}), there exists a minimum external hot and cold requirement for the entire process (Figure 3.13). The Composite Curves represent the heat and mass balance of the process.

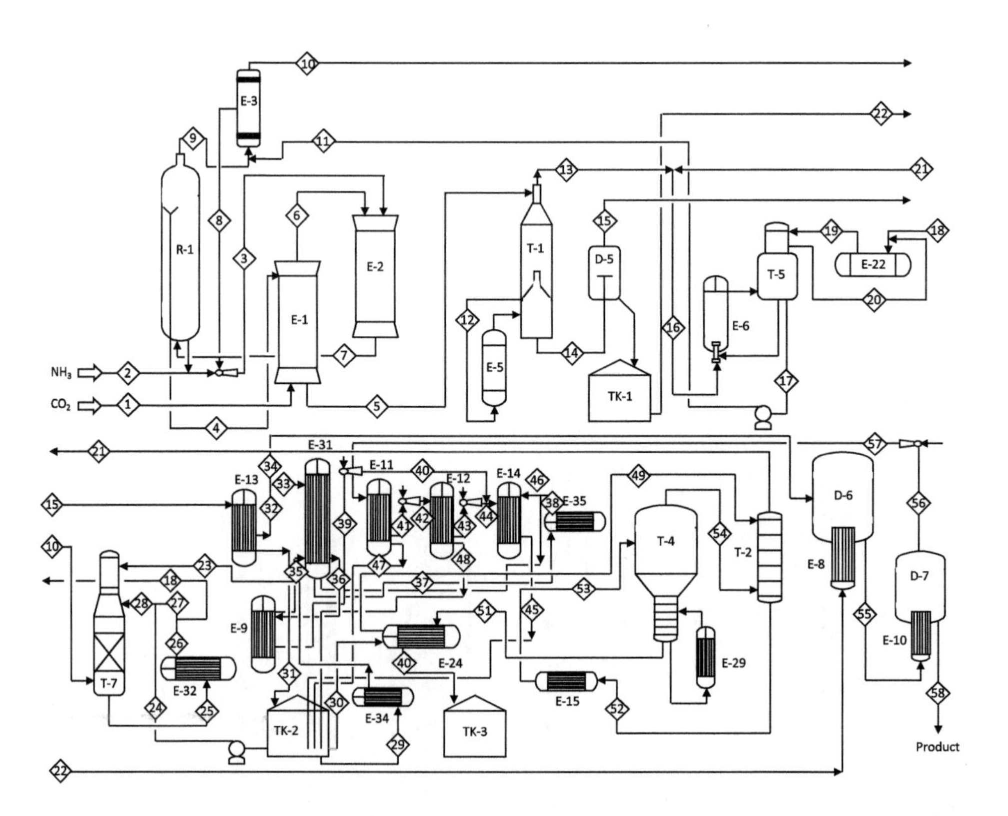

Figure 3.12: Flow diagram of a process plant.

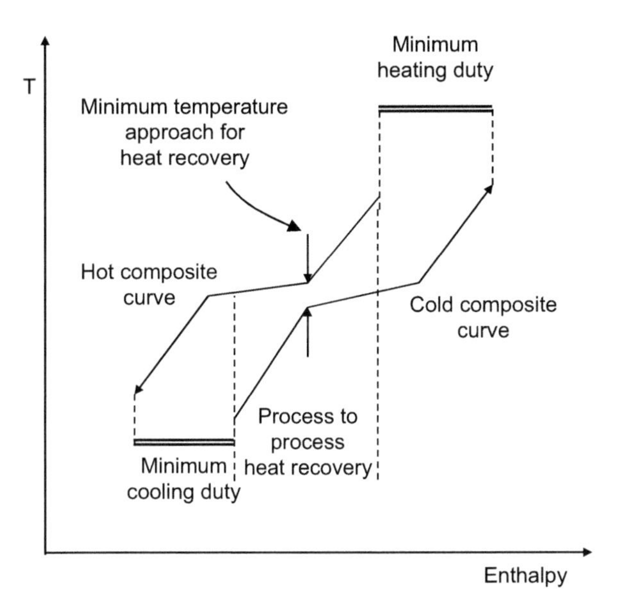

Figure 3.13: Composite Curves for process heat recovery.

The actual heat recovery network of the process plant of Figure 3.12 is shown in Figure 3.14. In this type of representation, process-to-process heat exchangers are represented as two circles connected by a vertical line linking a hot stream with a cold stream; coolers and heaters are displayed by a circle with the letter C or H, respectively. Some heat exchanger technologies such as the plate and fin and the plate and frame exhibit geometrical features that make them suitable for handling more than two different hot and cold streams in the same unit at the same time. A heat exchanger that can meet the thermal needs of more than two streams in the same unit is referred to as a multistream heat exchanger. Figure 3.15 shows the schematic of such a system. In principle, a multistream heat exchanger could fit the whole heat recovery network in a single structure. For design purposes, the Composite Curves provide a clear illustration of the entry and exit points of the various streams within a multistream exchanger. The sections identified as enthalpy intervals represented by a vertical line drawn at every point where the slope of the curves change is related to the inlet or outlet of a stream as shown in Figure 3.16. Here the diagram shows a four-stream problem: two hot streams (A and B) and two cold streams (C and D).

Compact heat exchangers of the plate and fin and plate and frame technology are suitable for multistream applications. The principles for the design of these types of units are provided in [39, 40]. Figure 3.17 illustrates the construction features of plate and fin exchangers extended to accommodate three hot and three cold streams. The stream population per enthalpy interval of a case study in Figure 3.18 shows the

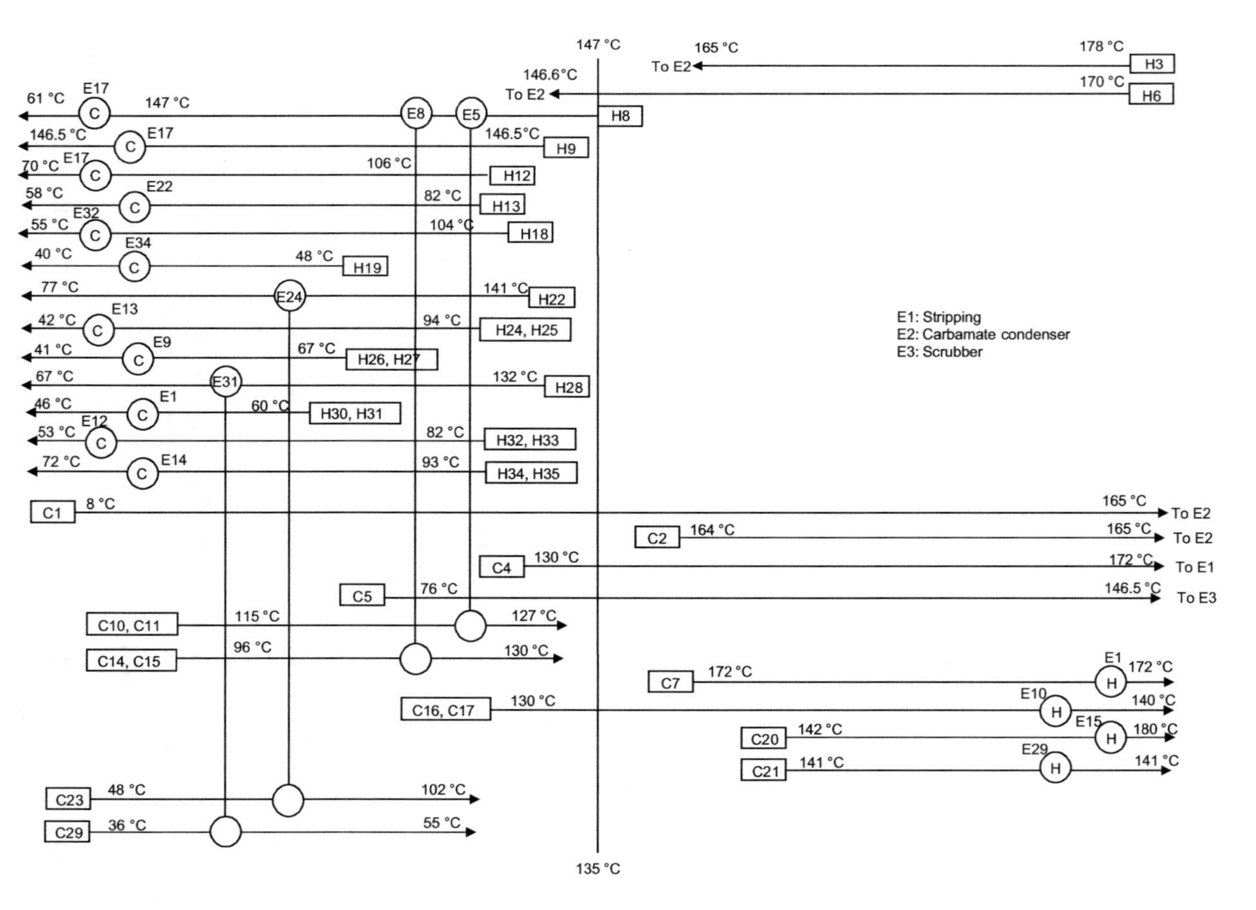

Figure 3.14: Graphical representation of a heat recovery network.

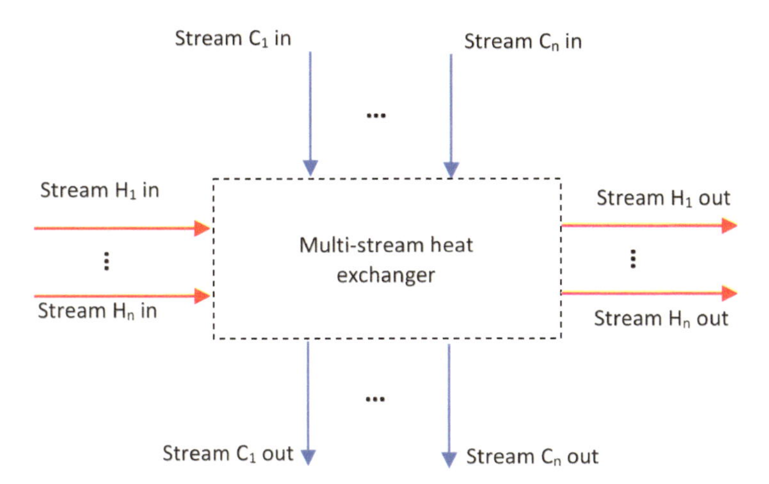

Figure 3.15: Schematic representation of a multistream heat exchanger.

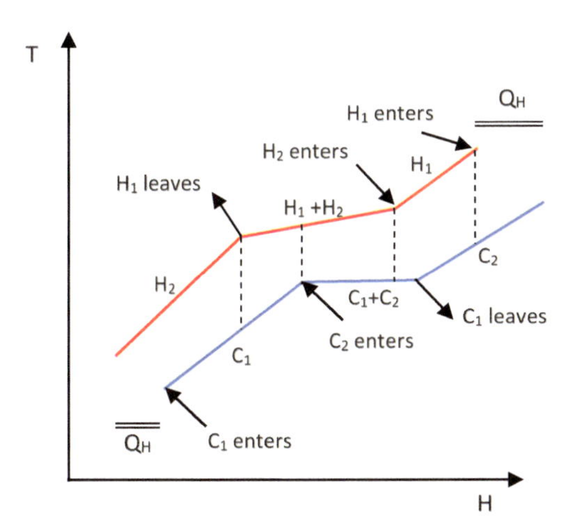

Figure 3.16: Enthalpy intervals in the Composite Curves illustrate the entry and exit point of the process streams in a multistream heat exchanger.

way separate blocks, each one per enthalpy interval, are formed and put together to integrate the whole multistream structure.

Plate and frame exchangers lend themselves for reducing the number of heat exchangers in networks (Figure 3.19). They are appropriate for accommodating up to three streams in the same frame (Figure 3.20). Applications with more than three streams are possible but the piping interconnections become rather intricate. Figure 3.21 shows the case of a six-stream application.

The first applications of multistream heat exchangers appeared in low-temperature processes such as the production of natural gas or air separation

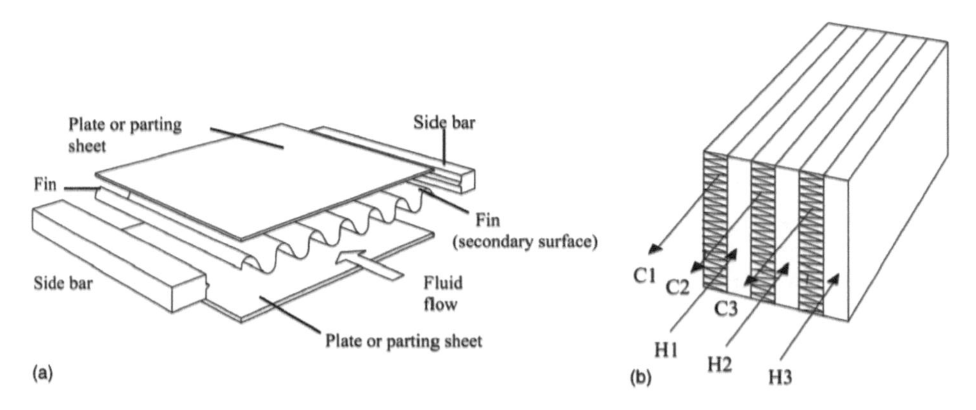

Figure 3.17: Plate and fin technology: (a) basic construction, (b) extension to multistream units.

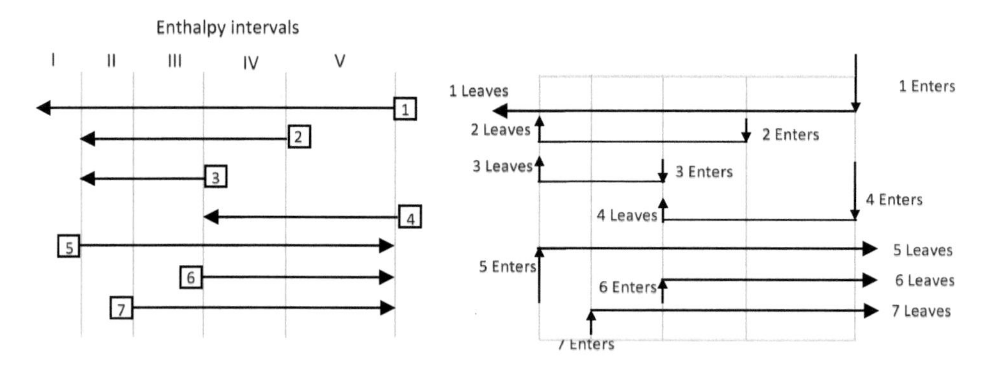

Figure 3.18: Use of enthalpy intervals from Composite Curves to determine the entry end exit points in a multistream heat exchanger.

processes [41]. In above temperature applications, the first attempts to move toward multistream units was on offshore oil platforms [42]. Three aspects are drawing the attention of research in this area: the design of the headers for appropriate flow distribution, the flow passage arrangement [43], and the potential operating problems multistream exchangers are bound to encounter [44]. To date, very few applications of multistream exchangers have been reported; however, the potential benefits in terms of installation costs and space are evident. Future plant construction processes should consider this heat transfer process option.

3.6 Conclusions

Heat transfer enhancement technology has been developed to achieve either of the two following goals: (1) obtain the smaller possible heat exchanger for a given heat

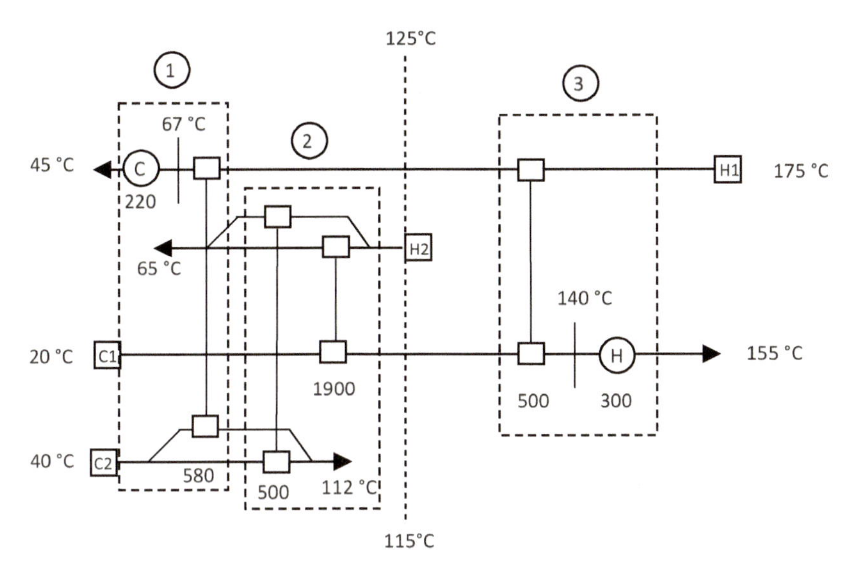

Figure 3.19: Use of multistream exchangers to reduce the number of units in a heat recovery network.

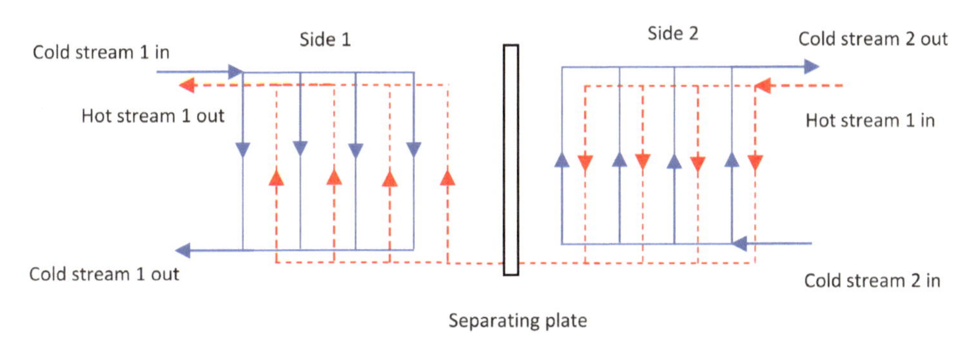

Figure 3.20: Plate and frame arrangement to fit a three-stream heat recovery application.

load and within the restriction of pressure drop or (2) obtain an increased heat duty within the limitations of pressure drop with the same installed equipment. From a physical point of view, heat transfer enhancement involves the removal of the laminar sublayer attached to the surface, where heat transfer takes place by conduction. This task can be achieved by passive mechanisms that include extended surfaces, fins, or turbulence promoters. Passive heat transfer enhancement techniques can be applied in grassroot design or in retrofit applications.

In single-phase heat transfer enhancement techniques, velocity is an important design variable. The relationship is: the higher the velocity, the higher the heat transfer coefficient. At the design stage, this parameter can be manipulated by appropriate choice of the exchanger cross-sectional area and number of passes.

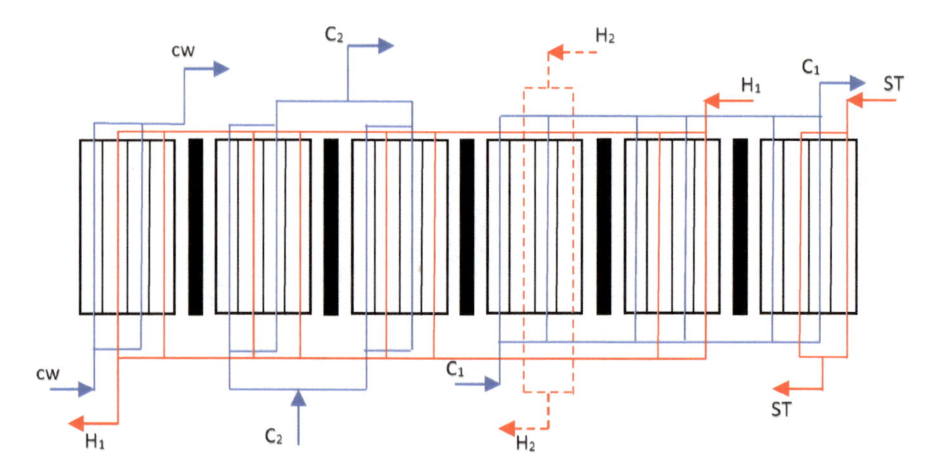

Figure 3.21: Extending the plate and frame geometry to a six-stream heat recovery problem.

Heat exchanger technology is moving in the direction of more compact surfaces. Such surfaces exhibit higher heat transfer rate with a moderate increase of pressure drop. Despite the thermohydraulic virtues of compact surfaces, most technologies are still limited to operate under moderate conditions of temperature and pressure. Such limitations can only be overcome with the development of new materials of construction.

Heat transfer enhancement can also be approached from a system´s point of view. A heat exchanger network is a set of heat exchangers that carries out the thermal duty of a whole process. Traditionally, it is formed by heat exchangers that perform a heat transfer duty between a single hot and single cold stream. New heat exchanger technology, such as the compact geometries, have the capability of fitting more than one hot and more than one cold stream into a single exchanger giving rise to multistream heat exchangers. This feature brings the possibility of performing the required thermal duty of a whole process in less number of units, subject to geographic and safety restrictions. The potential application of multistream heat exchangers is large. Design approaches seem to be mature enough to undertake this task.

Nomenclature

Ac	Free flow area
a	Coefficient in friction factor Eq. (3.11)
b	Plate spacing (mm); exponent in friction factor Eq. (3.11)
C_p	Heat capacity mass flow rate
D_i	Inner diameter (m)

D_s Outer diameter (m)
D_p Port diameter
d_h Hydraulic diameter (m)
h Heat transfer coefficient
F_t Ratio of original condition to new condition
f Friction factor
j Colburn factor
L Plate length (m)
m Mass flow rate; exponent of the Reynolds number in the expression for pressure drop
N Number of thermal plates
Nu Nusselt number
n Exponent of the Reynolds number in the Nusselt correlation
P_{hel} Helical pitch
Pr Prandtl number
Re Reynolds number
St Stanton number
VPI Volume performance index
v Fluid velocity (m/s)
W Plate width (m)
x Coefficient in Colburn factor Eq. (3.12)
y Exponent in Colburn factor Eq. (3.12)

Greek Letters

β Helical baffle angle; Chevron angle
δ Plate spacing
η Thermal enhancement factor
φ Area enlargement factor
ΔP Pressure drop (Pa)

Subscripts

p Bare tube
t Tube side
s Shell side, tube with inserts

Superscripts

N New condition
0 Original condition

References

[1] Gugulothua R, Somanchia N S, Kumar Reddya K V, Akkirajub K. A Review on Enhancement of Heat Transfer in Heat Exchanger with Different Inserts. Materials Today: Proceedings 2017, 4, 1045–1050.
[2] Kumar B, Srivastava G P, Kumar M, Kumar Patil A. A review of heat transfer and fluid flow mechanism in heat exchanger tube with inserts. Chemical Engineering & Processing: Process Intensification 2018,123, 126–137.
[3] Zhang J, Zhu X, Mondejar M E, Haglind F. A review of heat transfer enhancement techniques in plate heat exchangers. Renewable and Sustainable Energy Reviews 2019, 101, 305–328.
[4] Hesselgreaves J E. Compact Heat Exchanger. Selection, Design and Operation. First Edition, Pergamon: Oxford, UK, 2001.
[5] Goel V, Garg M O, Nautiyal H, Khurana S, Shukla M K. Heat transfer augmentation using twisted tape inserts: A review. Renewable and Sustainable Energy Reviews 2016, 63, 193–225.
[6] Smith R, Pan M, Butalov I. Chapter 32: Heat transfer enhancement in heat exchanger networks. In: Handbook of Process Integration: Minimization of Energy and Water Use, Waste and Emissions, ed. Jiri Klemes. Cambridge, UK, Woodhead Publishing Limited, 2103, 966–1037.
[7] Gundersen T. Chapter 8: Analysis and design of heat recovery systems for grassroot and retrofit situations, 2013. In: Handbook of Process Integration: Minimization of Energy and Water Use, Waste and Emissions, ed. Jiri Klemes. Cambridge, UK, Woodhead Publishing Limited, 2013, 262–309.
[8] Tabish A, Man-Hoe K. A comprehensive review on single phase heat transfer enhancement techniques in heat exchanger applications. Renewable and Sustainable Energy Reviews 2018,81, 813–839.
[9] Polley G T, Reyes-Athie C M, Gough M. Use of heat transfer enhancement in process integration. Heat Recovery Systems & Combined Heat and Power 1992, 12, 3, 191–202.
[10] Pan M, Bulatov I, Smith R. Improving heat recovery in retrofitting heat exchanger networks with heat transfer intensification, pressure drop constraint and fouling mitigation. Applied Energy 2016, 161, 611–626.
[11] Gut J A W, Pinto J M. Modeling of plate heat exchangers with generalized configurations. International Journal of Heat and Mass Transfer 2003, 46, 2571–2585.
[12] Prabhakara B R, Krishna P K, Sarit K D. Effect of flow distribution to the channels on the thermal performance of a plate heat exchanger. Chemical Engineering and Processing 2002, 41, 49–58.
[13] Wilhelmsson B. Consider spiral heat exchangers for fouling applications. Hydrocarbon Processing July 2005, 81–83.
[14] Minton P E. Designing spiral heat exchangers, Chemical Engineering May 1970, 4, 103–112.
[15] Martin H. Heat Exchangers. New York, USA. Hemisphere Publishing Corporation, 1992.
[16] Egner M W, Burmeister L C. Heat Transfer for Laminar Flow in Spiral Ducts of Rectangular Cross Section. Journal of Heat Transfer 2005, 27, 352–356.
[17] Kays W M, London A L. Compact Heat Exchangers. New York, USA. 3rd ed. McGraw-Hill, 1984.
[18] Picon-Núñez M, Polley G T, Torres-Reyes E, Gallegos-Muñoz A. Surface selection and design of plate-fin heat exchangers. Applied Thermal Engineering 1999, 19, 917–931.
[19] Sivarajan C, Rajasekaran B, Krishnamohan N. Numerical and experimental study of helix heat exchanger. International Journal of Engineering Research and Technology 2014, 2, 4, 8–13.
[20] Movassang S Z, Taher F N, Razmi K, Azar R T. Tube bundle replacement for segmental and helical shell and tube heat exchangers. Applied Thermal Engineering 2013, 51, 1162–1169.

[21] Zhang L, Xiaa Y, Jianga B, Xiaoa X, Yang X. Pilot experimental study on shell and tube heat exchangers with small-angles helical baffles. Chemical Engineering and Processing, 2013, 69,112–118.
[22] Chen G D, Zeng M, Wang Q W, Qi S Z. Numerical studies on combined parallel multiple shell-pass shell-and-tube heat exchangers with continuous helical baffles. Chemical Engineering Transactions 2010, 21, 229–234.
[23] Gao B, Bi Q, Nie Z, Wu J. Experimental study of effects of baffle helix angle on shell-side performance and shell-and-tube heat exchangers with discontinuous helical baffles. Exp Therm Fluid Sci 2015, 68, 48–57.
[24] Picón-Núñez M, García-Castillo J L, Alvarado-Briones B. Thermo-hydraulic design of single and multi-pass helical baffle heat exchangers. Applied Thermal Engineering 2016, 105, 25, 783–791.
[25] Drögemüller P. The use of HitRAN wire matrix elements to improve the thermal efficiency of tubular heat exchangers in single and two-phase flow. Chemie-Ingenieur-Technik 2015, 87, 3, 188–202.
[26] Bhuiya M M K, Ahamed J U, Chowdhury S U, Kalam M A. Heat transfer enhancement and development of correlation for turbulent flow through a tube with triple helical tape inserts. Int Commun Heat Mass Transf 2012, 39, 94–101.
[27] Murugesan P, Mayilsamy K, Suresh S. Heat transfer and friction factor studies in a circular tube fitted with twisted tape consisting of wire-nails. Chinese J Chem Eng 2010, 18, 1038–1042.
[28] Eiamsa-Ard S, Kongkaitpaiboon V, Nanan K. Thermohydraulics of turbulent flow through heat exchanger tubes fitted with circular-rings and twisted tapes. Chinese J Chem.Eng 2013, 21, 585–593.
[29] Murugesan P, Mayilsamy K, Suresh S, Srinivasan P S S. Heat transfer and pressure drop characteristics in a circular tube fitted with and without V-cut twisted tape insert. Int Commun Heat Mass Transf 2011, 38, 329–334.
[30] Wongcharee K, Eiamsa-ard S. Heat transfer enhancement by twisted tapes with alternate-axes and triangular, rectangular and trapezoidal wings. Chem Eng Process Process Intensif 2011, 50, 211–219.
[31] Skullong S, Promvonge P, Thianpong C, Jayranaiwachira N. Thermal behaviors in a round tube equipped with quadruple perforated-delta-winglet pairs. Applied Thermal Engineering 2017, 115, 229–243.
[32] Skullong S, Promvonge P, Thianpong C, Pimsarn, M. Heat transfer and turbulent flow friction in a round tube with staggered-winglet perforated-tapes. Int J Heat Mass Transf 2016, 95, 230–242.
[33] Chingtuaythong W, Promvonge P, Thianpong C, Pimsarn M. Heat transfer characterization in a tubular heat exchanger with V-shaped rings. Applied Thermal Engineering 2017, 110, 1164–1171.
[34] Promvonge P, Tamna S, Pimsarn M, Thianpong C. Thermal characterization in a circular tube fitted with inclined horseshoe baffles. Applied Thermal Engineering 2015, 75, 1147–1155.
[35] Bhuiya M M K, Chowdhury M S U, Saha M. Heat transfer and friction factor characteristics in turbulent flow through a tube fitted with perforated twisted tape inserts. Int Commun Heat Mass Transf 2013, 46, 49–57.
[36] Gunes S, Ozceyhan V, Buyukalaca O. Heat transfer enhancement in a tube with equilateral triangle cross sectioned coiled wire inserts. Exp Therm Fluid Sci 2010, 34, 684–691.
[37] Kemp I C. Pinch Analysis and Process Integration: A User Guide on Process Integration for the Efficient Use of Energy, 2nd ed. Oxford, UK, 2007.
[38] Klemes J J, Varbanov P S, Alwi R, Manan Z A. Process intensification and integration. Saving energy, water and resources. Ed. Walter de Gruyter GmbH, Berlin/Boston, 2014.

[39] Picón-Núñez, Polley G. T, Medina-Flores M. Thermal design of multi-stream heat exchangers. Applied Thermal Engineering 2002, 22, 1643–1660.
[40] Picón-Núñez M, Martínez-Rodríguez G, López-Robles J L. Alternative design approach for multi-pass and multi-stream plate heat exchangers for use in heat recovery systems. Heat Transfer Engineering 2006, 27, 6, 12–21.
[41] Suessmann W, Mansour A. Passage arrangement in plate–fin heat exchangers. XV International Congress of Refrigeration, Venice, 1979, 1, 421–429.
[42] Taylor M A. Plate–fin exchangers offshore- the background. The Chemical Engineer June 1990.
[43] Zhe W, Li Y. Layer pattern thermal design and optimization for multi-stream plate fin exchangers- A review. Renewable and sustainable energy reviews 2016, 53, 500–514.
[44] Picón Núñez M, López Robles J L. Flow passage arrangement and surface selection in multi-stream plate-fin heat exchangers. Heat Transfer Engineering 2005, 26, 9, 5–14.

Massimiliano Errico, Claudio Madeddu and Roberto Baratti

4 Reactive absorption of carbon dioxide: Modeling insights

Abstract: Intensified processes where reaction and separation are simultaneously carried out in the same equipment represent an attractive option to substitute the traditional reaction-separation sequences. Reactive absorption is already applied in different processes at industrial scale, but its modeling still represents an open topic where researchers are contributing to improve the understanding of the phenomena involved. Aspen Plus, one of the most powerful process simulators available on the market, can be used to model such kind of complex operation, but the user must be aware of the applicability and the limitations of the correlations implemented. In the present chapter, a model for the reactive absorption of carbon dioxide in monoethanolamine is discussed and validated using Aspen Plus. A particular analysis is given for the definition of the number of segments used to describe the column packing. Often overlooked, this parameter has a primary importance for the model output.

Keywords: Reactive absorption, intensified absorption, mathematical modeling

4.1 Introduction

Chemical engineering as a profession and as an independent study program arose from the clear necessity to transfer the knowledge gained at the laboratory scale into a competitive industrial dimension. Peppas [1] summarized different historical events that characterized the establishment of chemical engineering, emphasizing how it was a changing in the needs of the society to bring the developing of this new area. Taking 1888 as reference, the year when the MIT offered the first 4-year curriculum in chemical engineering, 130 years after, the evolution of the society is perfectly mirrored in the chemical engineering content. Process system engineering, process synthesis, process optimization, process design, sustainability, process intensification, are all new concepts introduced within chemical engineering to answer specific needs or to improve existing approaches. Among all, process intensification embraces most of the challenges of the modern process industry introducing a new approach for the design of the most used unit operations. A unique definition

Massimiliano Errico, Claudio Madeddu, University of Southern Denmark, Department of Chemical Engineering, Biotechnology and Environmental Technology, Odense M, Denmark
Roberto Baratti, Universitá degli Studi di Cagliari, Dipartimento di Ingegneria Meccanica Chimica e dei Materiali, Cagliari, Italy

https://doi.org/10.1515/9783110596120-004

of process intensification is not available, and the general definition proposed by Stankiewicz and Moulijn [2] as the development of *novel equipment and techniques that potentially could transform our concept of chemical plants and lead to compact, safe, energy-efficient, and environment-friendly sustainable processes* was modified and enriched by different authors. Ponce-Ortega et al. [3] defined process intensification as an activity characterized by one or more of the following principles: smaller equipment compared to those normally used, higher process throughput using the same equipment or process, reduced holdup or inventory for the same throughput, reduced utility consumption and feedstock, and increase of the process equipment performance.

In this context, intensified operations where reaction and separation are carried out simultaneously in a single unit are the leading light among all the intensified processes. Great emphasis was given to reactive distillation due to the possibility to reach a high reactant conversion, product selectivity, and the reduction in capital and operating costs. Reactive distillation applied to biodiesel production represents a clear breakthrough in overcoming the high production costs of acid- or alkali-catalyzed processes. Niju et al. [4] reached, with a laboratory-scale reactive distillation column, about 93.5% conversion of methyl esters using waste egg shells as catalyzer. Gomez-Castro et al. [5] considered the biodiesel production with supercritical methanol using a reactive distillation column obtaining promising results regarding the energy requirement, the production cost, and the pollutant emissions. Moreover, Bildea and Kiss [6] proved how, even with less degrees of freedom, the reactive distillation column can be controllable.

Reactive distillation is not the only intensified process that combines reaction and separation. Reactive absorption, or chemical absorption, also belongs to this category and its importance is growing rapidly due to its application in gas treatment and pollution control [7, 8]. Absorption, as most of the stage processes, is realized through the contact of a gas phase and a liquid phase. Depending on which phase is continuous and which one is disperse it is possible to identify different process alternatives. Among all, plate and packed columns are undoubtedly the most used ones [9].

The classical absorption process, named physical absorption, is based in the solubility of a gas in the liquid phase and is generally fostered at low temperature and high pressure. The physical absorption of carbon dioxide was reviewed by Zhenhong et al. [10] highlighting the necessity to develop intensified alternatives to reduce the equipment sizes. In this case, the approach followed is to develop new solvents like ionic liquids [11]. As an intensified alternative to the physical absorption, reactive absorption overcomes the limitation to operate at high pressure or the low gas solubility by a chemical reaction between the gas and the solvent. Considering the tight legislation regarding carbon dioxide emissions and the potential of the technology in its removal, reactive absorption is considered an alternative

of paramount importance [12]. The research activity on carbon dioxide removal by reactive absorption topic can be divided in four main branches:
1. Synthesis of process alternatives
2. Experimental comparison of different solvents
3. Kinetic studies
4. Process modeling

The study of different process alternatives regards the coupling of the reactive absorption with the solvent regeneration section and the generation of flowsheets aimed to reduce the energy consumption of the whole process. Jung et al. [13] proposed two new configurations based on rich vapor recompression and cold solvent split, while Madeddu et al. [14] studied the influence of the reflux on the stripper column. A very active research area is the solvent development since its choice is directly related to the energy consumption of the plant. The evaluation of the absorption rate, degradability, and environmental impact are some of the parameters taken into account during comparison of solvents [15, 16]. Together with the definition of new solvents, the interest in defining a reliable kinetic model emerges.

The review of the data available for the different solvents proposed in the literature is not the focus of this work, but an interesting review limited to alkanolamine solvents was given by Vaidya and Kenig [17]. Finally, the modeling part is the one directly related to the equipment design and represent an open topic in the literature. As evidenced by Van Duc Long et al. [7], reactive separation processes and intensified processes, in general, have been limited by the need of complex modeling. However, with intensive studies the impassable issues related, e.g., to the design of divided wall columns, have been addressed. Similarly, many studies have contributed to improve the actual knowledge on the modeling and design of reactive absorption operations. Eventually, it should be noted how only a reliable model allows the full achievement of the process intensification goals making possible the comparison of different process alternatives, the energy evaluation, and the capital costs estimation. Process modeling is the tool that holds together process intensification, process synthesis, process integration, and optimization and in this chapter its importance for the reactive absorption will be discussed considering carbon dioxide as an example.

4.2 Reactive absorption applications

Different reactive absorption processes are already implemented at industrial scale and some of them are briefly commented here.

4.2.1 Nitric acid production

Nitric acid is an important chemical mainly used for the production of fertilizers [18]. It can be produced by ammonia oxidation in the Ostwald process, summarized in the following five steps:
1. Oxidation of NH_3 to NO
2. Oxidation of NO to NO_2
3. Absorption of NO_x in water for the production of nitric acid
4. Tail gas purification.
5. Removal of NOx from the nitric acid

Reactive absorption is involved in the third step. In the column, an additional current of air is used to oxidize the NO not converted to NO_2. The absorption takes place in the trays and the oxidation in-between the trays. Typical dimensions of these towers are 2 m in diameter and 46 m height. Depending on the tower design, it is possible to obtain a solution of 55%–70% of acid [19].

Nitric acid can be also produced by absorption of NO_x as part of the treatment of industrial gas streams coming from combustion processes. In this case the modeling was proved to be fundamental to achieve the environmental restrictions imposed for pollutant emissions [20].

4.2.2 The SCOT process

The Shell Claus Off-gas Treating process, usually referred as SCOT process, was introduced as auxiliary to the Claus sulfur recovery process in order to reduce the sulfur emissions from the tail gas [21]. The process evolves in two steps: the reduction step where all sulfur values in the Claus off-gas are reduced to hydrogen sulfide and the absorption step where the hydrogen sulfide is absorbed in an amine solution. The absorber operates at atmospheric pressure and the off-gas usually contains 200–500 ppm of hydrogen sulfite.

4.2.3 Carbon dioxide capture and storage (CCS)

Motivated by the necessity to decrease the carbon dioxide emissions, mainly associated to power plants fed by fossil fuels, CCS technologies have catalyzed the interest of the research and industrial community. According to Scopus, in the year 2000, four articles were recorded under the topic CCS. This value became 735 in 2017. The post-combustion approach using reactive absorption is schematically reported in Figure 4.1.

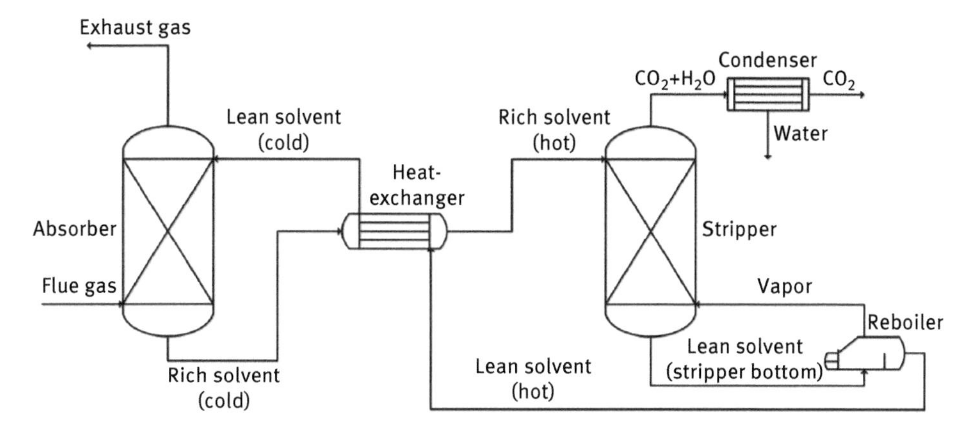

Figure 4.1: Simplified flowsheet for the CO_2 capture by reactive absorption-stripping.

The plant consists of two main sections: the absorption, where the CO_2 is transferred from the vapor/gas phase to the liquid one, and the stripping, where the solvent is regenerated. The two sections are interconnected by means of a cross heat exchanger. The CO_2-rich flue gas enters the bottom of the absorber where it flows countercurrent with the liquid solvent. From the top of the column, the exhaust vapor/gas exits and is sent to the stack. On the other hand, the CO_2-rich solvent exits the bottom of the absorber and, after being heated in the heat-exchanger, is sent to the top of the stripper, where it flows countercurrent with a vapor stream generated by the reboiler. A mixture containing mainly CO_2 and H_2O exits from the top of the column and is sent to the partial condenser where the CO_2 is concentrated in the gas phase while the water is recovered in the liquid phase. The regenerated solvent exiting the bottom of the stripper is firstly sent to the reboiler, where it is partially vaporized, and then to the heat-exchanger, where it supplies its sensible heat to the rich solvent. After the heat-exchanger, the lean solvent is recycled back to the top of the absorber.

A review of the CCS technologies was done by Leung et al. [22], while a more comprehensive description of different reactive absorption technologies was reported by Yildirim et al. [8].

4.3 CO_2 reactive absorption: Solvent selection, reaction scheme, and thermodynamic

Considering the CO_2 reactive absorption as target process, before examining the equations used for the modeling, the solvent used, the corresponding reactions, and the thermodynamic model are defined in this section.

4.3.1 Solvent selection

As stated in the introduction, the research on new solvents for the CO_2 absorption represents a very active topic due to the relation between the properties of the solvent and the operating and capital costs.

Peeters et al. [23] identify the chemical binding energy, the absorption rate, the solvent loading, the absorption and desorption temperatures, the solvent stability, and the solvent cost as the principal characteristics to be examined in the selection of a good solvent.

Historically, tertiary amine aqueous solutions have been introduced as reacting solvent in 1930 as patented by Bottoms [24] opening the study on the application of different alkanolamines. Among all the alkanolamines, aqueous solutions of monoethanolamine (MEA) are recognized as the most studied and used solvent. A very large sets of experimental data are available together with kinetic studies on the MEA-CO_2 reaction mechanism [25–34]. Due to the availability of experimental data for the model validation and for the maturity in the developing of kinetic models, MEA was chosen as solvent.

A review of alkanolamine properties toward CO_2 absorption was explored by Vaidya and Kenig [17].

4.3.2 Reaction scheme

The definition of the reaction scheme is essential for the correct description of the CO_2-MEA-H_2O system. A significant number of reaction sets have been proposed in the literature. Some authors have proposed a pure equilibrium reactions set [26, 35–38], while others have considered only one kinetic reaction between CO_2 and MEA [39–42], or two kinetic reactions involving CO_2 and MEA and OH^- ions [29, 43, 44].

However, the most used approach is to consider a hybrid set that takes into account both equilibrium and kinetic reactions [25, 27, 29, 31, 45–54]. This approach appears to be the most correct for two main reasons:

1. It has been demonstrated that the reactions involving CO_2 have kinetic limitations [55–58], therefore they cannot be described rigorously by a pure chemical equilibrium approach.
2. The presence of MEA and carbonate in water generates ions. Consequently, ionic equilibrium reactions are needed to describe the reactions involving electrolytes.

The hybrid approach including three ionic equilibrium reactions and the two kinetic reversible reactions has been here considered [45, 50].

- Self-ionization of water

$$2H_2O \leftrightharpoons H_3O^+ + OH^- \tag{4.1}$$

- Dissociation of MEA

$$MEA^+ + H_2O \leftrightharpoons H_3O^+ + MEA \tag{4.2}$$

- Dissociation of bicarbonate ion

$$HCO_3^- + H_2O \leftrightharpoons H_3O^+ + CO_3^- \tag{4.3}$$

- Formation of carbamate

$$CO_2 + MEA + H_2O \leftrightharpoons MEACOO^- + H_3O^+ \tag{4.4}$$

- Formation of bicarbonate ion

$$CO_2 + OH^- \leftrightharpoons HCO_3^- \tag{4.5}$$

The equilibrium reactions 4.1–4.3 were considered instantaneous, while the last two reactions 4.4 and 4.5, which have kinetic limitations, were assumed to be reversible.

For the mathematical description of the equilibrium reactions, the equilibrium constants in terms of temperature were computed by means of the standard Gibbs free-energy change, while activities were used as concentration basis for the equilibrium constants in terms of concentration. For what concerns the kinetic reversible reactions, the reaction rate was expressed by a second order power law expression, where the kinetic constant resulted from the Arrhenius equation and the concentration basis was expressed in terms of molarities:

- Formation of carbamate:

$$r_1 = k_{1,f} C_{CO_2}^L C_{MEA}^L - k_{1,r} C_{MEACOO^-}^L C_{H_3O^+}^L \tag{4.6}$$

- Formation of bicarbonate ion

$$r_2 = k_{2,f} C_{CO_2}^L C_{OH^-}^L - k_{2,r} C_{HCO_3^-}^L \tag{4.7}$$

The kinetic parameters for the formation of carbamate were taken from Hikita et al. [58], while for the formation of the bicarbonate ion the work of Pinsent et al. was considered [55]. The values are summarized in Table 4.1.

It should be noted that the reactions are limited to the liquid phase only.

Table 4.1: Kinetic parameters for reactions 4 and 5.

Reaction	Forward reaction		Reverse reaction	
	$k^0_{f,j}\left[\frac{kmol}{m^3s}\right]$	$E_{a,\,f_j}\left[\frac{cal}{mol}\right]$	$k^0_{r,j}\left[\frac{kmol}{m^3s}\right]$	$E_{a,\,r_j}\left[\frac{cal}{mol}\right]$
4	$9.77 \cdot 10^{10}$	9,855.8	$3.23 \cdot 10^{19}$	15,655
5	$4.32 \cdot 10^{13}$	13,249	$2.38 \cdot 10^{17}$	29,451

4.3.3 Definition of the thermodynamic model

The CO_2-MEA-H_2O system is an electrolyte system characterized by a non-ideal behavior due to the presence of ions [35, 59]. For this reason, the thermodynamic model for the liquid phase must be able to describe the interactions between electrolytes. In particular, the correct evaluation of the components activity coefficients is crucial, since these parameters are involved in the calculation of the vapor-liquid and gas-liquid equilibria, the equilibrium and kinetic reactions, and the calculation of the driving force for the interphase material transfer. Models with different level of detail have been used in the literature for the computation of the liquid phase thermodynamic properties. For example, the simple Kent-Eisenberg model, which fixes all the activities to unity (ideal behavior) was used by several authors [47, 60–63]. In the early 1980s, Deshmukh and Mather [64] and Chen and Evans [65] proposed two different models based on rigorous thermodynamic principles [66]. The first model, used by various authors [36, 66, 67], involves the Guggenheim equation [68] to compute the activity coefficients, while the latter, which can be found in a significant number of works [25, 26, 29, 35, 37, 43–46, 48–50, 59, 69, 70–74], is known as Electrolyte-NRTL model, and applies the Pitzer-Debye-Hückel theory together with the Electrolyte-NRTL equation for the evaluation of the activity coefficients. More recently, Hoff et al. [75] developed a model, used by different authors [27, 28, 76] that includes the non-idealities of the liquid phase by means of the salting-out correlation proposed by Van Krevelen et al. [77], obtaining results which are very similar to the Electrolyte-NRTL model. The CO_2 post-combustion capture process is operated at low pressures, for this reason, as reported by Freguia [78], no relevant non-idealities are present for what concerns the thermodynamics of the vapor/gas phase. Nevertheless, different equations of state (EoS) have been used in the literature for the evaluation of the fugacity coefficients of the components, such as the Peng-Robinson EoS [36, 47, 51, 67], the Redlich-Kwong EoS [35, 37, 45, 48, 50, 79], and the Soave-Redlich-Kwong EoS [49, 59, 78]. Furthermore, as carbon dioxide is a gas, the solubility of CO_2 in the amine aqueous solution is typically modeled by means of the Henry's law.

Here, the Electrolyte-NRTL model and the Redlich-Kwong EoS were used for the description of the thermodynamics of the liquid phase and the vapor/gas phase, respectively.

4.4 Process modeling: Equilibrium stage vs rate-based model

In the absorber the CO_2 is transferred from the vapor/gas to the liquid phase, where it reacts with the solvent to enhance the absorption process. Different phenomena are involved:

1. Non-ideal thermodynamics
2. Chemical equilibrium and kinetic reactions
3. Simultaneous material and energy transfer
4. Vapor-liquid and gas-liquid equilibria
5. Distribution of the components between the liquid and the gaseous phase

The definition of the type of model is the first step to accomplish to describe the process. The equilibrium-stage and rate-based models are compared here. An important clarification regards the terms used in the two models that are very often confused.

The introduction of the equilibrium stage model was attributed to Sorel in 1893 [80] and an equilibrium stage can be considered as a well mixed stage where the liquid and the vapor leaving the stage are in thermodynamic equilibrium with each other. It is easy to apply this concept to tray unit operations where each try corresponds to an equilibrium stage or a theoretical stage. For packed type columns where the equilibrium is not reached at any point, the packing can be divided in layers each of which can be considered equivalent to a theoretical stage leading to the definition of HETP or "height equivalent to one theoretical plate." Differently, the concept of segment is introduced to discretize the spatial domain of a packed column. Each segment contains both phases and the packing, it can be considered homogeneous in temperature and composition. Once the packing height is fixed, the number of segments used to discretize the spatial domain, influences the temperature and composition profiles correctness and it could also affect the global separation. At this point, it must be noted that the number of theoretical stages does not correspond to the number of segments used in rate-based models [83]. The number or equilibrium stages and the number of segments are two completely different concepts.

4.4.1 The equilibrium stage model

The equilibrium stage model is the simplest way to describe simultaneous material and energy transfer processes [82]. In the case of intimate contact between the phases and sufficient residence time, it can be assumed that the streams exiting each stage reach the thermodynamic equilibrium. The material and energy balances can be written according to the graphical representation of one stage reported in Figure 4.2. In the following equations the index i indicates the components, while index j indicates the stage.

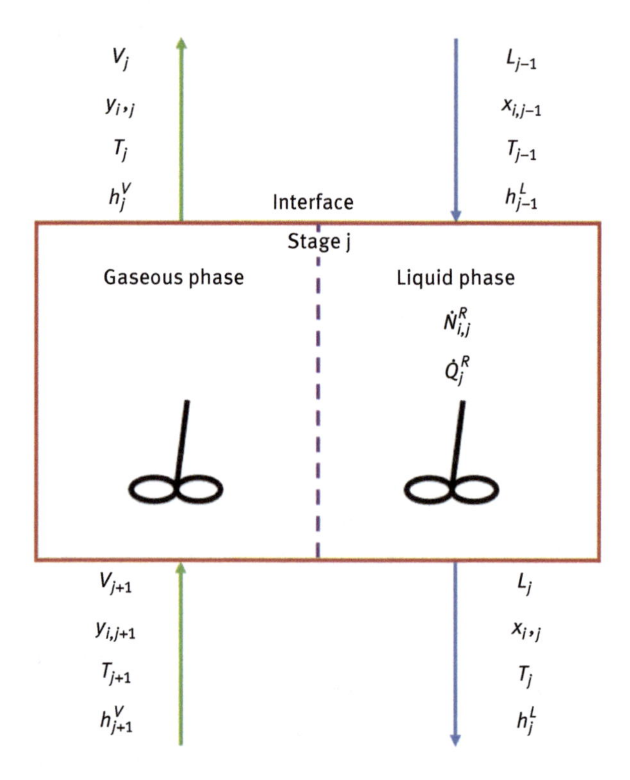

Figure 4.2: Equilibrium stage model.

Component material balance:

$$\frac{dM_{i,j}}{dt} = L_{j-1}x_{i,j-1} + V_{j+1}y_{i,j+1} - L_j x_{i,j} - V_j y_{i,j} + \dot{N}_{i,j}^R \tag{4.8}$$

Energy balance:

$$\frac{dU_j}{dt} = L_{j-1}h_{j-1}^L + V_{j+1}h_{j+1}^V - L_j h_j^L - V_j h_j^V + \dot{Q}_j^R \tag{4.9}$$

Vapor-liquid equilibrium:

$$P\ \phi_{i,j}\ y^{*}_{i,j} = P^{sat}_{i,j}\ \gamma_{i,j}\ x_{i,j} \tag{4.10}$$

Gas-liquid equilibrium:

$$P\ \phi_{i,j}\ y^{*}_{i,j} = He_{i,j}\gamma_{i,j}x_{i,j} \tag{4.11}$$

Summation equations:

$$\sum_{i=1}^{n_c^L} x_{i,j} = 1 \tag{4.12}$$

$$\sum_{i=1}^{n_c^V} y_{i,j} = 1 \tag{4.13}$$

Material and energy hold up:

$$M_{i,j} = \varepsilon\ Sdz\left(\psi_j^L C_j^L x_{i,j} + \psi_j^V C_j^V y_{i,j}\right) \tag{4.14}$$

$$U_{i,j} = \varepsilon\ Sdz\left(\psi_j^L C_j^L h_j^L + \psi_j^V C_j^V h_j^V\right) \tag{4.15}$$

The phase equilibrium assumption is a condition difficult to achieve due to the simultaneous presence of material transfer and chemical reactions. For this reason, some authors introduced an efficiency to take into account the deviation from the equilibrium. The most used expression for the efficiency is the one proposed by Murphree [83]. Values of the Murphree efficiency are typically very low when CO_2 reactive absorption is considered. For example, Walter and Sherwood [84] found an efficiency range between 0.65% and 4.2% using a glycerine aqueous solution as solvent in a plate column, while Afkhamipour and Mofarahi [85] used a range of efficiencies between 0.1 and 0.4 with a 2-amino-2-methyl-1-propanol solution as absorbent in a packed column. Using MEA as solvent, Øi [60, 86] defined a Murphree efficiency of 0.25 for a packed column. The equilibrium stage model combined with the Murphree efficiency was applied in several works [37, 60, 61, 86].

4.4.2 The rate-based model

The efficiency values reported for the equilibrium stage model give a clear indication on how the CO_2 absorption process is far from reaching the ideal equilibrium. For this reason, a more rigorous non-equilibrium model is necessary. In particular, this model must be able to take into account the material transfer limitations due to the presence of chemical reactions. For this purpose, a rate-based model was developed, which accounts for interfacial material and energy transfer between the gaseous and the liquid phases, the reaction kinetics and the electrolytic interactions. The main assumption of this approach is that phase equilibrium occurs at the vapor/gas-liquid interface only and that material and energy transport in the two phases are described separately. Furthermore, the first implication of this assumption is the fact that two distinct balances must be written for the two phases. Different authors compared the equilibrium stage and the rate-based model. Among all, Afkhamipour and Mofarahi [85] stated: *the rate-based model gives a better prediction of the temperature and concentration profiles compared to the equilibrium-stage model*, in line with Neveux et al. [51] that reported: *the use of a rate-based model seems mandatory to take into consideration all phenomena*.

4.4.2.1 Interphase material transfer

The peculiarity of the rate-based model is the quantification of the material transport across the gaseous-liquid interface. Three main theories have been proposed:
1. The two-film theory by Lewis and Whitman [87]
2. The penetration theory by Higbie [88]
3. The surface renewal model by Danckwerts [89]

Although applications of the penetration theory and the surface renewal model can be find in the literature for CO_2 absorption [27, 56, 90], the two film theory is by far the most used model in this field [25, 26, 29, 37, 39, 41, 43–45, 50, 51, 53, 54, 74, 76, 91–98], due to the fact that it is simple and a significant number of correlations for the parameters evaluation are present in the literature [99].

A graphical representation of the two-film theory in the presence of chemical reaction in the liquid phase is reported in Figure 4.3.

The model is based on the following assumption: all the resistance to material and energy transfer is concentrated in two thin films close to the gaseous-liquid interface, where phase equilibrium exists. In particular, as showed in Figure 4.3, the spatial domain can be divided in four parts:
1. Gaseous bulk
2. Gaseous film
3. Liquid film
4. Liquid bulk

In the absorption process, the CO_2 is transferred from the gaseous bulk to the interface through the gaseous film. Then CO_2 is absorbed in the liquid, where it reacts with the solvent. Finally, the remaining CO_2 dissolved in the liquid transferred to the liquid bulk.

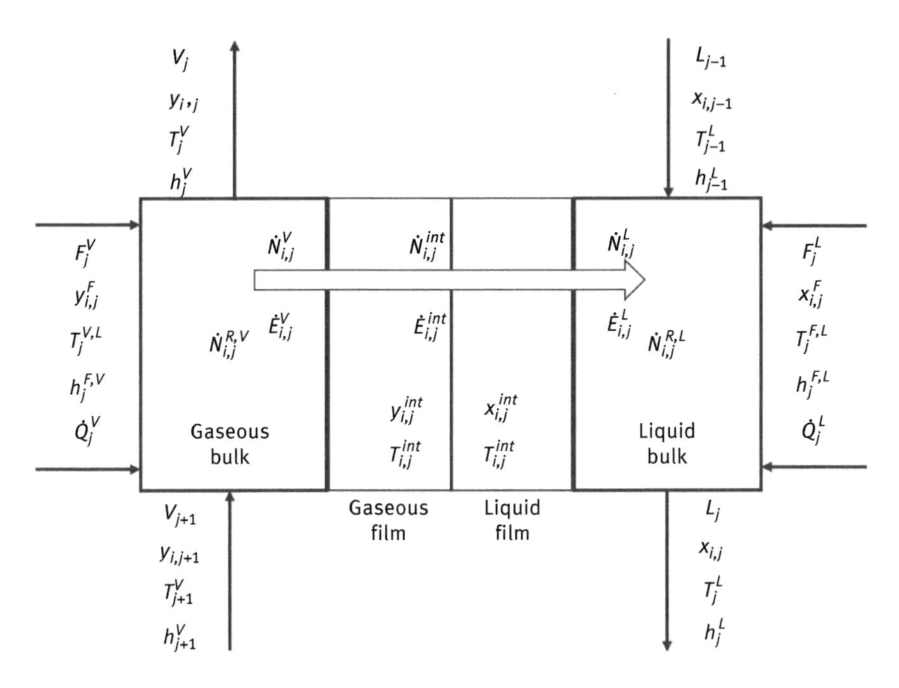

Figure 4.3: Rate-based segment represented using the two-film theory.

4.4.2.2 Modeling the bulk

The *RadFrac* model – Rate-Based mode, like all the Aspen Plus blocks, is constituted by algebraic equations. For this reason, reactive columns are modeled as a series of n-CSTRs. Material and energy balances for the two phases bulk are reported in Eqs. 4.16–4.21 with reference to Figure 4.3 where index i indicates the component, while index j indicates the segment:

- Material balance in the liquid bulk

$$F_j^L x_{i,j}^F + L_{j-1} x_{i,j-1} + \dot{N}_{i,j}^L + \dot{N}_{i,j}^{R,L} - L_j x_{i,j} = 0 \tag{4.16}$$

- Material balance in the gaseous bulk

$$F_j^V y_{i,j}^F + V_{j+1} y_{i,j+1} + \dot{N}_{i,j}^V + \dot{N}_{i,j}^{R,V} - V_j x_{i,j} = 0 \tag{4.17}$$

- Energy balance in the liquid bulk

$$F_j^L h_j^{F,L} + L_{j-1} h_{j-1}^L + \dot{Q}_j^L + \dot{E}_j^L - L_j h_j^L = 0 \tag{4.18}$$

- Energy balance in the gaseous bulk

$$F_j^V h_j^{F,V} + V_{j+1} h_{j+1}^V + \dot{Q}_j^V + \dot{E}_j^V - V_j h_j^V = 0 \tag{4.19}$$

– Summation equation for both phases

$$\sum_{i=1}^{n_c^L} x_{i,j} = 1 \tag{4.20}$$

$$\sum_{i=1}^{n_c^V} y_{i,j} = 1 \tag{4.21}$$

For the evaluation of the bulk conditions in each segment, five different flow models are available in the *RadFrac* model. In particular, bulk properties can be assumed to be equal to the outlet conditions (Mixed flow), obtaining an ideal CSTR, or they can be computed as an average between the inlet and the outlet conditions (Countercurrent flow), or as a combination of average conditions for one phase and outlet conditions for the other one (VPlug flow, VPlugP flow, and LPlug flow). Figure 4.4 schematically resumes the different flow models.

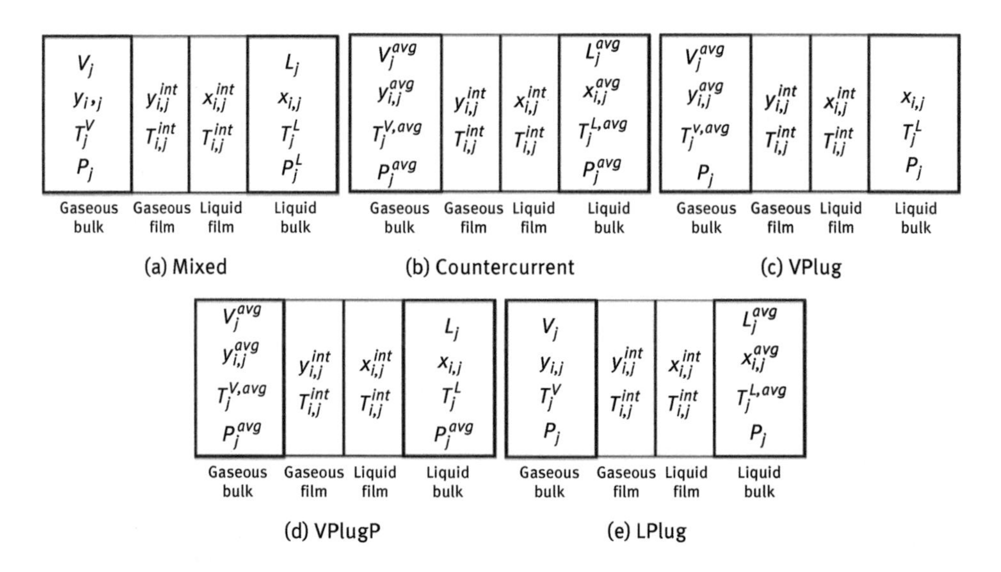

Figure 4.4: Flow models available in the Aspen Plus-*RadFrac*.

4.4.2.3 Modeling the film

To model the film, the *RadFrac* model offers different options to establish which resistances are to be considered:

1. *Nofilm*: there is no distinction between bulk and film, the entire phase is completely mixed and treated like an equilibrium stage;

2. *Film*: only resistance to material diffusion is considered and no reactions are present;
3. *Filmrxn*: both resistance to diffusion and reactions are present. The film is not discretized;
4. *Discrxn*: both resistance to diffusion and reactions are present, but this time the film is discretized. This option is suggested whenever the reactions lead to significant changes in the concentration profiles in the film.

As the gaseous phase does not involve chemical reactions, the option *film* can be chosen. For what concerns the liquid film, reactions involving CO_2 are typically very fast [29, 45, 50, 93] and lead to steep profiles close to the interface. For this reason, the option *Discrxn* is recommended to model the liquid film.

The choice of the *Discrxn* option implicates that the liquid film is discretized into a certain number of points and gives the possibility to choose how to arrange these points. Since the chemical reactions involving CO_2 are fast, the profiles in the liquid film are expected to be significantly steep close to the interface, until the variables reach their equilibrium value. For this reason, the discretization points must be concentrated close to the interface in order to describe these profiles correctly. To do so, the *geometric* discretization option was used. In this way, the discretization points are non-equidistant. In the *RadFrac* model – Rate-Based mode three parameters must be defined to setup the film discretization:

- *Reaction Condition Factor (RCF)*. This parameter weights the interface and bulk compositions and temperature in the computation of the reaction rates within the film. A large value of this parameter increases the influence of the bulk conditions on the computation of the film reaction rate. Considering the molar fraction for the i-th component in the liquid film, its relation with the *RCF* is evidenced in Eq. 4.22:

$$x_i^f = RCFx_i + (1 - RCF)x_i^{int} \tag{4.22}$$

  For the CO_2 capture with MEA process, the film reactions are very fast; consequently the bulk conditions must have a higher weight on the film modeling.
- *Number of discretization points in the film*. It represents the number of non-equidistant points in which the liquid film is divided for the evaluation of the temperature and composition profiles.
- *Film Discretization Ratio (FDR)*. This parameter is related to the distance between the discretization points in the film. It is defined, according to Eq. 4.23, as the ratio between adjacent film segments moving in the direction of the interface:

$$FDR = \frac{\delta_k}{\delta_{k+1}} \tag{4.23}$$

  Where δ_k is the thickness of the k-th segment.

The film profiles are very steep close to the interface and they approach the equilibrium value close to the bulk. Therefore, it is preferable to use a small number of points placing most of them close to the interface, rather than use a high number of equidistant points [93]. However, there must be a compromise between the number of points and the film discretization ratio. Even if it is necessary to define a *FDR* higher than one in order to concentrate the internal points near the interface, a high value can lead to numerical problems due to the small discretization steps close to the interface.

4.5 Rate-based parameters evaluation

A further aspect that demonstrates the higher level of detail of the rate-based model compared to the equilibrium stage one is the need to evaluate the following parameters:
- Wetted surface area
- Material transfer coefficients
- Heat transfer coefficients
- Fractional liquid hold-up

Different correlations are available for the calculation of these parameters based on the packing type, i.e., random or structured packing. A brief literature review is given in the following subsections.

4.5.1 Wetted surface area

One characteristic parameter of the packing is the dry specific area, i.e., the packing surface available per cubic meter of packing. When the liquid irrigates the packing, it covers a fraction of this available surface, which is referred to as wetted surface area. This parameter indicates the contact surface between the gaseous and the liquid phase. Typically, it is a function of the liquid properties and the packing geometry and its expression is usually given with reference to the three dimensionless numbers of Reynolds, Freud, and Weber.

Several correlations have been developed for the evaluation of the wetted surface area, depending on the packing type, as they can be used for:
- Random packing only: Onda et al. [100]
- Structured packing only: Bravo et al. [101, 102], Henriques de Brito et al. [103], Tsai et al. [104, 105]
- All kinds of packing: Billet and Schultes [106], Hanley and Chen [107]

4.5.2 Material transfer and heat transfer coefficients

For the description of the interphase material and energy transfer, the correspondent coefficients were introduced and defined as the ratio between the diffusion coefficient (material transfer) or the thermal conductivity (energy transfer) and the film thickness as reported in Eqs. 4.24 and 4.25, respectively:

$$k_M^p = \frac{D^p}{\delta^p} \tag{4.24}$$

$$h_T^p = \frac{k_T^p}{\delta^p} \tag{4.25}$$

where p indicates the phase.

Since the estimation of the film thickness is difficult, these coefficients are typically evaluated by means of empirical correlations. The coefficients are functions of the phase properties and the geometric features of the packing. Depending on the packing type, the correlations for the estimation of the material transfer coefficients can be applied for:

- Random packing only: Onda et al. [100]
- Structured packing only: Bravo et al. [101, 102], Rocha et al. [108]
- All kinds of packing: Billet and Schultes [106], Hanley and Chen [107]

Once the computation of the material transfer coefficient is performed, the most used approach for the evaluation of the heat transfer coefficients is the Chilton-Colburn analogy [109]. According to the theory, the mechanisms through which material and heat transfer happen are identical. Then, once the transfer coefficient is known for one transfer type, the evaluation of the remaining transfer coefficient is straightforward.

4.5.3 Fractional liquid hold-up

Once the packing is placed in the column, the volume-free portion for the liquid and the gaseous phase is identified by the packing void fraction. When the column operates, a percentage of this volume is occupied by the liquid, while the gaseous phase occupies the remaining part. In order to properly evaluate the control volumes in which the balances must be written, it is necessary to evaluate the fractional liquid hold-up, i.e., the volume fraction occupied by the liquid. Once this parameter is known, the volume fraction occupied by the gas/vapor evaluation is immediate, since the sum of the two fractions must be equal to unity. Even in this case, several relationships are present based on the packing type. Differently from

the wetted surface area and the material transfer coefficients, no random packing only correlations are available. The fractional liquid hold-up can be evaluated for:
- Structured packing only: Suess and Spiegel [110], Bravo et al. [102], Rocha et al. [111]
- All kinds of packing: Stichlmair et al. [112], Billet and Schultes [106, 113, 114]

4.6 Fluid dynamics analysis

The definition of the equations used for the parameters estimation is not enough to complete the rate-based model. Another fundamental aspect that must be taken into account regards the system fluid dynamics. Writing the material and energy balance for a single segment, two equivalent approaches can be followed:
1. A differential form, which leads to an ideal Plug-Flow reactor model without axial dispersion
2. A discrete form, which leads to a Plug-Flow reactor modeled as a series of n-CSTR reactors

In the *RadFrac* – Rate-based mode block, reactive columns are modeled as a series of n-CSTRs according to the second approach.

The PFR model without axial dispersion, although it is by far the most used model to describe the fluid dynamics of the absorption column, is an assumption which must be validated in order to be definitely acceptable. In particular, it must be verified that the column has a fluid dynamic behavior that resembles that of an ideal plug-flow. In general, two factors can make the PFR assumption fall:
1. The axial diffusion/dispersion (microscale phenomenon)
2. The back-mixing due to the countercurrent (macroscale phenomenon)

If only one of these factors had an important effect on the process, the PFR assumption would not be valid anymore. The methods to analyze the influence of the axial diffusion/dispersion and the back-mixing due to the countercurrent are discussed in the following sections.

4.6.1 Axial diffusion/dispersion: Peclet number definition

In order to analyze the possible effect of the axial diffusion/dispersion on the system fluid dynamics, it is worth to evaluate the Peclet number. This dimensionless group is defined as reported in Eq. 4.26:

$$Pe = \frac{rate\ of\ transport\ by\ convection}{rate\ of\ transport\ by\ diffusion/dispersion} = \frac{FL_c}{\varepsilon\,\psi SCD} \tag{4.26}$$

where L_c represents the characteristic length.

In general, high values of the Peclet number indicate a column behavior close to that of an ideal plug-flow. On the other hand, if the value of the dimensionless group is small, the axial diffusion/dispersion has an effect on the process that could not be neglected.

The Peclet number can be defined for both the phases either for what concerns the material transport and the energy transport. Moreover, two characteristic lengths can be considered, i.e., the column height and the packing equivalent diameter. The former provides information on the overall column behavior, while the latter considers the effect of the diffusion/dispersion at a local level, around the packing.

For what concerns the material transport, the Peclet number is defined for all the components and the mixture and for both the phases, in Eqs. 4.27-a, b with reference to the packing height and in Eqs. 4.28-a, b with reference to the packing equivalent diameter:

$$Pe^V_{M,H_i} = \frac{VH}{\varepsilon\,\psi^V SC^V D^V_{i,mix}} \tag{4.27-a}$$

$$Pe^L_{M,H_i} = \frac{LH}{\varepsilon\,\psi^L SC^L D^L_{i,mix}} \tag{4.27-b}$$

$$Pe^V_{M,deq_i} = \frac{Vd_{eq}}{\varepsilon\,\psi^V SC^V D^V_{i,mix}} \tag{4.28-a}$$

$$Pe^L_{M,deq_i} = \frac{Ld_{eq}}{\varepsilon\,\psi^L SC^L D^L_{i,mix}} \tag{4.28-b}$$

Similarly, the thermal Peclet number is defined as following for both the phases and with reference to the packing height in Eqs. 4.29-a,b and in Eqs. 4.30-a,b with reference to packing equivalent diameter:

$$Pe^V_{T,H_i} = \frac{c^V_{P,mix} VH}{\varepsilon\,\psi^V Sk^V_T} \tag{4.29-a}$$

$$Pe^L_{T,H_i} = \frac{c^L_{P,mix} LH}{\varepsilon\,\psi^L Sk^L_T} \tag{4.29-b}$$

$$Pe^V_{T,d_{eq}} = \frac{c^V_{P,mix} Vd_{eq}}{\varepsilon\,\psi^V Sk^V_T} \tag{4.30-a}$$

$$Pe^{L}_{T,d_{eq}} = \frac{c^{L}_{P,mix} L d_{eq}}{\epsilon \psi^{L} S k^{L}_{T}} \tag{4.30-b}$$

The expression of the Peclet number is obtained from the dimensionless form of the material and energy balances considering the diffusion/dispersion term. Therefore, to evaluate the dimensionless group, an appropriate reference condition is required. The easiest choice is to consider the feed conditions (top for the liquid phase and bottom for the gaseous one) as reference for the physical properties, in order to make the Peclet number analysis independent from the system solution, since these data are known before performing any simulation.

4.6.2 Back-mixing due to the countercurrent effect

The second factor which affects the ideal plug-flow assumption is represented by the back-mixing due to the countercurrent. This macroscale phenomenon is not taken into account by the Peclet number, which contains the axial diffusion/dispersion, a microscale phenomenon. The back-mixing is implicitly included in the material and energy balances because of the countercurrent streams arrangement and to investigate its possible effect on the process it is necessary to obtain the correct numerical solution of the resulting system of equations.

4.6.3 The analysis of the number of segments

The discussion on the modeling of the system fluid dynamics up to now dealt with the physics only, but it is strictly linked to a fundamental mathematical aspect which is the solution of the resulting system of equations obtained from the model development.

When dealing with a system of algebraic equations, a crucial step to find the solution is the correct definition of the number of points in the axial domain, i.e., the packing height, where this system is to be solved. In other words, it must be defined an appropriate number of segments for the discretization of the packing height. Very often this parameter has been too easily defined in the literature, and its influence on the model has not been appropriately discussed. For example, Mores et al. [47], Kucka et al. [93], and Mac Dowell et al. [97] focused their research on the mathematical modeling of the CO_2-MEA absorption system, validating their models using the experimental data reported in the work of Tontiwachwuthikul et al. [40]. They all considered the same absorber with a packing height of 6.55 m. The first research group discretized the packing height using 10 segments, the second one 15, and the third 25. The same discussion can be made with reference to several authors who validated their model using the experimental data from the work of Dugas [115], who made an experimental campaign using a pilot-plant facility at the University of Texas at Austin. Plaza et al. [43], Kvamsdal and Rochelle [44], Lawal et al. [37], and Zhang et al. [29] all used a different number of segments of discretization, although they all dealt with the same column.

A list of some relevant works where the rate-based method was used to model the CO_2-MEA absorber is reported in Table 4.2.

Table 4.2: Literature review for the number of segments applied in the rate-based model for the CO_2-MEA absorption system.

Reference	Segments	Height [m]	Diameter [m]
39	39	3.89	0.15
116	20	4.25	0.125
43	12	6.10	0.427
47	10	6.55	0.100
93	15	6.55	0.100
97	25	6.55	0.100
44	30	6.10	0.427
30	15	6.10	0.427
29	20	6.10	0.427
92	15	8.00	1.680
73	24	12.00	0.150
25	40	17.00	1.100

From the analysis of Table 4.2 it can be noticed that there is no correspondence between the number of segments and the column dimensions. Even when the same column height is considered, there is still a variation in the number of segments considered. This disagreement deserves to be explored in detail. The definition of a correct number of segments for the discretization of the axial domain is strictly related to the possible influence that the axial dispersion and the back-mixing due to the countercurrent have on the process.

In particular:

- For large values of the Peclet number the axial dispersion effect can be neglected. In this case the system fluid dynamics resembles that of an ideal plug-flow and, consequently, a high number of segments would be needed to correctly describe the process.

- Due to the countercurrent streams arrangement, even if the axial dispersion had no influence, the plug-flow assumption would fall if the back-mixing generated by the countercurrent played an important role on the process. To investigate the possible effect of the back-mixing, it is necessary to obtain the correct numerical solution of the system. This means that the number of segments must be increased until an asymptotic behavior is reached [117–119], i.e., until the differences between two consecutive simulations with different number of segments are negligible. If the number of segments for the discretization of the axial domain necessary to obtain the correct solution is sufficiently high, it can be concluded that the back-mixing due to the countercurrent has no important effect on the process.

These considerations proved how the definition of a proper number of segments is of fundamental importance to obtain a robust and correct model from a numerical point of view.

4.7 Model validation

Once the principles and the different correlations used in the rate-based model have been defined, the model needs to be validated with experimental data. In this section the parameters defined in Aspen Plus are discussed together with their influence in the model output.

Two experimental pilot-plant facilities with different peculiarities were chosen for the validation. The first case considered was the laboratory-scale absorption plant designed by Tontiwachwuthikul et al. [40, 120], while the second one was the CESAR (CO_2 Enhanced Separation and Recovery) large-scale absorption/desorption system described by Razi et al. [25]. The two systems were chosen because of their differences in the dimensions (laboratory- vs large-scale) and in the packing type (random vs structured). The two facilities are briefly described in the following two sections.

4.7.1 Laboratory-scale pilot-plant

The column reported in the works of Tontiwachwuthikul et al. [40, 120] is packed with 12.7 mm ceramic Berl saddles divided in six sections separated by redistributors. The column and packing features are reported in Table 4.3.

Each section is equipped with sample points for the measurement of the liquid temperature and the CO_2 concentration in the gaseous phase. In particular, the Run 20 of which the feed characterization is reported in Table 4.4, was chosen in this work to validate the model.

The experimental data available are reported in Table 4.5, where the height 0 m corresponds to the column bottom.

Table 4.3: Lab-scale plant column and packing features.

Variable	Value
Packing height [m]	6.55
Column diameter [m]	0.1
Void fraction [m^3/m^3]	0.62
Dry specific surface area [m^2/m^3]	465

Table 4.4: Feed characterization for the selected run of the lab-scale plant.

	Flue gas	Lean solvent
Temperature [K]	288.15	292.15
Molar flow [mol/s]	0.14	1.04
CO_2 [mol frac]	0.192	0
MEA [mol frac]	0	0.0497
H_2O [mol frac]	0.1	0.9503
N_2 [mol frac]	0.708	0
Pressure [kPa]	103.15	103.15

Table 4.5: Experimental data for the selected run of the lab-scale plant.

Sample	H [m]	T^L [K]	y_{CO_2} [mol frac]
1	0.00	321.15	0.192
2	1.05	330.15	0.177
3	2.15	320.15	0.142
4	3.25	305.15	0.077
5	4.35	295.15	0.028
6	5.45	293.15	0.006
7	6.55	292.15	0.000

4.7.2 Large-scale pilot-plant

The CESAR pilot plant described in the work of Razi et al. [25] is a large-scale plant which treats part of the flue gas coming from the Dong Esbjerg power station (power plant fed with 400 MWe pulverized bituminous coal) extracted directly from the desulfurization section without any pretreatment. In this case the packing height is divided into four identical sections containing structured Mellapak 2X. After the absorption zone, a water-wash section is present at the top of the column to recover the volatilized MEA from the gas phase. The recovered solvent is then recycled back to the top of the absorption section after cooling. In the simulations the water-wash section was not included, since no information on the recycle stream were reported.

The column and packing features are reported in Table 4.6. Each section is equipped with sample points for the measurement of the temperature. One run, Run 1-A2, of which the feed characterization is reported in Table 4.7, was chosen to validate the model. The experimental data available for the run are reported in Table 4.8, where the height 0 m corresponds to the column bottom.

Table 4.6: Large-scale plant column and packing features.

Variable	Value
Packing height [m]	17
Column diameter [m]	1.1
Void fraction [m^3/m^3]	0.99
Dry specific surface area [m^2/m^3]	205

4.7.3 Peclet number analysis

Before setting the simulations, the Peclet number analysis was performed for all the runs. The Peclet number was evaluated for both the material (single components and mixture) and energy transport in each phase and with reference to both the packing height and the packing equivalent diameter using Aspen Custom Modeler. The inlet conditions (column top for the liquid phase, column bottom for the gaseous phase) were used as reference conditions. The results of the computations are reported in Tables 4.9 and 4.10.

As it is possible to notice from the analysis of Tables 4.9 and 4.10, the values of the Peclet number are quite large. Table 4.9 show the results for the lab-scale plant.

Table 4.7: Feed characterization for the selected run of the large-scale plan.

	Flue gas	Lean solvent
Temperature [K]	326.92	332.57
Molar flow [mol/s]	52.33	214.55
CO_2 [mol frac]	0.12	0.0263
MEA [mol frac]	0	0.102
H_2O [mol frac]	0.12	0.8717
N_2 [mol frac]	0.76	0
Pressure [kPa]	106.391	101.325

Table 4.8: Experimental data for the selected run of the large-scale plan.

Sample	H [m]	T [K]
1	0.00	326.92
2	4.25	333.47
3	8.50	339.95
4	12.75	346.55
5	17	332.57

When the liquid phase is considered and the column height is used as characteristic length, the material Peclet number values are in the order of 10^9. If the packing equivalent diameter is used, this value is reduced to about 10^6. For what concerns the gaseous phase, the material Peclet number is in the order of 10^5 and 10^2 when the column height and the packing equivalent diameter are considered, respectively. On the other hand, the thermal Peclet number for the liquid phase is in the range of 10^6 when the packing height is used and 10^3 with the packing equivalent diameter as characteristic length. With regards to the gaseous phase, the thermal Peclet number is in the order of 10^5 with the packing height as characteristic length, while the same value is reduced to 10^2 when the packing equivalent diameter is considered.

For what concerns the large-scale plant, as expected, the values of the Peclet number shown in Table 4.10 are higher compared to the lab-scale plant due to the

Table 4.9: Peclet number evaluation for run T20.

Phase	Liquid		Gaseous	
Characteristic length	H	d_{eq}	H	d_{eq}
$Pe_{M,i}$				
CO_2	$3.54 \cdot 10^8$	$2.65 \cdot 10^5$	$3.05 \cdot 10^5$	228.03
CO_3^{2-}	$3.84 \cdot 10^8$	$2.87 \cdot 10^5$	–	–
H_2O	$3.04 \cdot 10^9$	$2.27 \cdot 10^6$	$2.32 \cdot 10^5$	173.91
H_3O^+	$3.80 \cdot 10^7$	$2.84 \cdot 10^4$	–	–
HCO_3^-	$2.44 \cdot 10^8$	$1.83 \cdot 10^5$	–	–
MEA	$3.32 \cdot 10^8$	$2.49 \cdot 10^5$	$5.19 \cdot 10^5$	388.64
MEA^+	$2.66 \cdot 10^8$	$1.99 \cdot 10^5$	–	–
$MEACOO^-$	$2.66 \cdot 10^8$	$1.99 \cdot 10^5$	–	–
N_2	$3.52 \cdot 10^8$	$2.63 \cdot 10^5$	$2.75 \cdot 10^5$	205.82
OH^-	$6.71 \cdot 10^7$	$5.02 \cdot 10^4$	–	–
Mixture	$2.15 \cdot 10^9$	$1.61 \cdot 10^6$	$2.75 \cdot 10^5$	$2.05 \cdot 10^2$
Pe_T				
Mixture	$3.10 \cdot 10^6$	2,319.71	$2.92 \cdot 10^5$	$2.19 \cdot 10^2$

higher column dimensions and molar flows involved. For the liquid phase, the material Peclet number is in the order of 10^9 when the column heigh is considered and 10^6 for the packing equivalent diameter. For the gaseous phase, values in the range of 10^6 with the column height and 10^3 considering the packing equivalent diameter are observed. The liquid thermal Peclet number is in the order of 10^7 and 10^4 considering the column height and the packing equivalent diameter, respectively. For the gaseous phase, similarly to the material Peclet number, values of the thermal Peclet number in the order of 10^6 considering the column height to 10^3 considering the packing equivalent diameter are obtained. Since the lowest value of the dimensionless group was in the order of 10^2, it could be concluded that the axial diffusion/dispersion had no effect on the process and that the column fluid dynamics resembled that of an ideal plug-flow. Since the axial diffusion/dispersion effect could be neglected, from a mathematical point of view a sufficiently high number of segments for the discretization of the axial domain was required to achieve a correct representation of the system.

Table 4.10: Peclet number evaluation for run 1-A2.

Phase	Liquid		Gaseous	
Characteristic length	H	d_{eq}	H	d_{eq}
$Pe_{M,i}$				
CO_2	$9.72 \cdot 10^8$	$1.14 \cdot 10^5$	$1.30 \cdot 10^6$	1,525.2
CO_3^{2-}	$2.68 \cdot 10^9$	$3.15 \cdot 10^6$	–	–
H_2O	$4.03 \cdot 10^9$	$4.74 \cdot 10^6$	$9.49 \cdot 10^5$	1,116.78
H_3O^+	$2.65 \cdot 10^8$	$3.12 \cdot 10^5$	–	–
HCO_3^-	$1.70 \cdot 10^9$	$2.00 \cdot 10^6$	–	–
MEA	$1.00 \cdot 10^9$	$1.18 \cdot 10^6$	$2.13 \cdot 10^6$	2,500.37
MEA^+	$1.86 \cdot 10^9$	$2.18 \cdot 10^6$	–	–
$MEACOO^-$	$1.86 \cdot 10^9$	$2.18 \cdot 10^6$	–	–
N_2	$9.66 \cdot 10^8$	$1.14 \cdot 10^6$	$1.11 \cdot 10^6$	1,300.85
OH^-	$4.68 \cdot 10^8$	$3.90 \cdot 10^6$	–	–
Mixture	$3.31 \cdot 10^9$	$3.90 \cdot 10^6$	$1.10 \cdot 10^6$	1,298.1
Pe_T				
Mixture	$2.03 \cdot 10^7$	$2.39 \cdot 10^4$	$1.19 \cdot 10^6$	$1.40 \cdot 10^3$

4.7.4 Component definition

The first step in building a model in Aspen Plus is the definition of the components involved in the Aspen Properties package. In the case of the CO_2 post-combustion capture with MEA aqueous solutions the process involves two different phases, the vapor/gas and the liquid one, which are characterized by different components. In particular, while the gaseous phase contains only gases, with the exception of water vapor, the liquid phase contains all the ions involved in the CO_2-MEA-H_2O system according to the reaction scheme described in Section 3.2.

Table 4.11 resumes the components for the two phases. The CO_2-MEA-H_2O system is characterized by the presence of ions, for a total of nine components in the liquid phase. The ions are a convenient way to represent the process by means of a mathematical model, but from an experimental point of view, the evaluation of their actual composition is not advantageous. Typically, the composition measurements in the liquid phase are given in terms of apparent carbon dioxide, monoethanolamine, and water concentrations. The apparent compositions are related

Table 4.11: Model components.

Gaseous phase		Liquid phase	
Component	**Name**	**Component**	**Name**
CO_2	Carbon dioxide	CO_2	Free carbon dioxide
H_2O	Water	${CO_3}^{2-}$	Carbonate ion
MEA	Monoethanolamine	H_2O	Free water
N_2	Nitrogen	H_3O^+	Hydronium ion
O_2	Oxygen	${HCO_3}^-$	Bicarbonate ion
		MEA	Free monoethanolamine
		MEA^+	Protonated monoethanolamine
		$MEACOO^-$	Carbamate ion
		OH^-	Hydroxide ion

to the actual ions compositions by means of the Eqs. 4.31–4.33, in terms of molar fractions:

$$x_{CO_2}^{app} = x_{CO_2} + x_{CO_3^{2-}} + x_{HCO_3^-} + x_{MEACOO^-} \tag{4.31}$$

$$x_{MEA}^{app} = x_{MEA} + x_{MEA^+} + x_{MEACOO^-} \tag{4.32}$$

$$x_{H_2O}^{app} = x_{H_2O} + x_{H_3O^+} + x_{HCO_3^-} + x_{OH^-} \tag{4.33}$$

Since nine compositions are computed for modeling purposes, six more equations are required to square the system. These equations are represented by the ionic equilibrium relations and the electroneutrality condition reported in Eqs. 4.34–4.39:

$$2H_2O \rightleftharpoons H_3O^+ + OH^- \tag{4.34-a}$$

$$K_{eq1} = \frac{a_{H_3O^+}\, a_{OH^-}}{a_{H_2O}^2} \tag{4.34-b}$$

$$CO_2 + 2H_2O \rightleftharpoons H_3O^+ + HCO_3^- \tag{4.35-a}$$

$$K_{eq2} = \frac{a_{H_3O^+}\, a_{HCO_3^-}}{a_{CO_2}\, a_{H_2O}^2} \tag{4.35-b}$$

$$HCO_3^- + H_2O \rightleftharpoons H_3O^+ + CO_3^{2-} \tag{4.36-a}$$

$$K_{eq3} = \frac{a_{H_3O^+} \ a_{CO_3^{2-}}}{a_{HCO_3^-} \ a_{H_2O}} \tag{4.36-b}$$

$$MEA^+ + H_2O \rightleftharpoons H_3O^+ + MEA \tag{4.37-a}$$

$$K_{eq4} = \frac{a_{H_3O^+} \ a_{MEA}}{a_{MEA^+} \ a_{H_2O}} \tag{4.37-b}$$

$$MEACOO^- + H_2O \rightleftharpoons HCO_3^- + MEA \tag{4.38-a}$$

$$K_{eq5} = \frac{a_{HCO_3^-} \ a_{MEA}}{a_{MEACOO^-} \ a_{H_2O}} \tag{4.38-b}$$

$$\sum_{i=1}^{N_C^L} x_i \tilde{z}_i = 0 \tag{4.39}$$

These equations are used by Aspen Plus in the simulations to characterize all the liquid streams composition. It must be noted that these reactions are not the same used to describe the reactions that happen inside the columns.

4.7.5 Run T20: rate-based mode set-up

As discussed in Subsection 4.2.3, since the option *Discrxn* was activated in the *RadFrac* model for the discretization of the liquid film, three parameters had to be defined for the setup of the rate-based model. In particular:

- The *Reaction Condition Factor* was set to 0.9 in order to give more weight to the bulk conditions in the evaluation of the film reaction rates, due to the high rate of the reactions in the liquid film. This value is in agreement with the work of Zhang et al. [29].
- After a significant number of simulations, it was found that 5 non-equidistant discretization points in the liquid film were sufficient for the correct description of the profiles on the basis of the comparison between the model results and the experimental data. This value is in agreement with the work of Kucka et al. [93].
- The *Film Discretization Ratio* was fixed to 10, value that was found to be a good compromise between the placing of the discretization points and the discretization steps for the numerical solution of the system.

Moreover, in the case of the random packing, since Zhang et al. [29] demonstrated that the correlation by Onda et al. [100] underestimated the wetted surface area, the *Interfacial Area Factor* available in the simulator to correct the evaluation of the parameter was set to 1.2. This factor was coherent with the dry specific area for the ceramic Berl saddles reported by Mores et al. [47]. For what concerns the fluid dynamics, the Mixed flow model was used in the simulations, since in this way it was possible to model the column as a series of CSTRs. A comparison between the Mixed flow model and the other flow models present in the software is reported in the Subsection 4.7.5.3.

4.7.5.1 Analysis of the number of segments

The absorber was simulated varying the number of segments until two consecutive profiles were overlapped. The liquid temperature and the CO_2 in the vapor phase profiles are reported in Figure 4.5.

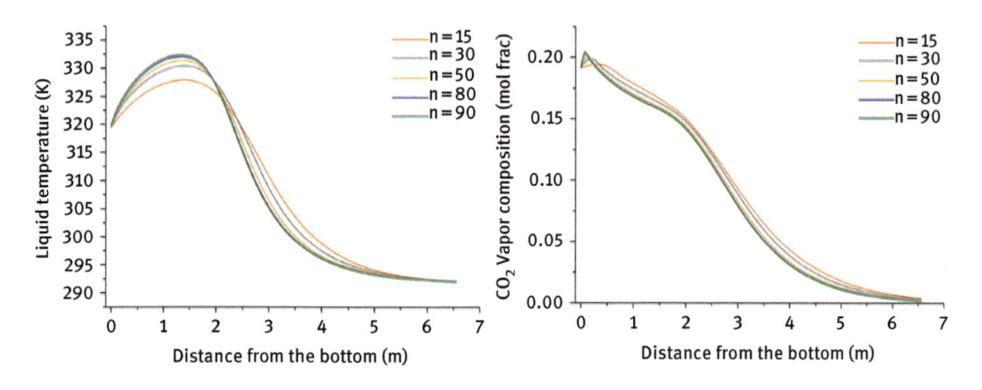

Figure 4.5: Run T20, variation with the number of segments: liquid temperature profile (left) and CO_2 vapor composition profile (right).

From the analysis of Figure 4.5 it is possible to notice that the profiles with 80 and 90 segments are overlapped. Then 90 segments are sufficient to obtain the correct system solution. This result proves the correctness of the previous Peclet number analysis and it leads to the conclusion that the back-mixing effect can be neglected. It must be noticed that only with an appropriate number of segments it is possible to identify a small concentration bulge in the bottom of the column, which is caused by the partial preponderance of the reverse reaction of carbamate to CO_2 (Eq. 4.4).

A number of segment analyses were needed to obtain a model which was robust and correct from a numerical point of view. Only at this point it was possible to make the comparison between the 90 segments model and the experimental data, which is reported in Figure 4.6 for both the liquid temperature and the CO_2 vapor composition.

From Figure 4.6 it is possible to observe the good agreement between the model and the experimental data profiles and this result was obtained with a model correct from a numerical point of view.

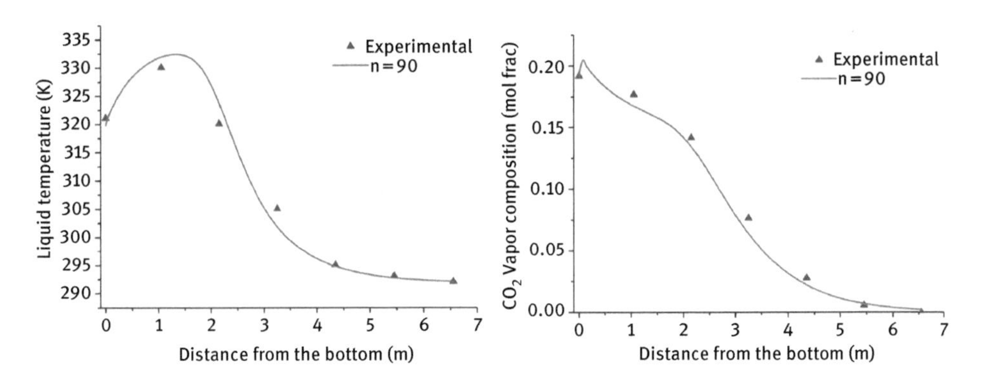

Figure 4.6: Experimental data comparison: liquid profile (left), CO_2 vapor composition profile (right).

From the point of view of the column performance, which are represented by the CO_2 removal and the loading in the outlet liquid stream, the results are reported in Table 4.12.

Table 4.12: Absorber performance variation with the number of segments.

Performance	Experimental	Number of segments				
		15	30	50	80	90
CO_2 removal %	100	98.2	98.8	99	99.1	99.1
Loading out	0.514	0.507	0.511	0.515	0.516	0.516

It can be noticed that when the number of segments is increased higher values of both the CO_2 removal and the loading are obtained. This fact can be explained analyzing the variation of the CO_2 interphase molar flow rate profile with the number of segments, reported in Figure 4.7.

It is possible to notice from Figure 4.7 that with the increase in the number of segments the simulator is always able to evaluate a higher CO_2 transfer from the gaseous to the liquid phase since, due the more detailed discretization of the axial domain, the calculations are performed in more points.

From these first results, the need of a sufficiently high number of segments to discretize the axial domain was highlighted. This necessity could be explained also by the fact that along the column several transitions take place between the absorption and the desorption process regimes and between water evaporation and condensation [118]. Using an inadequate number of segments, the net fluxes, especially for water, between liquid and vapor could be under/over-estimated, as shown in Figure 4.8, leading to different temperature and composition profiles. It must be noticed that in

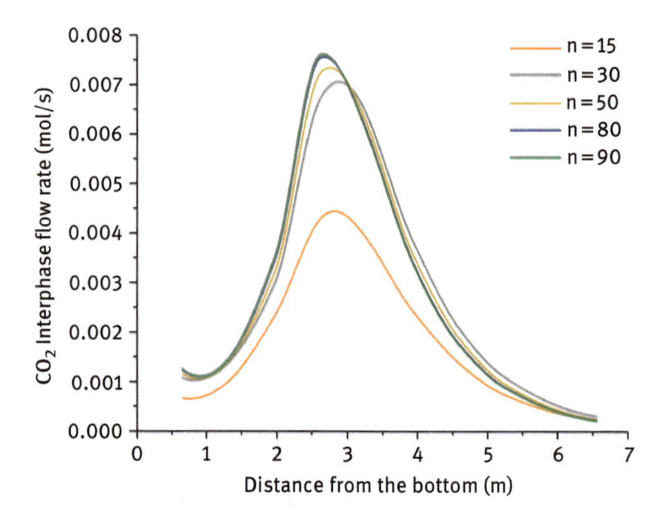

Figure 4.7: CO_2 interphase molar flow rate variation with the number of segments.

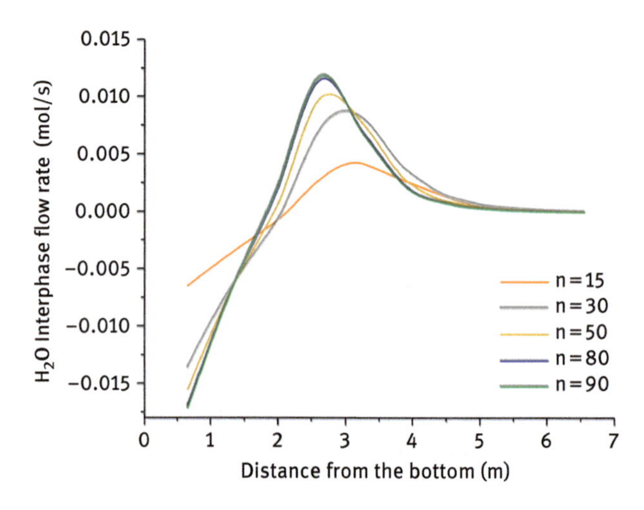

Figure 4.8: H_2O interphase molar flow rate profile variation with the number of segments.

Figure 4.8 a negative flow rate indicates a transfer from the liquid to the vapor phase (evaporation), while a positive flow rate indicates a transfer from the vapor to the liquid phase (condensation).

4.7.5.2 Kinetic parameters calibration

After the identification of an appropriate number of segments for the discretization of the axial domain, it is possible to assure that the proposed model is numerically robust and it is in agreement with the fluid dynamics predicted by the Peclet number.

At this point, it is possible to improve the model focusing on the physical parameters involved. After a literature review and comparing different models, it was possible to conclude that some parameters are affected by uncertainty. In particular, the attention was concentrated on the kinetic parameters of the forward carbamate formation reaction (Eq. 4.4), of which Table 4.13 reports the range of variation found in the literature.

Table 4.13: Kinetic reaction parameters.

$k^0_{f,j}$ $\left[\frac{kmol}{m^3 s}\right]$	E_{a,f_j} $\left[\frac{cal}{mol}\right]$	Reference
$4.495 \cdot 10^{11}$	10,733.22	Kucka et al. [93]
$9.77 \cdot 10^{10}$	9,855.8	Hikita et al. [58]
$1.17 \cdot 10^{6}$	1,797.1	Kvamsdal & Rochelle [44]

In general, it is known that to evaluate the influence of uncertain parameters a sensitivity analysis is required [121]. Moreover, as reported by Rodriguez-Aragon and Lopez-Fidalgo [122], the kinetic parameters are essential in the modeling of the phenomena and the most accurate estimation will produce the best results in the use of the model. Different studies [123–125] dealt with the uncertainty related to the parameters of the Arrhenius equation and in the present work the pre-exponential factors of both the forward and reverse carbamate reactions were calibrated assuming the correctness of the equilibrium constant of the overall reaction. The pre-exponential factor was optimized minimizing the standard error (*SE*) between the experimental data and the model values. The *SE* is defined, according to Eq. 4.40, as the square root of the mean squared error (*MSE*):

$$SE = \sqrt{MSE} \tag{4.40}$$

The optimization led to a 30% reduction of the pre-exponential factor of the forward reaction. Remembering that the equilibrium constant must be always respected for the reaction, the pre-exponential factor for the reverse reaction was evaluated consequently using the Eq. 4.41:

$$K_{eq} = \frac{k_f}{k_r} \tag{4.41}$$

The new set of kinetic parameters is reported in Table 4.14.

The optimized parameters of the reaction, together with the parameters of the bicarbonate reaction (Eq. 4.5) were varied of ±10% in order to verify the robustness of the system. All the possible combinations were considered, demonstrating the robustness of the system for small variations of the modified kinetic parameters set, since no significant variations are observed in the profiles.

Table 4.14: Modified kinetic parameters.

	Forward reaction		Reverse reaction	
	$k^0_{f,j}$ $\left[\frac{kmol}{m^3 s}\right]$	E_{a,f_j} $\left[\frac{cal}{mol}\right]$	$k^0_{r,j}$ $\left[\frac{kmol}{m^3 s}\right]$	E_{a,r_j} $\left[\frac{cal}{mol}\right]$
Hikita et al. [58]	$9.77 \cdot 10^{10}$	9,855.8	$3.23 \cdot 10^{19}$	15,655
Proposed model	$6.839 \cdot 10^{10}$	9,855.8	$2.261 \cdot 10^{19}$	15,655

Figure 4.9 reports the comparison between the absorber profiles with the initial set of kinetic parameters from Hikita et al. [58] and the modification proposed.

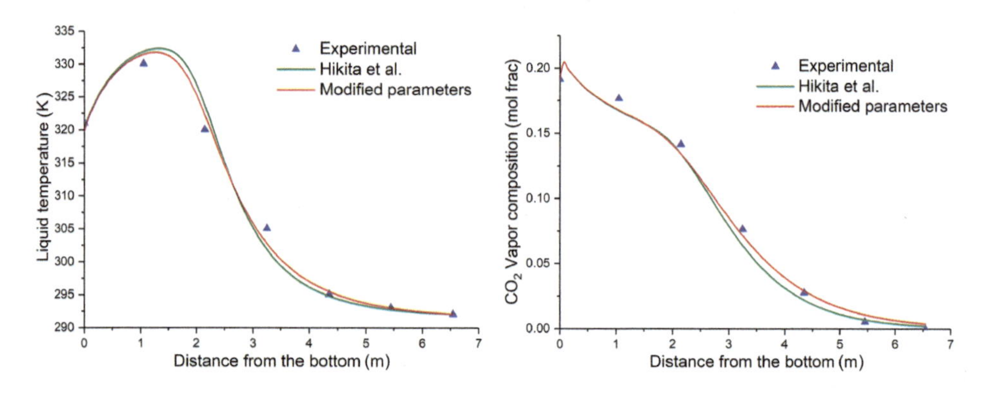

Figure 4.9: Liquid temperature profile (left) and CO_2 vapor composition profile (right) for the two sets of kinetic parameters.

An effective improvement can be immediately observed in the description of the experimental data for both the liquid temperature and the CO_2 molar fraction.

In order to quantify this improvement, the standard error was evaluated and reported in Table 4.15.

Table 4.15: Standard error for the two sets of kinetic parameters.

	SE_T	SE_x
Hikita et al. [58]	2.03	$7.12 \cdot 10^{-3}$
Proposed model	1.44	$5.5 \cdot 10^{-3}$

A reduction of more than 0.5 K was obtained for the temperature, while a slightly reduction of about $1.5 \cdot 10^{-3}$ was obtained for the CO_2 composition.

Table 4.16: Performance of the absorber with the two sets of kinetic parameters.

	Experimental	Hikita et al. [60]	Proposed model
CO_2 removal %	100	99.1	98.4
Loading out	0.511	0.516	0.512

Following the modification of the kinetic parameters, as expected, a slightly reduction of both the CO_2 removal and loading were obtained, as shown in Table 4.16. Anyway, the results remained coherent with the experimental values.

4.7.5.3 Influence of the different flow models

In Subsection 4.2.2 the five flow models included in the *RadFrac* model for the evaluation of the bulk properties were presented. All the models were tested to investigate their possible influence on the simulations. The results of this analysis related to Run T20 for both the liquid temperature and the CO_2 vapor composition showed no significant differences using the different flow models. The values of the standard error together with the performance of the absorber are reported in Table 4.17.

Table 4.17: Standard error and absorber performance using the different flow models.

	Mixed	Countercurrent	VPlug	VPlug-Pavg	LPlug
SE_T	1.44	1.77	1.65	1.65	1.47
SE_x	$5.5 \cdot 10^{-3}$	$7.4 \cdot 10^{-3}$	$6.42 \cdot 10^{-3}$	$6.42 \cdot 10^{-3}$	$6.35 \cdot 10^{-3}$
CO_2 removal %	98.4	98.7	98.6	98.6	98.5
Loading out	0.512	0.513	0.513	0.513	0.512

The results from Table 4.17 indicate that the minimum *SE* is obtained using the Mixed model; on the other hand, the performance of the absorber remain practically constant varying the flow models.

4.7.6 Run 1-A2: extension to the large-scale plant

With the purpose of proving the general validity of the proposed method, the developed model was tested on a large-scale facility, equipped with structured packing. Since the column dimensions and the molar flows involved were higher compared to the lab-scale pilot-plant, in the light of the Peclet number values obtained and reported in Table 4.10, a higher number of segments was expected in this case. The model with the modified kinetic parameters was directly applied to find the proper number of segments. For this plant, only temperature measurements were available and the phase was not specified. For this reason, Figure 4.10 reports the profiles for the liquid and vapor/gas temperature with different number of segments.

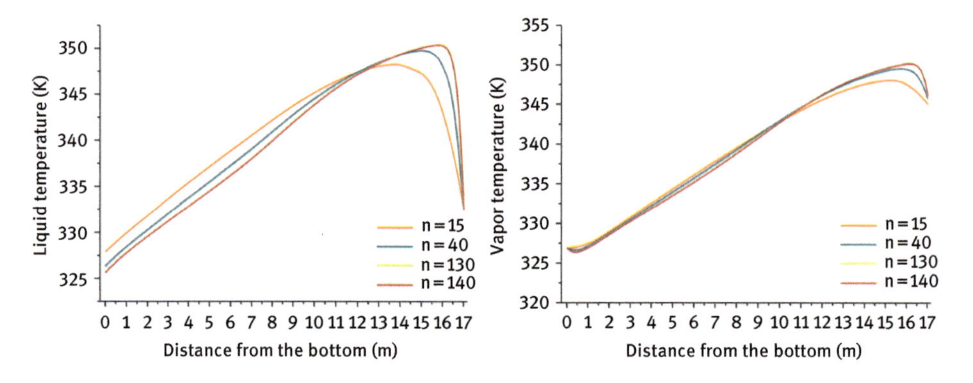

Figure 4.10: Run 1-A2 liquid temperature profile (left) and vapor temperature profile (right) variation with the number of segments.

As it is possible to notice, 140 segments are needed in this case to obtain a numerically robust solution of the system. After the evaluation of a proper number of segments, the comparison between the model and the experimental data is reported in Figure 4.11.

A good agreement is found between the model with 140 segment and the experimental data.

From the point of view of the column performance, only the percentage CO_2 removal was available in the work of Razi et al. [25]. The comparison between the experimental values and the model estimations is reported in Table 4.18.

In the light of these results, it was validated again the necessity of an appropriate number of segments to obtain a numerically correct and robust solution of the resulting system of algebraic equation.

Using a too low number of segments can lead to incorrect evaluations especially if the model is extended to describe the dynamic behavior of the absorber.

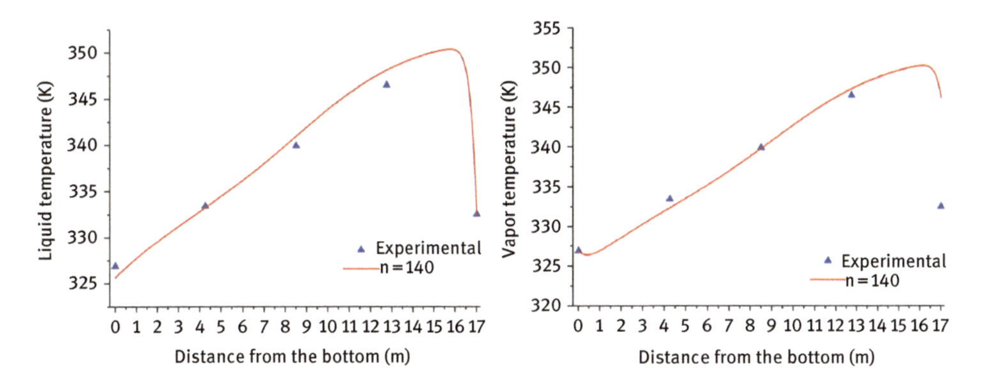

Figure 4.11: Comparison between liquid temperature profile (left) and vapor composition profile (right) with the experimental data.

Table 4.18: Run 1-A2 absorber performance.

Performance	Experimental	Number of segments		
		15	40	140
CO_2 removal %	90	85.2	86.6	87.2

4.8 Conclusion

Process intensification is a powerful tool that can lead to the definition of new ways of production characterized by lower capital and operating costs. Units obtained by combining chemical reaction and separation are a classic example of intensification and reactive absorption belongs to this category. Even if applied at industrial scale, reactive absorption design still represents a challenge because of the complexity of the phenomena involved. The definition of reliable models is the key to bring process intensification closer to the industrial world. In this chapter the modeling of the absorption of CO_2 into monoethanolamine is discussed starting from the general theory and its transposition into the process simulator Aspen Plus. It was proved how the knowledge of the transport phenomena is essential in defining a useful model. Moreover, it was proved how the number of segments has a paramount influence in the definition of the model output. This parameter was very often underestimated in the literature and sometimes confused with the number of stages leading to conceptual errors. An analysis based on the Peclet number was proposed and discussed to choose the correct number of segments. The model was validated considering experimental data from both lab-scale and large-scale plants. The procedure described and the model obtained are important tools in guiding the designer to the definition of reliable models for intensified processes involving chemical absorption.

Bibliography

[1] Peppas NA. The first century of chemical engineering. Distillations 2008, fall.
[2] Stankiewicz A, Moulijn JA. Process intensification: transforming chemical engineering. Chemical Engineering Progress 2000, 96, 22–34.
[3] Ponce-Ortega JM, Al-Thubaiti MM, El-Halwagi MM. Process intensification: new understanding and systematic approach. Chem Eng Process 2012, 53, 63–75.
[4] Niju S, Meer KM, Beguma S, Anantharaman N. Continuous flow reactive distillation process for biodiesel production using waste egg shells as heterogeneous catalysts. RCS Advances 2014, 4, 54109–54114.
[5] Gomez-Castro FI, Rico-Ramirez V, Segovia-Hernandez JG, Hernandez-Castro S, El-Halwagi MM. Simulation study on biodiesel production by reactive distillation with methanol at high pressure and temperature: Impact on costs and pollutant emissions. Computers and Chemical Engineering 2013, 52, 204–215.
[6] Bildea CS, Kiss AA. Dynamics and control of a biodiesel process by reactive absorption. Chemical Engineering Research and Design 2011, 89, 187–196.
[7] Van Duc Long N, Minh LQ, Ahmad F, Luis P, Lee M. Intensified distillation –based separations processes: recent developments and perspectives. Chem Eng Technol 2016, 39, 2183–2195.
[8] Yildirim O, Kiss AA, Hüser N, Leßmann K, Kenig EY. Reactive absorption in chemical process industry: A review on current activities. Chemical Engineering Journal 2012, 213, 371–391.
[9] Kenig EY, Gorak A. Reactive Absorption. In: Sundmacher K., Kienle A., Seidel-Morgenstern A., ed. Integrated Chemical Processes: Synthesis, Operation, Analysis, and Control. Wiley, 2005, 265–311.
[10] Zhenhong B, Kokkeong L, Mohdshariff A. Physical absorption of CO_2: A review. Advanced Material Research 2014, 917, 134–143.
[11] de Riva J, Suarez-Reyes J, Moreno D, Díaz I, Ferro V, Palomar J. Ionic liquids for post-combustion CO_2 capture by physical absorption: Thermodynamic, kinetic and process analysis. International Journal of Greenhouses Gas Control 2017, 61, 61–70.
[12] Koronaki IP, Prentza L, Papaefthimiou V. Modeling of CO_2 capture via chemical absorption processes – An extensive literature review. Renewable and Sustainable Energy Reviews 2015, 50, 547–566.
[13] Jung J, Jeong YS, Lee U, Lim Y, Han C. New configuration of the CO_2 capture process using aqueous monoethanolamine for coal-fired power plants. Industrial & Engineering Chemistry Research 2015, 54, 3865–3878.
[14] Madeddu C, Errico M, Baratti R. Process analysis for the carbon dioxide chemical absorption–regeneration system. Applied Energy 2018, 2015, 532–542.
[15] Hoff KA, da Silva EF, Kim I, Grimstvedt A, Ma'mun S. Solvent development in post combustion CO_2 capture-Selection criteria and optimization of solvent performance, cost and environmental impact. Energy Procedia 2013, 37, 292–299.
[16] Aronu UE, Hoff KA, Svendsen HF. CO_2 capture solvent selection by combined absorption–desorption analysis. Chemical Engineering Research and Design 2011, 89, 1197–1203.
[17] Vaidya PD, Kenig EY. CO_2-alkanolamine reaction kinetics: A review of recent studies. Chemical Engineering & Technology 2007, 30, 1467–1474.
[18] Clarke SI, Mazzafro WJ. Nitric acid. In: Kirk-Othmer Encyclopedia of Chemical Technology, 2005 John Wiley & Son, Inc.
[19] Martin MM, Nitric acid. In: Martin MM., ed. Industrial Chemical Process Analysis and Design. Elsevier, 2017, 200–345.
[20] Hüpen B, Kenig EY. Rigorous modelling of NOx absorption in tray and packed columns. Chemical Engineering Science 2005, 60, 6462–6471.

[21] Swaim CD Jr. The Shell Claus offgas treating (SCOT) process. Advances in Chemistry 1975, 139, 111–119.
[22] Leung DYC, Caramanna G, Maroto-Valer MM. An overview of current status of carbon dioxide capture and storage technologies. Renewable and Sustainable Energy Reviews 2014, 39, 426–443.
[23] Peeters ANM, Faaij APC, Turkenburg WC. Techno-economic analysis of natural gas combined cycles with post-combustion CO_2 absorption, including a detailed evaluation of the development potential. International Journal of Greenhouse Gas Control 2007, 1, 396–417.
[24] Bottoms RR. Process for separating acid gases. 1930, US Patent 1,783,901
[25] Razi N, Svendsen HF, Bolland O. Validation of mass transfer correlations for CO_2 absorption with MEA using pilot data. International Journal of Greenhouse Gas Control 2013, 19, 478–491.
[26] Biliyok C, Lawal A, Wang M, Seibert F. Dynamic modelling, validation and analysis of post-combustion chemical absorption CO_2 capture plant. International Journal of Greenhouse Gas Control 2012, 9, 428–445.
[27] Tobiensen FA, Svendsen HF. Experimental validation of a rigorous absorber model for CO_2 postcombustion capture. AIChE Journal 2007, 53, 846–865.
[28] Tobiensen FA, Juliussen O, Svendsen HF. Experimental validation of a rigorous desorber model for CO_2 post-combustion capture. Chemical Engineering Science 2008, 63, 2641–2656.
[29] Zang Y, Chen H, Chen C-C, Plaza JM, Dugas R, Rochelle GT. Rate-based process modeling study of CO_2 capture with aqueous monoethanolamine solution. Ind Eng Chem Res 2009, 48, 9233–9246.
[30] Putta KR, Svendsen HF, Knuutila HK. Kinetic of CO_2 absorption in to aqueous MEA solutions near equilibrium. Energy Procedia 2017, 114, 1576–1583.
[31] Aboudheir A, Tontiwachwuthikul P, Chakma A, Idem R. Kinetics of the reactive absorption of carbon dioxide in high CO_2-loaded, concentrated aqueous monoethanolamine solutions. Chemical Engineering Science 2003, 58, 5195–5210.
[32] Savage DW, Kim CJ. Chemical kinetics of carbon dioxide reactions with diethanolamine and diisopropanolamine in aqueous solutions. AIChE Journal 1985, 31, 296–301.
[33] McCann N, Phan D, Wang X, Conway W, Burns R, Attalla M, Puxty G, Maeder M. Kinetics and mechanism of carbamate formation from CO_2(aq), carbonate species, and monoethanolamine in aqueous solution. J Phys Chem A 2009, 113, 5022–5029.
[34] Putta KR, Pinto DDD, Svendsen HF, Knuutila HK. CO_2 absorption into loaded aqueous MEA solutions: kinetics assessment using penetration theory. International Journal of Greenhouse Gas Control. 2016, 53, 338–353.
[35] Austgen DM, Rochelle GT, Peng X, Chen C. Model of vapor-liquid equilibria for aqueous acid gas-alkanolamine systems using the electrolyte-NRTL equation. Ind Eng Chem Res 1989, 28, 1060–1073.
[36] Weiland RH, Chakravarty T, Mather AE. Solubility of carbon dioxide and hydrogen sulfide in aqueous alkanolamine. Ind Eng Chem Res 1993, 32, 1419–1430.
[37] Lawal A, Wang M, Stephenson P, Yeung H. Dynamic modelling of CO_2 absorption for post combustion capture in coal-fired power plants. Fuel 2009, 88, 2455–2462.
[38] Lin Y, Pan T-H, Wong DS-H, Jang SS, Chi YW, Yeh C-H. Plantwide control of CO_2 capture by absorption and stripping using monoethanolamine solution. Ind Eng Chem Res 2011, 50, 1338–1345.
[39] Gaspar J, Cormos A-M. Dynamic modeling and validation of absorber and desorber columns for post-combustion CO_2 capture. Comput Chem Eng 2011, 35, 2044–2052.

[40] Tontiwachwuthikul P, Meisen A, Lim J. CO_2 absorption by NaOH, monoethanolamine and 2-amino-2-methyl-1-propanol solutions in a packed column. Chem Eng Sci 1992, 47, 381–390.

[41] Khan FM, Krishnamoorti V, Mahmud T. Modelling reactive absorption of CO_2 in packed columns for post-combustion carbon capture applications. Chem Eng Res Des 2011, 89, 1600–1608.

[42] Jayarathna SA, Lie B, Melaaen MC. NEQ rate based modeling of an absorption column for post combustion CO_2 capturing. Energy Procedia 2011, 4, 1797–804.

[43] Plaza JM, Wagener DV, Rochelle GT. Modeling CO_2 capture with aqueous monoethanolamine. Energy Procedia 2009, 1, 161–166.

[44] Kvamsdal HM, Rochelle GT. Effects of the temperature bulge in CO_2 absorption from flue gas by aqueous monoethanolamine. Ind Eng Chem Res 2008, 47, 867–875.

[45] Errico M, Madeddu C, Pinna D, Baratti R. Model calibration for the carbon dioxide-amine absorption system. Appl Energy 2016, 183, 958–968.

[46] Abu-Zahra MRM, Schneiders HJ, Niedered JPM, Feron PHM, Versteeg GF. CO_2 capture from power plants –Part I. A parametric study of the technical performance based on monoethanolamine. Int J Greenhouse Gas Control 2007, 1, 37–46.

[47] Mores P, Scenna N, Mussati S. A rate based model of a packed column for CO_2 absorption using aqueous monoethanolamine solution. Int J Greenhouse Gas Control 2012, 6, 21–36.

[48] Strube R, Manfrida G. CO_2 capture in coal-fired power plants – Impact on plant performance. Int J Greenhous Gas Control 2011, 5, 710–726.

[49] Moioli S, Pellegrini LA, Gamba S. Simulation of CO_2 capture by MEA scrubbing with a rate-based model. Procedia Eng 2012, 42, 1651–1661.

[50] Madeddu C, Errico M, Baratti R. Rigorous modeling of a CO_2-MEA stripping system. Chem Eng Trans 2017, 57, 451–456.

[51] Neveux T, Moullec YL, Corriou J-P, Favre E. Modeling CO_2 capture in amine solvents: Prediction of performance and insights on limiting phenomena. Ind Eng Chem Res 2013, 52, 4266–4279.

[52] Mores P, Scenna N, Mussati S. CO_2 capture using monoethanolamine (MEA) aqueous solution: Modeling and optimization of the solvent regeneration and CO_2 desorption process. Energy 2012, 45, 1042–1058.

[53] Faramarzi L, Kontogeorgis GM, Michelsen ML, Thomsen K, Stenby EH. Absorber model for CO_2 capture by monoethanolamine. Ind Eng Chem Res 2010, 49, 3751–3759.

[54] Meldon JH, Morales-Cabrera JA. Analysis of carbon dioxide absorption in and stripping from aqueous monoethanolamine. Chem Eng Journ 2011, 171, 753–759.

[55] Pinsent BRW, Pearson L, Roughton FJW. The kinetics of combination of carbon dioxide with hydroxide ions. Trans Faraday Soc 1956, 52, 1512–1520.

[56] Clarke JKA. Kinetics of absorption of carbon dioxide in monoethanolamine solutions at short contact times. Ind Eng Chem Res 1964, 313, 239–245.

[57] Astarita G, Savage DW, Bisio A. Gas treating with chemical solvents. John Wiley & Sons, 1983.

[58] Hikita H, Asai S, Ishikawa H, Honda M. The kinetic of reaction of carbon dioxide with monoethanolamine, diethanolamine, triethanolamine by a rapid mixing model. Chem Eng Journ 1977, 1, 7–12.

[59] Hilliard MD. A predictive thermodynamic model for an aqueous blend of potassium carbonate, piperazine and monoethanolamine for carbon dioxide capture from flue gas. PhD thesis, The University of Texas at Austin, 2008.

[60] Øi LE. Aspen HYSYS simulation of CO_2 removal by amine absorption from a gas based power plant. SIMS 2007Conference Gøtheborg 2007, 71–81.

[61] Mores P, Scenna N, Mussati S. Post-combustion CO_2 capture process: Equilibrium stage mathematical model of the chemical absorption of CO_2 into monoethanolamine (MEA) aqueous solution. Chem Eng Res Des 2011, 89, 1587–1599.

[62] Jayarathna SA, Lie B, Melaaen MC. Dynamic modeling of the absorber of a post combustion CO_2 capture plant: Modeling and simulations. Comput Chem Eng 2013, 53, 178–189.
[63] Jayarathna SA, Lie B, Melaaen MC. Amine based CO_2 capture plant: Dynamic modeling and simulations. Int J Greenhous Gas Control 2013, 14, 282–290.
[64] Deshmukh RD, Mather AE. A mathematical model for equilibrium solubility of hydrogen sulfide and carbon dioxide in aqueous alkanolamine solutions. Chem Eng Sci 1981, 36, 355–362.
[65] Chen C-C, Evans LB. A local composition model for the excess Gibb energy of aqueous electrolyte systems. AIChE J 1986, 32, 444–454.
[66] Benamor A, Aroua MK. Modeling of CO_2 solubility and carbamate concentration in DEA, MDEA and their mixtures using the Deshmukh-Mather model. Fluid Phase Equilibr 2005, 231, 150–162.
[67] Jou F-Y, Mather AE, Otto FD. The solubility of CO_2 in a 30 mass percent monoethanolamine solution. Can J Chem Eng 1995, 73, 140–147.
[68] Guggenheim EA, Stokes RH. Activity coefficient of 2:1 and 1:2 electrolytes in aqueous solution from isopiestic data. Trans Faraday Soc 1958, 54, 1646–1649.
[69] Alie C, Backham L, Croiset E, Douglas PL. Simulation of CO_2 capture using MEA scrubbing: a flowsheet decomposition method. Energy Convers Manage 2005, 46, 475–487.
[70] Ziaii S, Rochelle GT, Edgar TF. Dynamic modeling to minimize energy use for CO_2 capture in power plants by aqueous monoethanolamine. Ind Eng Chem Res 2009, 48, 6105–6111.
[71] Oexmann J, Kather A. Post-combustion CO_2 capture in coal fired power plants: comparison of integrated chemical absorption process with piperazine promoted potassium carbonate and MEA. Energy Procedia 2009, 1, 799–806.
[72] Van Wagener DH, Rochelle GT. Stripper configurations for CO_2 capture by aqueous monoethanolamine. Chem Eng Res Des 2011, 89, 1639–1646.
[73] Posh S, Haider M. Dynamic modeling of CO_2 absorber from coal-fired power plant into an aqueous monoethanolamine solution. Chem Eng Res Des 2013, 91, 977–987.
[74] Luo X, Wang M. Improving prediction accuracy of a rate-based model of an MEA-based carbon capture process for large-scale commercial deployment. Engineering 2017, 3, 232–243.
[75] KA Hoff, Juliussen O, Falk-Pedersen O, Svendsen HF. Modeling and experimental study of carbon dioxide absorption in aqueous alkanolamine solutions using a membrane contactor. Ind Eng Chem Res 2004, 43, 4908–4921.
[76] Harun N, Nittaya T, Douglas PL, Croiset E, Ricardez-Sandoval LA. Dynamic simulation of MEA absorption process for CO_2 capture from power plants. Int J Greenhouse Gas Control 2012, 10, 295–309.
[77] van Krevelen DW, Hoftijzer PJ, Huntjens FJ. Composition and vapor pressure of aqueous solution of ammonia, carbon dioxide and hydrogen sulphide. Rec Trav Chim 1949, 68, 191–216.
[78] Freguia S. Modeling of CO_2 removal from flue gases with monoethanolamine. Master's thesis, The University of Texas at Austin, 2002.
[79] Lawal A, Wang M. Stephenson P, Koumpouras G, Yeung H. Dynamic modeling and analysis of post-combustion CO_2 chemical absorption process for coal-fired power plants. Fuel 2010, 89, 2791–2801.
[80] Sorel E. La rectification de l´alcohol. Gauthier Villars et Fils G. Masson, 1893.
[81] Seader JD, Henley HJ, Roper DK. Separation process principles. 3rd ed. John Wiley & Son, Inc., 2013.
[82] Seader JD, Henley EJ, Roper DK Separation Process Principles: Chemical and Biochemical Operations. John Wiley & Sons, Inc., 2010.
[83] Murphree EV. Rectifying Column Calculations with Particular Reference to N Component Mixtures. Ind Eng Chem 1925, 17, 747–750.

[84] Walter JF, Sherwood TK. Gas Absorption in Bubble-Cap Columns. Ind Eng Chem 1941, 33, 493–501.
[85] Afkhamipour M, Mofarahi M. Comparison of rate-based and equilibrium-stage models of a packed column for post-combustion CO_2 capture using 2-amino-2-methyl-1-propanol (AMP) solution. Int. J. Greenhouse Gas Control 2013, 15, 186–199.
[86] Øi LE. Comparison of Aspen HYSYS and Aspen Plus simulation of CO_2 absorption into MEA from atmospheric gas. Energy Procedia 2012, 23, 360–369.
[87] Lewis WK, Whitman WG. Principles of gas absorption. Ind Eng Chem 1924, 16, 1215–1220.
[88] Higbie R. The rate of absorption of a pure gas into a still liquid during short periods of exposure. Trans Am Inst Chem Engrs 1935, 31, 365–392.
[89] Danckwerts PV. Significance of liquid-film coefficients in gas absorption. Ind Eng Chem 1951, 43, 1460–1467.
[90] Leder F. The absorption of CO_2 into chemically reactive solutions at high temperatures. Chem Eng Sci 1971, 26, 1381–1390.
[91] Kvamsdal HM, Jakobsen JP, Hoff KA. Dynamic modeling and simulation of a CO_2 absorber column for post-combustion CO_2 capture. Chem Eng Process Intensif 2009, 49, 135–144.
[92] Pacheco MA, Rochelle GT. Rate-based modeling of reactive absorption and CO_2 and H_2S into aqueous amine methyldiethanolamine. Ind Chem Eng Res 1998, 37, 4107–4117.
[93] Kucka L, Muller I, Kenig EY, Gorak A. On the modelling and simulation of sour gas absorption by aqueous amine solutions. Chem Eng Sci 2003, 58, 3571–3578.
[94] Cormos A-M, Gaspar J. Assessment of mass transfer and hydraulic aspects of CO_2 absorption in packed columns. Int J Greenhouse Gas Control 2012, 6, 201–209.
[95] Gaspar J, Cormos A-M. Dynamic modeling and absorption capacity assessment of CO_2 capture process. Int J Greenhous Gas Control 2012, 8, 45–55.
[96] Kvamsdal HM, Hillestad M. Selection of model parameters correlations in a rate-based CO_2 absorber model aimed for process simulation. Int J Greenhouse Gas Control 2012, 11, 11–20.
[97] Mac Dowell N, Samsatli NJ, Shah N. Dynamic modelling and analysis of an amine-based post-combustion CO_2 capture absorption column. Int J Greenhouse Gas Control 2013, 12, 247–258.
[98] Nittaya T, Douglas PL, Croiset E, Ricardez-Sandoval LA. Dynamic modeling and evaluation of an industrial-scale CO_2 capture plan using monoethanolamine absorption processes. Ind Eng Chem Res 2014, 53, 11411–11426.
[99] Noeres C, Kenig EY, Gorak A. Modelling of reactive separation processes: reactive absorption and reactive distillation. Chem Eng Process Intensif 2003, 42, 157–178.
[100] Onda K, Takeuchi H, Okumoto Y. Mass transfer coefficient between gas and liquid phases in packed columns. J Chem Eng Jpn 1968, 1, 56–62.
[101] Bravo JL, Rocha JA, Fair JR. Mass transfer in gauze packings. Hydrocarb Processes 1985, 64, 91–95.
[102] Bravo JL, Rocha JA, Fair JR. A comprehensive model for the performance of columns containing structured packings. IChEME Symp S 1992, 128, A439-53.
[103] Henriques de Brito M, von Stockar U, Menendez Bangerter A, Bomio P, Laso M. Effective mass-transfer area in pilot plan column equipped with structured packing and with ceramic rings. Ind Eng Chem Res 1994, 33, 647–656.
[104] Tsai RE, Seibert AF, Eldridge RB, Rochelle GT. Influence of viscosity and surface tension on the effective mass transfer area of structured packing. Energy Procedia 2009, 1, 1197–1204.
[105] Tsai RE, Seibert AF, Eldridge EB, Rochelle GT. A dimensionless model for predicting the mass-transfer area of structured packing. AIChE J 2011, 57, 1173–1184.
[106] Billet R, Schultes M. Predicting mass transfer in packed columns. Chem Eng Technol 1993, 16, 1–9.

[107] Hanley B, Chen C-C. New mass-transfer correlations for packed towers. AIChE J 2012, 58, 132–152.
[108] Rocha JA, Bravo JL, Fair JR. Distillation columns containing structured packings: A comprehensive model for their performance. 2. Mass transfer model. Ind Eng Chem Res 1996, 35, 1660–1667.
[109] Chilton TH, Colburn AP. Mass transfer (absorption) coefficients. Prediction from data on heat transfer and fluid friction. Ind Chem Eng Chem 1934, 26, 1183–1187.
[110] Suess P, Spiegel L. Hold-up of Mellapak structured packing. Chem Eng Process Intensif 1992, 2, 119–124.
[111] Rocha JA, Bravo JL, Fair JR. Distillation columns containing structured packings: a comprehensive model for their performance. 1. Hydraulic models. Ind Eng Chem Res 1993, 32, 641–651.
[112] Stichlmair J, Bravo JL, Fair JR. General model for prediction of pressure drop and capacity of countercurrent gas/liquid packed columns. Gas Sep Purif 1989, 3, 19–28.
[113] Billet R, Schultes M. Influence of phase ratio on packing efficiency in columns for mass transfer processes. Chin J Chem Eng 1997, 5, 117–126.
[114] Billet R, Schultes M. Prediction of mass transfer columns with dumped and arranged packings: Updated summary of the calculation method of Billet and Schultes. Chem Eng Res Des 1999, 77, 498–504.
[115] Dugas RE. Pilot plant study on carbon dioxide capture by aqueous monoethanolamine. Master´s thesis, The University of Texas at Austin, 2006.
[116] Zhang Y, Chen C-C. Modeling CO_2 absorption and desorption by aqueous monoethanolamine solution with Aspen rate-based model. Energy Procedia 2013, 37, 1584–1596.
[117] Davis ME. Numerical methods and modelling for chemical engineers. New York, John Wiley & Sons, 1984.
[118] Kenig EY, Schneider R, Gorak A. Rigorous dynamic modelling of complex reactive absorption processes. Chem Eng Sci 1999, 54, 5195–5203.
[119] Schneider R, Kenig EY, Gorak A. Complex reactive absorption processes: model optimization and dynamic column simulation. Comput Aided Chem Eng 2001, 9, 285–290.
[120] Tontiwachwuthikul P, Meisen A, Lim CJ. Novel pilot plant technique for sizing gas absorbers with chemical reactions. Can J Chem Eng 1989, 67, 602–607.
[121] Edgar TF, Himmelblau DM, Lasdon LS. Optimization of chemical processes. McGraw Hill, 2001.
[122] Rodriguez-Aragon LJ, Lopez-Fidalgo J. Optimal design for the Arrhenius equation. Chemometr Intell Lab 2005, 77, 131–138.
[123] van der Spek M, Ramirez A, Faaij A. Improving uncertainty evaluation of process models by using pedigree analysis. A case study on CO_2 capture with monoethanolamine. Comput Chem Eng 2016, 85, 1–15.
[124] Nagy T, Valko E, Sedyo I, Zsely IGy, Pilling MJ, Turayi T. Uncertainty of the rate parameters of several important elementary reactions of the H_2 and syngas combustion systems. Combust Flame 2015, 162, 2059–2076.
[125] Schwaab M, Pinto JC. Optimum reference temperature for the reparameterization of the Arrhenius equation. Part 1: Problems involving one kinetic constant. Chem Eng Sci 2007, 62, 2750–2764.

Notation

Roman symbols

Symbol	Unit	Description
a	$[-]$	Activity
C	$\left[\frac{kmol}{m^3}\right]$	Molar concentration
c_P	$\left[\frac{kJ}{kmol\ K}\right]$	Specific heat at constant pressure
D	$\left[\frac{m^2}{s}\right]$	Diffusion coefficient
d_{eq}	$[m]$	Packing equivalent diameter
$\dot{E}$	$\left[\frac{kJ}{s}\right]$	Interphase energy flow rate
F	$\left[\frac{kmol}{s}\right]$	Molar flow rate
FDR	$[-]$	Film discretization ratio
H	$[m]$	Packing height
He	$[bar]$	Henry constant
h	$\left[\frac{kJ}{kmol}\right]$	Specific enthalpy
h_T	$\left[\frac{W}{m^2\ K}\right]$	Heat transfer coefficient
K_{eq}	$[-]$	Equilibrium constant
k	$\left[\frac{m^3}{kmol\ s}\right]$	Kinetic constant
k_M	$\left[\frac{m}{s}\right]$	Material transfer coefficient
k_T	$\left[\frac{W}{m\ K}\right]$	Thermal conductivity
L	$\left[\frac{kmol}{s}\right]$	Liquid molar flow rate
L_c	$[m]$	Characteristic length
M	$[kmol]$	Molar hold-up
MSE	$[-]$	Mean squared error
$\dot{N}$	$\left[\frac{kmol}{s}\right]$	Interphase molar flow rate
$\dot{N}^R$	$\left[\frac{kmol}{s}\right]$	Reaction molar flow rate

Symbol	Unit	Description
P	$[bar]$	Pressure
P^{sat}	$[bar]$	Vapor pressure
Pe	$[-]$	Peclet number
$\dot{Q}$	$\left[\frac{kJ}{s}\right]$	Heat flow rate input to a stage/segment
$\dot{Q}^R$	$\left[\frac{kJ}{s}\right]$	Reaction heat flow rate
RCF	$[-]$	Reaction condition factor
R	$\left[\frac{kmol}{m^3 s}\right]$	Reaction rate
S	$[m^2]$	Column cross-sectional area
SE	$[-]$	Standard error
t	$[s]$	Time
U	$[kJ]$	Energy hold-up
V	$\left[\frac{kmol}{s}\right]$	Gas/vapor molar flow rate
x	$[mol\ frac]$	Liquid component molar fraction
y	$[mol\ frac]$	Gas/vapor component molar fraction
y*	$[mol\ frac]$	Equilibrium gas/vapor component molar fraction
z	$[m]$	Axial coordinate
$\tilde{z}$	$[-]$	Component electric charge number

Greek symbols

Symbol	Unit	Description
γ	$[-]$	Activity coefficient
δ	$[m]$	Film thickness
ε	$\left[\frac{m^3}{m^3}\right]$	Void fraction
ϕ	$[-]$	Fugacity coefficient
ψ	$\left[\frac{m^3}{m^3}\right]$	Fractional hold-up

Subscripts

Symbol	Description
f	Forward reaction
i	i-th component
j	j-th stage or segment
k	k-th component
M	Material
mix	Mixture
r	Reverse reaction
T	Thermal

Superscripts

Symbol	Description
app	Apparent
F	Feed
L	Liquid phase
p	Phase
V	Vapor/gas phase

Indexes extremes

Symbol	Description
n_c^L	Number of components in the liquid phase
n_c^V	Number of components in the gas/vapor phase

Abel Briones-Ramírez and Claudia Gutiérrez-Antonio

5 Optimal design methodology for homogeneous azeotropic distillation columns

Abstract: Research pertaining to azeotropic distillation has received periodic attention over the years; however, it has attracted more interest recently, particularly in the design area. The Boundary Value Method was the first design methodology reported for azeotropic distillation columns, which was the basis for the development of other design methodologies. However, all these methodologies have in common the fact that they verify the feasibility of the split of the components of the mixture in a graphic way. This issue limits the application of the procedure to ternary mixtures or pseudo-ternary mixtures. In this work, we propose an optimal design methodology for the separation of multicomponent azeotropic mixtures to overcome this limitation. The main advantage of this methodology is the use of an analytical procedure to evaluate the feasibility of the split, and calculate, at the same time, the minimum reflux ratio. The method is tested with different azeotropic mixtures and the results are compared with those obtained by a multiobjective genetic algorithm with constraints, linked to an Aspen Plus processes simulator. Results show that designs obtained with the proposed procedure are similar to those generated through stochastic optimization, but with less time and computational resources.

Keywords: azeotropic distillation, optimal design, multicomponent mixtures, homogeneous azeotropic mixtures, shortcut design methodology

5.1 Introduction

Nowadays, the design of new processes in chemical engineering takes into account multiple objectives; usually, these objectives include the minimization of total annual costs and environmental issues, along with the maximization of profit and products generation. Thus, we have to design optimal processes. This approach in chemical engineering emerges, mainly, as a consequence of the increase in the price of fuels and raw materials, the climate change problem, and the availability of the appropriate tools to perform the intensification of the process. Thus, the goal of the task of

Abel Briones-Ramírez, Departamento de Investigación y Desarrollo, Exxerpro Solutions, Av. Del Av. del Sol 1B Local 4B Col. El Sol, Querétaro, México
Claudia Gutiérrez-Antonio, Facultad de Química, Universidad Autónoma de Querétaro, Av. Cerro de las Campanas Col. Las Campanas s/n, Querétaro, México

https://doi.org/10.1515/9783110596120-005

designing equipment or processes implies to obtain the optimal design: perform the required task (separation, mixing, or reaction) with reduced costs of capital and operation. If the process has minimum operation costs, as a consequence of reduced energy consumption, there is also minimal emission of greenhouse gases associated with the generation of the required energy.

Among the industrial separation processes, distillation is one of the most used. Usually, the thermodynamic efficiency of a distillation column is around 10%; therefore, the increase of the efficiency of these systems receives a lot of attention [1–8]. Nevertheless, most of these efforts focus on non-azeotropic mixtures. In spite of this, the separation of azeotropic mixtures through distillation is widely used in petrochemical, chemical, and biochemical industries for the production of many important chemicals required in our daily life [9]. Over time, azeotropic distillation has received periodic attention, usually as a consequence of world events that required products obtained through this kind of distillation, like benzene in the World War II, for instance [9]. However, in recent years, research in azeotropic distillation has increased, with special emphasis on the design area; it is worth mentioning that in 1985, Doherty et al. [10, 11] proposed the first design method for azeotropic distillation columns.

The intricate nature of azeotropic mixtures makes the design procedure more complicated than for non-azeotropic mixtures. Unlike non-azeotropic mixtures, the composition space of azeotropic mixtures has regions, as a consequence of boundary distillation that occurs due to the presence of azeotropes. The existence of different regions restricts the feasibility of the separations, since compositions of feed and products must be located in the same region of the composition space. Therefore, an important step, before the design, is the verification of the feasibility of the split; since we can waste a lot of time trying to design a column with an infeasible split. Therefore, Stichlmair and Fair [12] proposed a verification procedure, where the feasibility zones are calculated using the distillation boundary, a residue curve through feed composition, and the lines of direct and indirect material balances, as shown in Figure 5.1.

From Figure 5.1 we observe that the feasibility it is easily verified for ternary mixtures, but this is complex for the multicomponent ones. Once the feasibility of the split is verified, then we apply a design methodology. Several shortcut design methodologies were developed for this purpose, the first one being the Boundary Value Method (BVM) [10, 11]. These methodologies are described next.

The BVM considers material balances in form of differential equations to calculate the number of stages, in each section of the column [10, 11]; these equations are solved from outside to inside the column, and the feed stage is located in the intersection of the profiles of both sections. However, in many cases the composition resulting from the intersection of operation profiles is different from the feed composition [13]. Moreover, the BVM uses residue curves to make calculations at total reflux, which are operation lines at total reflux in packed columns, while in staged columns this condition is represented by distillation lines [12]. In addition,

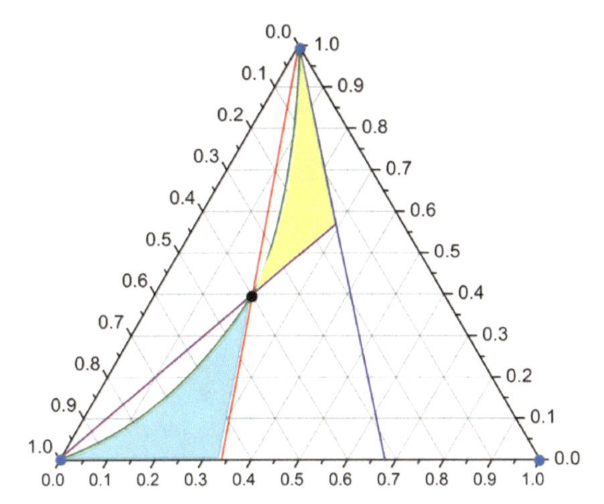

Figure 5.1: Feasibility zone for a feed composition of an azeotropic mixture. The triangular diagram includes the indirect mass balance (purple line), direct mass balance (red line), distillation line through the feed composition (green curve), and the distillation boundary (blue line). Yellow and green areas represent the feasible zones for distillate and bottom compositions, respectively. Blue circles represent pure components while black circle indicates the feed composition.

the BVM assumes that at the minimum reflux condition there is collinearity between the feed, the feed pinch point, and the saddle pinch point. Nevertheless, this collinearity is just an approximation in azeotropic mixtures.

In 1990, Julka and Doherty [14] proposed the Zero-volume Criterion (ZVC) which was an extension to multicomponent mixtures of the BVM. The ZVC method considers that at the minimum reflux ratio there is collinearity between the feed composition and the C-1 pinch points located in a hyperplane, where C is the number of components. Nevertheless, this method works properly only in direct and indirect splits, since the required pinch points to determine the minimum reflux ratio must be known a priori. Since this method was conceived as an extension of the BVM, it uses the criterion of collinearity at the minimum reflux ratio condition; however, this criterion is just an approximation in azeotropic mixtures.

Köhler et al. [15] presented the Minimum Angle Criterion (MAC) method in 1991. This method considers that at the minimum reflux ratio the angle between the feed composition and a pinch point of each section of the column must be minimum. Then, at the minimum reflux ratio condition the minimal angle is zero; this consideration is equivalent to the collinearity criterion used in the BVM and the ZVC. In addition, the minimal angle criterion is empiric and it does not have a physical explanation for more than three components.

Later, Pöllmann et al. [16] developed the Eigenvalue Criterion (EC), which was a hybrid of the BVM and ZVC methods. In the EC method, the profiles are calculated just after the pinch point, in the direction of the unstable eigenvector of the pinch point, using a tray-to-tray calculation. The minimum reflux ratio is the smallest value that allows the intersection of the profile of the rectifying section with a pinch point of the stripping section. Nevertheless, the computational effort increases when there is more than one unstable eigenvector in the pinch point.

In 1998, Bausa et al. [17] proposed the Rectification Bodies Method (RBM). They defined a rectification body as a triangle delimited by all possible trajectories of pinch points, from products to feed, in the stripping or rectifying section. For each trajectory, the entropy generated is calculated to verify that all the trajectories are thermodynamically consistent. In the RBM, the minimum reflux ratio is the value that allows the rectifying and stripping bodies to touch each other in one point. Nevertheless, according to Petlyuk et al. [18] this method is not always effective in all possible types of pinch branches.

Later, Thong and Jobson [19] proposed a method to design distillation columns for the separation of multicomponent azeotropic mixtures. The procedure employs a feasibility matrix to identify the combination of the design parameters of the distillation column. The authors claim that the accuracy of the design method could be improved by the application of an iterative convergence procedure [19].

On the other hand, Liu et al. [20] developed a method that considered that a distillation region with one additional saddle node has compartments, which are considered as ideal regions. Therefore, in these compartments the Fenske-Underwood-Gilliland method was applied [21], considering azeotropes as pseudo components. In their paper, they reported that the minimum reflux ratio and number of stages calculated with their method had deviations until 22% and 30%, in respect to the values calculated in the HYSIS simulator.

In 2007, Gutiérrez-Antonio et al. [13] presented a design method based on distillation lines, which used algebraic material balances solved from the extremes to inside the column. The feed stage is located using as a criterion the minimum difference in composition, between each composition in the operation profile in both sections and the feed composition. Therefore, the assumption of collinearity is not used, and this overcomes the limitations of the previous design methodologies [10–12, 14–17, 20]. However, the calculation of the minimum reflux ratio is restricted to ternary mixtures, since the intersection of the operation profiles is verified graphically. Nevertheless, in some cases the design of the distillation column includes pinch zones, which increase the capital costs without improving the separation.

Moreover, in the literature there are other design methods for azeotropic distillation columns based on optimization strategies, which are briefly described next.

In 2008, Lucia et al. [22] presented a method that used the approach of the shortest stripping line to minimize the energy requirements in distillation. The design method is a global optimization algorithm, which represents a mixed integer nonlinear programming problem. Several examples are tested with this strategy, and the results show that the minimum energy requirements are found for any number of components. In spite of the fact that this procedure is reliable and useful for multicomponent mixtures, it requires deep knowledge in mathematical programming and considerable computational resources to get the optimal design.

In the same line of optimization strategies, in 2009 an optimization strategy was proposed for the design of distillation sequences for homogeneous azeotropic mixtures [23]. In this strategy, two stages are considered. In the first one, shortcut models are used to generate the initial values for the optimization process of the second stage. In the second stage, the problem is reformulated and solved as a nonlinear continuous optimization problem. However, the use of this strategy requires deep knowledge in mathematical programming techniques.

Later in 2012, a method was presented for the conceptual design of homogenous distillation systems [24]. This method includes a software tool that uses ∞/∞ – analysis along with piece-wise linearized invariant manifolds of vapor–liquid equilibria. According to the authors, the key is the replacement of the original, highly nonlinear process model by piece-wise linear submodels [24]; this simplification reduces the complexity of the mathematical model.

Recently, Skiborowski et al. [25] presented a reformulation of the feed angle method to have a shortcut-based process optimization. The improvement considered elements of the feed angle method and the zero volume criterion [14]; this resulted in a shortcut strategy that is easily incorporated into an optimization strategy.

In general, all the aforementioned methodologies are useful to design azeotropic distillation columns; from all of them, we identify two main approaches: shortcut methodologies and optimization strategies. In the first category, the BVM and its derivations assume a collinearity criterion to calculate the minimum reflux, which is just an approximation in azeotropic mixtures. The method based on a minimum difference in composition [13] overcomes the previous limitations; however, the calculation of the minimum reflux ratio is restricted to ternary mixtures, and in some cases the design of the distillation column includes pinch zones. In the second category, the optimization algorithms are reliable and useful for multicomponent mixtures, but its use requires considerable computer resources. Moreover, it is important to mention that the procedure to verify the feasibility of the split has the disadvantage of being a graphical method [12]; and, as a consequence, its use is limited to ternary mixtures. It would be desirable to have a design methodology useful for multicomponent mixtures with an analytical verification of the feasibility, which generates optimal designs.

In this work, we propose an optimal methodology to design distillation columns for the separation of homogeneous azeotropic multicomponent mixtures; this methodology considers material balances and the concept of minimum difference in composition based on a previous contribution [13]. The number of stages is calculated with algebraic material balances, in each section of the column, which are solved from outside to inside, in order to satisfy the global material balance. The minimum reflux ratio is calculated through an analytical procedure, and it is useful for multicomponent mixtures. The methodology also includes a criterion to optimize the structure of the distillation column through the elimination of the pinch zones, if there are any; it is important to mention that the elimination of pinch zones does not increase the energy consumption of the separation. The proposed methodology generates optimal designs with minimum number of total stages and a reduced heat duty due to the optimal location of the feed stage; thus, the designs have a reduction in capital cost and better energy efficiency, in comparison with those generated with other methodologies. In this manner, this methodology lies in the process intensification area, which is defined as the process development that involves reduction in equipment sizes that lead to improvements in reactions kinetics, better energy efficiency, reduction in capital cost, and improvements in process safety [26]. The methodology is applied to different azeotropic mixtures and their results are compared with those obtained from a multiobjective genetic algorithm with constraints [27]. Results show that the new method generates similar designs to those from the optimization algorithm, but it requires less time and computational resources.

5.2 Optimal design methodology

In this section, we describe the optimal design methodology along with the calculation of the design variables of a simple distillation column, Figure 5.2.

First, we present the set of equations to calculate the number of stages. Next, we show the proposed strategy to calculate the minimum reflux ratio, which also verifies the feasibility of the split. Finally, we use the concept of minimum difference in composition to locate the feed stage.

5.2.1 Calculation of number of stages

The methodology employs distillation lines, since staged columns are considered [12]. The method consists of algebraic material balances, which are solved from outside to inside the distillation column. The equations for rectifying (Eq. 5.1) and stripping (Eq. 5.2) sections are:

$$y_{n+1,i} = \frac{r}{r+1} x_{n,i} + \frac{1}{r+1} z_{D,i} \tag{5.1}$$

$$x_{n+1,i} = \frac{s}{s+1} y_{n,i} + \frac{1}{s+1} z_{B,i} \tag{5.2}$$

$$\sum_{i=1}^{C} x_i = 1; \sum_{i=1}^{C} y_i = 1 \tag{5.3}$$

Where:

$x_{j,i}$ Liquid phase composition of component *i* in stage *j*

$y_{j,i}$ Vapor phase composition of component *i* in stage *j*

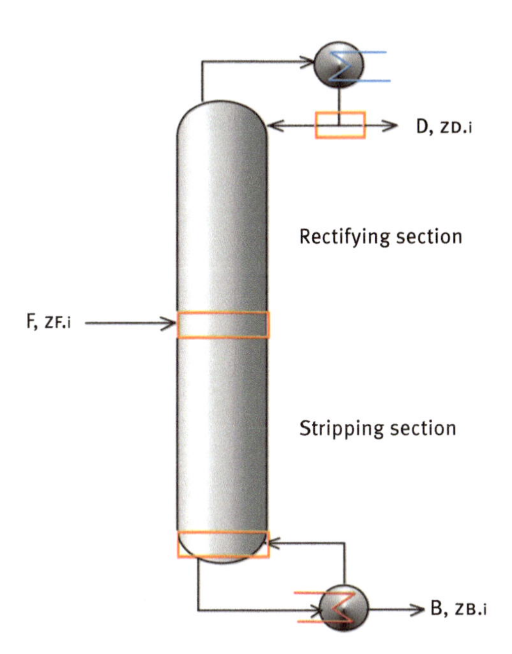

Figure 5.2: Design variables of a simple distillation column. F, D, and B represent the feed, distillate, and bottoms streams of the distillation column, while $Z_{F,i}$, $Z_{D,i}$, and $Z_{B,i}$ are the composition of the components of the feed, distillate, and bottoms streams, respectively. Rectifying section comprises the stages between the feed stage and the top of the column, while the stages included between the feed stage and the bottoms of the column represent the stripping section.

r Reflux ratio
s Reboil ratio
$z_{D,i}$ Composition of component *i* in distillate product, it can be liquid or vapor phase
$z_{B,i}$ Composition of component *i* in bottoms product, it can be liquid or vapor phase
C Number of components
n Number of stages

The Eq. 5.3 represents the summation equations for liquid and vapor phases; therefore, Eqs. 5.1 and 5.2 are applied to *i=1* to C-1 components. Considering the thermal condition, *q*, of the feed stream, the reboil ratio is calculated with Eq. 5.4:

$$s = (r + q)\left[\frac{z_{B,1} - z_{F,1}}{z_{F,1} - z_{D,1}}\right] + (q - 1) \tag{5.4}$$

Where:
$Z_{F,i}$ Composition of component *i* in feed stream, it can be liquid or vapor phase

Moreover, the feed, bottom, and top product compositions must be collinear to satisfy the global material balance. Thus, the fulfillment of the global material balance for multicomponent mixtures implies that the bottom compositions for components are calculated with the next equation:

$$z_{B,l} = z_{F,l} + (z_{F,l} - z_{D,l}) \left[\frac{s - q + 1}{r + q} \right] \quad (5.5)$$

$$l = 2, \ldots, C - 1 \quad (5.6)$$

Once the degrees of freedom are fixed, we calculate the reboil ratio with Eq. 5.4, and the bottoms compositions of l components with Eqs. 5.5 and 5.6. For instance, for C=3 two specifications of distillate fraction and one of bottoms composition must be provided; thus, the missing bottom compositions are calculated with Eqs. 5.5 and 5.6. Finally, we compute the rectifying and stripping profiles with Eqs. 5.1–5.3, along with thermodynamic equations for the estimation of the phase equilibrium, for a given operating reflux ratio. The operating reflux ratio is calculated based on the minimum reflux ratio, in a procedure that is explained next.

5.2.2 Calculation of the minimum reflux ratio

According to Towler et al., the intersection of composition profiles anywhere in the composition space is a necessary and sufficient condition to establish the feasibility of the split [28]. This necessary and sufficient condition is employed to calculate the minimum reflux ratio.

The minimum reflux ratio occurs when the operation profiles intersect for the first time. In order to find the minimal value for the reflux ratio we propose an iterative process, where the reflux ratio is varied in small increments (it has been verified that increments of 0.01 are sufficiently small). For each reflux ratio, the operation profiles are calculated (Eqs. 5.1–5.6) and compared to find their first intersection; at this point, the reflux ratio is minimum. This implies that we have to guarantee the intersection profiles; however, considering that the compositions of the operation profiles are not continuous, it is clear that establishing a tolerance criterion does not guarantee the intersection. In addition, there is a wide variety of azeotropic maps and, as a consequence, different configurations in the operation profiles. Then, in order to ensure the intersection of the operation profiles, we construct lines for each two consecutive composition points of each operating profile; these points were generated previously (Eqs. 5.1–5.6). Once the lines are constructed, we verify the intersection of each line of the rectifying profile against each line of the stripping profile. The expressions for the lines in rectifying (Eq. 5.7) and stripping (Eq. 5.8) sections are the following parametric equations:

$$R_{m \to m+1} = R_m + \alpha(R_{m+1} - R_m) \quad (5.7)$$

$$0 \le \alpha \le 1 \tag{5.8}$$

$$S_{k \to k+1} = S_k + \beta(S_{k+1} - S_k) \tag{5.9}$$

$$0 \le \beta \le 1 \tag{5.10}$$

$$R_m + \alpha(R_{m+1} - R_m) = S_k + \beta(S_{k+1} - S_k) \tag{5.11}$$

$$0 \le \alpha, \beta \le 1 \tag{5.12}$$

Where:

$R_{m \to m+1}$ Line in the rectifying section that includes the compositions m and $m+1$

$S_{k \to k+1}$ Line in the stripping section that includes the compositions k and $k+1$

α, β Parameters of the equation from 0 to 1

The parameters α and β are introduced to consider that the intersection can take place in any part of the line, including the extremes (for instance, compositions m and $m+1$, in the case of the rectifying section). So, if there are two parameters α and β such that $R_m = S_k$, then there is intersection between the operation profiles. As result, we have a simple linear system with two unknown parameters and C equations, one for each component (Eqs. 5.11–5.12). The pseudo code of the minimum reflux ratio algorithm is presented in Figure 5.3.

It is important to mention that the procedure to calculate the minimum reflux ratio also verifies the feasibility of the split in multicomponent mixtures; if there is no minimum reflux found in a very wide range of values then the separation is not feasible. Once the minimum reflux ratio is found, then the operating reflux ratio is fixed with heuristic rules for azeotropic mixtures of Douglas ($1.1R_{min}$) [29] or Doherty and Malone ($1.5R_{min}$) [1]. It is important to mention that this methodology allows the calculation of the minimum reflux ratio and the feasibility of the split in multicomponent ideal, nonideal, and azeotropic mixtures.

5.2.3 Location of the feed stage

The position of the feed stage is necessary to have the complete design of the column. The feed stage is located where there is a minimum difference between compositions in rectifying and stripping profiles and the feed stream. In this way, the feed stream is introduced to the column in the stage where the composition is more similar and, as a consequence, the disturbances are minimal and the design is more efficient. The two operation profiles of the column are considered in the calculation of the difference in composition, so we can know the number of stages in each section and the location of the feed stage. The composition difference for each section of the column is determined with the following equations:

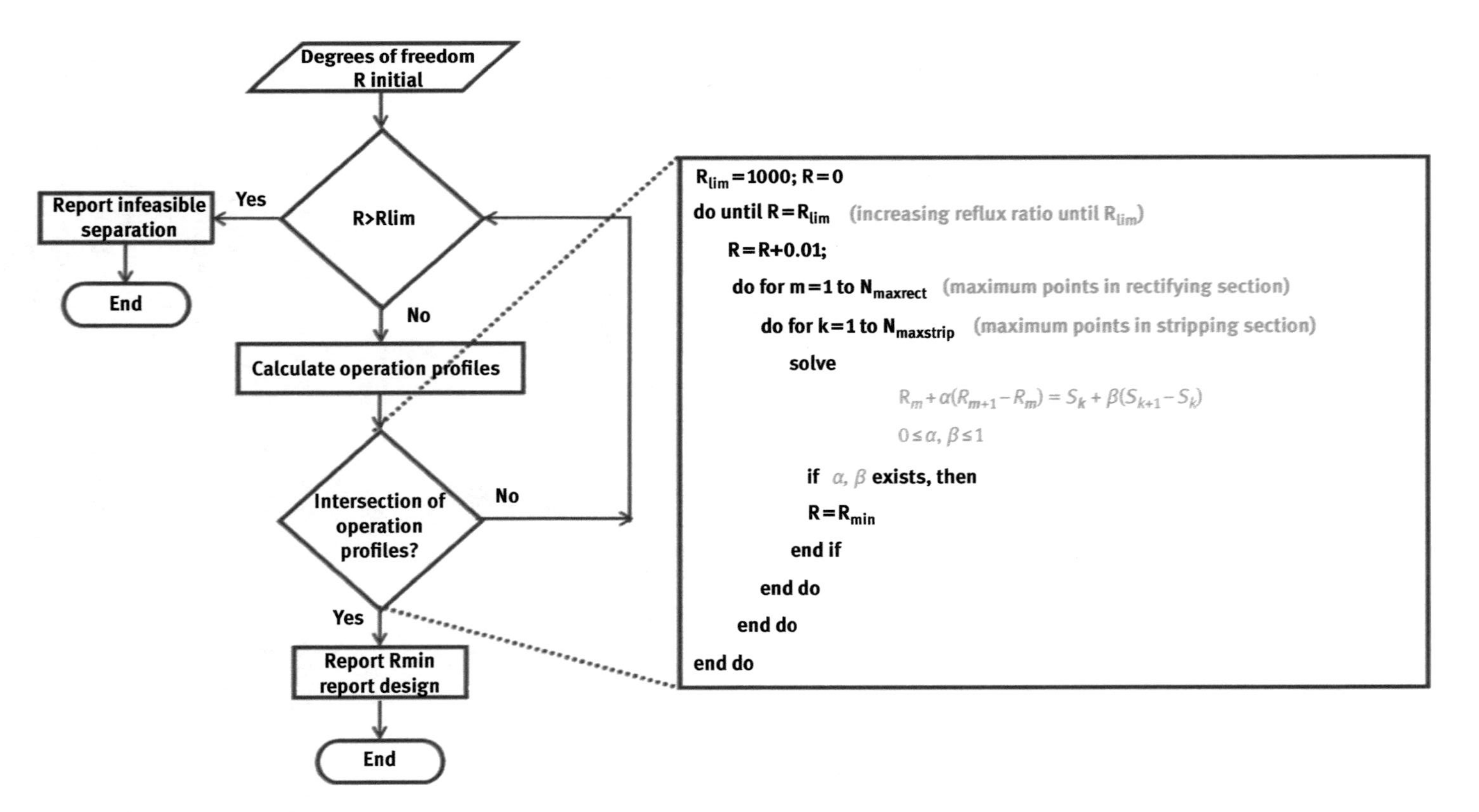

Figure 5.3: Pseudo code of the minimum reflux ratio algorithm.

$$d_R = \sqrt{\sum_{i=1}^{c} (z_{F,i} - x_{NR,i})^2 + \sum_{i=1}^{c} \left(z_{F,i} - y_{NR,i}\right)^2} \tag{5.13}$$

$$d_S = \sqrt{\sum_{i=1}^{c} (z_{F,i} - x_{NS,i})^2 + \sum_{i=1}^{c} \left(z_{F,i} - y_{NS,i}\right)^2} \tag{5.14}$$

Where:

d_R Composition difference between the feed and stage NR in the rectifying section
d_S Composition difference between the feed and stage NS in the stripping section

Once the complete design is obtained, then we apply the criterion shown in Eq. 5.15 to the operation profiles to eliminate the pinch zones, if there are any:

$$\left|\frac{d^2 x_i}{dn^2}\right| \le 1.0{*}10^{-4} \tag{5.15}$$

The differential equation (Eq. 5.15) is calculated as composition differences of the operation profiles. Therefore, the optimal location of the feed stage along with the refinement of the final design generates a better structure for the separation. Moreover, the algorithm for the minimum reflux ratio also identifies the minimum energy consumption, which is used for the design of the distillation column. The most remarkable feature is that this methodology just considers material balances, and very simple equations.

5.3 Cases of study

The proposed design method is applied to four azeotropic mixtures, which are shown in Table 5.1; in all cases, the feed stream is introduced as saturated liquid to the distillation column.

Table 5.1: Selected mixtures as study cases.

Mixture	Components (A-B-C)
M1	acetone – isopropanol – water
M2	acetone – chloroform – benzene
M3	ethanol – water – ethylene glycol
M4	methanol – isopropanol – water

The operating pressure of the distillation columns for all mixtures is 14.7 psia, and due to this the vapor phase is considered ideal. The NRTL solution model describes the nonideality in the liquid phase [30]. Adjusted parameters for all mixtures are taken from Dechema Collection [31], at 14.7 psia. With these parameters, boiling point surfaces are generated for each mixture to observe its thermodynamic nature.

The designs obtained with the proposed methodology are compared with a multiobjective genetic algorithm with constraints [27], coupled to Aspen Plus; this design tool performs the simultaneous minimization of number of stages and heat duty, considering the complete models for material and energy balances along with the phase equilibrium calculations. Usually, the total annual cost is used as a single objective function, which includes capital and operating costs. However, this function considers the equipment cost, energy, parameters of actualization and projection that depend on time and location [27]. Thus, we choose as objectives of the optimization strategy the number of stages and heat duty, which are variables in competition in a distillation column; the first one is associated with capital costs, and the second one directly affects the energy consumption [27]. As manipulated variables we select the reflux ratio, number of stages, and location of the feed stage. The multiobjective genetic algorithm is implemented in MATLAB. A complete description of the optimization strategy is presented in [27], while the explanation of the link between Matlab and Aspen Plus is exposed in [32]. As parameters of the evolutionary strategy, we selected 30 generations of 250 individuals each for all mixtures. The parameters of the multiobjective genetic algorithm are obtained through a tuning process, where typically graphs are constructed to show the evolution of the objectives versus the generation number, for a fixed number of individuals.

5.4 Results

In this section, we present the designs obtained from the application of the proposed optimal design methodology and the stochastic strategy for all the study cases. All studied mixtures present one distillation boundary, but with different curvatures. From here, we name shortcut design to that obtained from the proposed methodology, and evolutionary design to that obtained from the stochastic strategy. The number of stages is counted from top to bottom, and since total condensers are considered, they are not counted as a stage. Moreover, in Section 5.6 some exercises are included. In these exercises, the objective is the use of the proposed methodology and the comparison with the results presented in this section.

5.4.1 Mixture acetone-isopropanol-water

The mixture acetone-isopropanol-water, M1, presents one azeotropic point, composed by isopropanol (X_{isop}: 0.6738) and water at 80.13 °C, at atmospheric pressure.

This azeotropic point generates a straight distillation boundary, with a slight curvature near the acetone node, Figure 5.4.

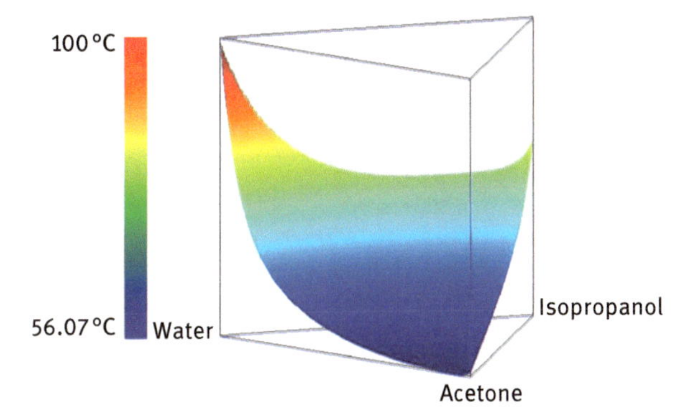

Figure 5.4: Boiling point surface for mixture acetone-isopropanol-water. The triangular diagram includes the composition of the ternary mixture and the pure components, which are located in the vertex. The boiling temperature is represented by the vertical axe, where blue color represent the minor boiling point temperature and red color the highest one.

The specifications of feed and product compositions for this mixture are shown in Table 5.2; from Table 5.2 we observe that the separation is performed in the region delimited by the nodes of acetone, water, and the azeotropic point.

Table 5.2: Feed and products specifications for the selected study cases.

Component	Feed	Distillate	Bottom	Component	Feed	Distillate	Bottom
Acetone	0.63	0.93	0.02	Ethanol	0.33	0.5026	1e-6
Isopropanol	0.07	0.02	0.1717	Water	0.33	0.4973	9.999e-3
Water	0.30	0.05	0.8083	Ethylene glycol	0.33	1e-04	0.99
Acetone	0.50	0.997	0.30	Methanol	0.33	0.99	1e-5
Chloroform	0.20	0.0015	0.2798	Isopropanol	0.33	0.006	0.492
Benzene	0.30	0.0015	0.4202	Water	0.34	0.004	0.508

After applying the proposed methodology and the evolutionary strategy with the parameters mentioned before, we obtained the designs shown in Table 5.3.

The shortcut design considers 18 total stages, and the feed stage is located in stage 13; this design is obtained considering the compositions specified in Table 5.2. The shortcut design is simulated in Aspen Plus, considering the complete energy and material balances, and the resulting compositions and heat duty are shown in the same line of Table 5.3. We observe that the purities are 2% under the specifications;

Table 5.3: Shortcut and evolutionary designs for the selected study cases.

	Reflux ratio	Feed stage	Number of stages	X_D	X_B	Heat Duty, Btu/h
	acetone – isopropanol – water (M1)					
Shortcut design	0.66385	13	18	0.9116	0.0574	1,561,735.1
Adjusted shortcut design	0.9667	13	18	0.9300	0.0200	1,824,709.8
Evolutionary design	1.25	13	18	0.9391	0.0015	1,854,668.6
	acetone – chloroform – benzene (M2)					
Shortcut design	2.75	46	49	0.9970	0.3000	1,379,681.9
Adjusted shortcut design	2.75	46	49	0.9970	0.3000	1,379,681.9
Evolutionary design	2.78	46	49	0.9970	0.3000	1,390,319.2
	ethanol – water – ethylene glycol (M3)					
Shortcut design	0.08668	11	18	4.82e-20	0.9900	1,593,957.8
Adjusted shortcut design	0.03368	11	18	1.02e-6	0.9900	1,533,173.0
Evolutionary design	0.05143	11	18	4.73e-7	0.9901	1,542,826.9
	methanol – isopropanol – water (M4)					
Shortcut design	7.6245	10	72	0.9900	6.12e-5	4,360,829.1
Aadjusted shortcut design	7.6245	10	72	0.9900	6.12e-5	4,360,829.1
Evolutionary design	3.8710	23	73	0.9900	6.13e-5	2,467,453.1

however, with an increase of 45% of the value of the reflux ratio the purities are reached, along with an increment of 16% in the energy consumption (these results are shown in the row adjusted shortcut design of Table 5.3). The shortcut design is compared with an evolutionary design resulting from the application of a multiobjective optimization, considering the simultaneous minimization of the reflux ratio and number of stages. As a result, a set of optimal designs are generated, which are known as the Pareto Front. Practically, the Pareto Front includes a set of designs from the minimum number of stages to the minimum reflux ratio, and all designs between these points; in this case, the Pareto Front consists of more than 100 designs with different numbers of total stages. Therefore, we select one of these designs to make the comparison with the proposed methodology. Each one of the designs included in the Pareto Front is optimal, but they represent different trades-offs between the optimization objectives. Thus, from all designs of the Pareto Front, we choose the one with the same number of total stages as the shortcut design, and we take this design to make the comparison. In this way, we can analyze the similarity in the resulting structure and the heat duty of the distillation column with both procedures, along with the fulfillment of the specified compositions. From Table 5.3, it is observed that the location of the feed stage in the evolutionary design is the same as in the shortcut design; the main difference is in the reflux ratio (88% higher in the evolutionary design), and even the heat duty is greater (18% higher in the evolutionary design) than the one observed for the adjusted shortcut design. However, in the evolutionary design the purities are slightly higher (0.97%) than the specification.

The time required to obtain the design with the shortcut methodology is in the order of seconds, instead of the hours required to perform the multiobjective optimization and get the evolutionary design. Therefore, there is a considerable reduction in computational time when the shortcut methodology is employed, without losing the optimality of the design of the distillation column.

5.4.2 Mixture acetone-chloroform-benzene

The mixture acetone-chloroform-benzene, M2, presents one azeotropic point, which is binary and composed of chloroform and acetone (X_{acet}: 0.3396) at 64.46 °C, at atmospheric pressure. This azeotropic point generates a curved distillation boundary, and modifies the curvature of the boiling point surface, Figure 5.5.

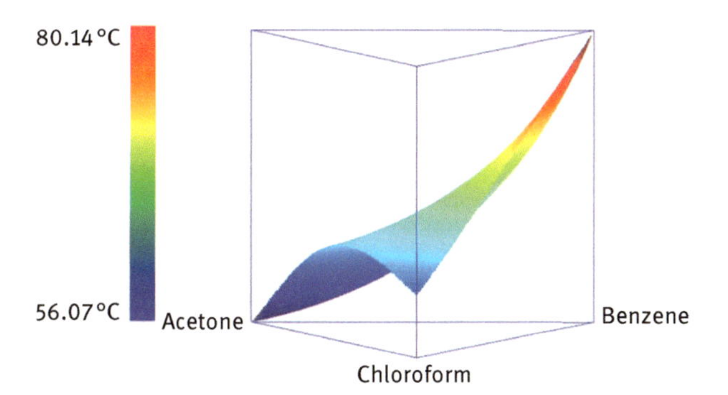

Figure 5.5: Boiling point surface for mixture acetone-chloroform-benzene. The triangular diagram includes the composition of the ternary mixture and the pure components, which are located in the vertex. The boiling temperature is represented in the vertical axe, where blue color represent the minor boiling point temperature and red color the highest one.

The specifications of feed and product compositions for this mixture are shown in Table 5.2, which was presented previously. Thus, we generate shortcut and evolutionary designs through the application of the proposed methodology and the evolutionary strategy, whose results are shown in Table 5.3.

The shortcut design includes 49 total stages, and the feed stage is located in stage 46; this design is obtained considering the compositions specified in Table 5.2. The shortcut design is simulated in Aspen Plus, and the resulting compositions and heat duty are in the row of the same name of Table 5.3. We observe that the purities reach the specifications; thus, the same purities of the shortcut design are presented in the row adjusted shortcut design of Table 5.3. Similar to the previous case, we perform the multiobjective optimization to obtain the Pareto Front, considering the simultaneous minimization of the reflux ratio and number of stages. From all designs of the Pareto Front, we choose the one with the same number of stages as the shortcut design to perform the comparison. As can be seen in Table 5.3, the evolutionary design has the same total number of stages and location of the feed stage in the distillation column as the one generated with the proposed methodology. In the reflux ratio value, the difference is small (1% higher in the evolutionary design), and in this case the heat duty is slightly greater (0.77 %) than the one observed for the adjusted shortcut design.

5.4.3 Mixture ethanol-water-ethylene glycol

The mixture ethanol-water-ethylene glycol, M3, presents one azeotropic point, integrated by ethanol (X_{etha}: 0.9187) and water at 78.09 °C, at atmospheric pressure. This azeotropic point generates a slightly curved distillation boundary, and the boiling point surface is almost flat, Figure 5.6.

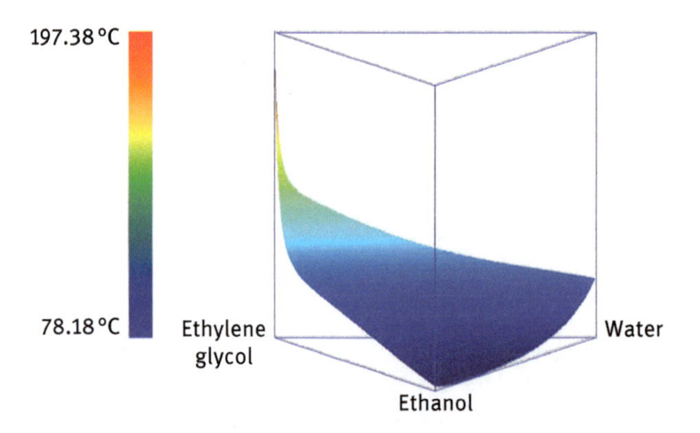

Figure 5.6: Boiling point surface for mixture ethanol-water-ethylene glycol. The triangular diagram includes the composition of the ternary mixture and the pure components, which are located in the vertex. The boiling temperature is represented in the vertical axe, where blue color represent the minor boiling point temperature and red color the highest one.

The specifications of feed and product compositions for this mixture are shown in Table 5.2, which was presented previously. Applying the proposed methodology and the evolutionary strategy, we obtain the designs shown in Table 5.3.

The shortcut design considers 18 total stages, and the feed stage is located in stage 11; this design is obtained considering the compositions specified in Table 5.2.

The shortcut design is simulated in Aspen Plus, in the Radfrac module, and the resulting compositions and heat duty are in the row of the same name of the Table 5.3. As we observe, the compositions reach the purities specifications, and it is even possible to decrease the reflux ratio by 2.5 times without losing the design specifications; this information is presented in the row adjusted shortcut design of Table 5.3. To compare the shortcut design, we perform the multiobjective optimization to generate the Pareto Front. From all designs of the Pareto Front, we choose the one with the same number of stages as the shortcut design to perform the comparison. As can be seen in Table 5.3, the evolutionary design has the same total number of stages and location of the feed stage as the one generated with the proposed methodology. In the reflux ratio, the difference is considerable (68% less in the evolutionary design), and in this case the heat duty is slightly less (3.3%) than the one observed for the adjusted shortcut design.

5.4.4 Mixture methanol-isopropanol-water

The mixture methanol-isopropanol-water, M4, presents one azeotropic point, integrated by isopropanol (X_{isop}: 0.6738) and water at 80.13 °C, at atmospheric pressure. This azeotropic point generates a straight distillation boundary, with a slight curvature near the acetone node; this surface is similar to the one presented in Figure 5.4.

The specifications of feed and product compositions for this mixture are shown in Table 5.2, presented previously. Applying the proposed methodology and the evolutionary strategy, we obtained the designs shown in Table 5.3.

The shortcut design considers 72 total stages, and the feed stage is located in stage 10; this design is obtained considering the compositions specified in Table 5.2. This shortcut design is simulated in Aspen Plus, considering the complete energy and material balances, and the resulting purities and heat duty are in the row of the same name of the Table 5.3. As we observe, the purities reach the specifications without any adjustment; due to this, the same values are presented in the row adjusted shortcut design of Table 5.3. In order to evaluate the shortcut design we perform the multiobjective optimization, considering the simultaneous minimization of the reflux ratio and number of stages. From all designs of the Pareto Front, we choose the one with the same number of stages as the shortcut design, and we take this design to perform the comparison. From Table 5.3 we observe that the evolutionary design has a different number of feed stage (one stage of difference) in comparison to the one generated with the proposed methodology. This modification decreases at half the reflux ratio, and as a consequence the heat duty of the distillation column is 76% less.

From the previous study cases, we observe that the designs obtained with the shortcut procedure are optimal, since they are quite similar to those obtained through a mixed integer, nonlinear and rigorous optimization strategy. This optimality is due to the location of the feed stage according to the concept of minimum difference in composition. In addition, the simple criterion of the second derivate helps to eliminate the pinch zones that could appear. The most important aspect is that all these operations involve the use of material balances and simple equations. Moreover, the algorithm to calculate the minimum reflux ratio is appropriate to verify the feasibility of the split in multicomponent mixtures, since if the intersection of profiles does not occur, then the separation is unfeasible.

5.5 Conclusions

An optimal design procedure for azeotropic distillation columns is presented. The procedure considers algebraic material balances, which are solved from outside to inside the column. The intersection of the composition profiles is verified to guarantee the feasibility of the split; however, the feed stage is located at the minimum composition difference between the stages of each operation profile and the feed stream. In this way, the feed stage is better located, generating fewer disturbances, and making the design more efficient. In addition, we propose an analytical procedure to calculate the reflux ratio and verify the feasibility at the same time, in multicomponent mixtures; in addition, this methodology only considers material balances. Moreover, the methodology includes the simple criterion to eliminate pinch zones, in case these exist. Finally, it is important to mention that the resulting designs from the application of the design methodology have a reduction in capital cost and better energy efficiency, in comparison with the designs generated with other methodologies.

Exercises

1. Design the distillation column to perform the separation of the mixture methyl acetate-methanol-ethyl acetate with an equimolar feed composition. Assume that the feed stream is fed as saturated liquid, and the desirable recoveries of the key components are 90%. Make the design with the methodology proposed in this chapter and perform a comparison with the BVM. What differences do you find in both designs?
2. Repeat exercise 1, but assume that the feed stream is saturated vapor. What differences do you find in the structure and heat duty of the distillation columns?
3. Design the distillation column to perform the separation of the mixture acetone-isopropanol-water with an equimolar feed composition. Assume that the feed stream is fed as saturated liquid, and the desirable recoveries of the key components are 99%. Make the design with the methodology proposed in this

chapter and perform a comparison with the results obtained for the same mixture in this chapter. What differences do you find in both designs?

4. Repeat exercise 3, but assume that the recoveries of key components are 90%. What differences do you find in the structure and heat duty of the distillation columns?

Acknowledgments: Financial support provided by FOMIX Gobierno del Estado de Querétaro-CONACyT, through grant 279753, for the development of this project is gratefully acknowledged.

References

[1] Olujić Ž, Jödecke M, Shilkin A, Schuch G, Kaibel B. Equipment improvement trends in distillation. Chemical Engineering and Processing: Process Intensification 2009, 48(6), 1089–1104.
[2] Caballero J. Thermally Coupled Distillation. Computer Aided Chemical Engineering 2009, 27, 59–64.
[3] Yildirim Ö, Kiss AA, Kenig EY. Dividing wall columns in chemical process industry: A review on current activities. Separation and Purification Technology 2011, 80(3), 403–417.
[4] Rong B-G, Errico M. Synthesis of intensified simple column configurations for multicomponent distillations. Chemical Engineering and Processing: Process Intensification 2012, 62, 1–17.
[5] Pătruţ C, Bîldea CS, Liţă I, Kiss AA. Cyclic distillation – Design, control and applications. Separation and Purification Technology 2014, 125(7), 326–336.
[6] Kiss AA, Olujić Ž. A review on process intensification in internally heat-integrated distillation columns. Chemical Engineering and Processing: Process Intensification 2014, 86, 125–144.
[7] Jana AK. Advances in heat pump assisted distillation column: A review. Energy Conversion and Management 2014, 77, 287–297.
[8] Aneesh V, Antony R, Paramsivan G, Selvaraju N. Distillation technology and need of simultaneous design and control: A review. Chemical Engineering and Processing: Process Intensification 2016, 104, 219–242.
[9] Doherty MF, Malone MF. Conceptual Design of Distillation Systems. NY, USA, Wiley-VCH, 1998.
[10] Van Dongen DB, Doherty MF. Design and synthesis of homogeneous azeotropic distillations. 1. Problem formulation for a single column. Industrial & Engineering Chemistry Fundamentals 1985, 24(4), 454–463.
[11] Levy SG, Van Dongen DB, Doherty MF. Design and synthesis of homogeneous azeotropic distillations. 2. Minimum reflux calculations for nonideal and azeotropic columns. Industrial & Engineering Chemistry Fundamentals 1985, 24(4), 463–474.
[12] Stichlmair JG, Fair JR. Distillation: principles and practice. NY, USA, Wiley-VCH, 1998.
[13] Gutiérrez-Antonio C, Jiménez-Gutiérrez A. Method for the Design of Azeotropic Distillation Columns. Industrial & Engineering Chemistry Research 2007, 46(20), 6635–6644.
[14] Julka V, Doherty MF. Geometric Behavior and Minimum Flows for Nonideal Multicomponent Distillation. Chemical Engineering Science 1990, 45(7), 1801–1822.
[15] Köhler J, Aguirre P, Blass E. Minimum Reflux Calculations for Nonideal Mixtures Using the Reversible Distillation Model. Chemical Engineering Science 1991, 46(12), 3007–3021.

[16] Pöllmann P, Glanz SB, Blaβ E. Calculating Minimum Reflux of Nonideal Multicomponent Distillation Using Eigenvalue Theory. Computers and Chemical Engineering 1994, 18(Suppl.), S49–S53.
[17] Bausa J, Watzdorf Rv, Marquardt W. Shortcut Methods for Nonideal Multicomponent Distillation:1. Simple Columns. AIChE Journal 1998, 44(10), 2181–2198.
[18] Petlyuk F, Danilov R, Skouras S, Skogestad S. Identification and analysis of possible splits for azeotropic mixtures – 1. Method for column sections. Chemical Engineering Science 2011, 66, 2512–2522.
[19] Thong D Y-C, Jobson M. Multicomponent homogeneous azeotropic distillation 2. Column design. Chemical Engineering Science 2001, 56(14), 4393–4416.
[20] Liu G, Jobson N, Smith R, Wahnschafft OM. Shortcut Design Method for Columns Separating Azeotropic Mixtures. Industrial and Engineering Chemistry Fundamentals 2004, 43(14), 3908–3923.
[21] Henley EJ, Seader JD. Equilibrium-Stage Separation Operations in Chemical Engineering. NY, USA, Wiley, 1981.
[22] Lucia A, Amale A, Taylor R. Distillation Pinch Points and more. Computers and Chemical Engineering 2008, 33, 1342–1364.
[23] Kraemer K, Kossack S, Marquardt W. Efficient Optimization-Based Design of Distillation Processes for Homogeneous Azeotropic Mixtures. Industrial and Engineering Chemistry Research 2009, 48, 6749–6764.
[24] Ryll O, Blagov S, Hasse H. ∞/∞ – Analysis of homogeneous distillation process. Chemical Engineering Science 2012, 84, 315–332.
[25] Skiborowski M, Recker S, Marquardt W. Shortcut-based optimization of distillation – based processes by a novel reformulation of the feed angle method. Chemical Engineering Research and Design 2018, 132, 135–148.
[26] Reay D, Ramshaw C, Harvey A. Process intensification. IChemE, Rugby, 2008.
[27] Gutiérrez-Antonio C, Briones-Ramírez A. Pareto front of Petlyuk sequences using a multiobjective genetic algorithm with constraints. Computers and Chemical Engineering 2009, 33, 454–464.
[28] Towler GP, Thong DYC, Castillo FJL. Homogeneous Azeotropic Distillation. 1. Design Procedure for Single-Feed Columns at Nontotal Reflux. Industrial and Engineering Chemistry Fundamentals 1998, 37, 987–997.
[29] Douglas JM. Conceptual Design of Chemical Processes. NY, USA, Mc Graw-Hill Inc., 1998.
[30] Gutiérrez-Antonio C, Iglesias-Silva GA, Jiménez-Gutiérrez A. Effect of Different Thermodynamic Models on the Design of Homogeneous Azeotropic Distillation Columns. Chemical Engineering Communications 2008, 195(9),1059–1075.
[31] Gmehling J, Onken U. Vapor-Liquid Equilibrium Data Collection, Dechema, Chemistry Data Series, DECHEMA, Frankfurt/Main, 1977.
[32] Briones-Ramírez A, Gutiérrez-Antonio C. Multiobjective optimization of chemical processes with complete model using Matlab and Aspen Plus. Computación y Sistemas 2018, 22(4), 1157–1170.

Nima Nazemzadeh, Isuru A. Udugama, Rasmus Fjordbak Nielsen, Kristian Meyer, Eduardo S. Perez-Cisneros, Mauricio Sales-Cruz, Jakob Kjøbsted Huusom, Jens Abildskov and Seyed Soheil Mansouri

6 Graphical tools for designing intensified distillation processes: Methods and applications

Abstract: This chapter will focus on application of graphical tools in designing intensified distillation units. The use of intensified distillation units in industry has been steadily rising due to their economic and environment benefits. The objective of this chapter is to illustrate how established shortcut graphical tools can be easily adopted to design and develop intensified distillation columns. The outstanding feature of graphical methods is their intuitiveness that makes them more convenient to use than mathematical models. As such, graphical tools can help save time and computational costs in comparison to the use of mathematical model-based optimization. To this end, this chapter will cover the application of extended McCabe-Thiele and Ponchon-Savarit on heat-integrated distillation columns (HIDiC) and the application of driving force approach on cyclic and reactive distillation. For each type of the abovementioned intensified columns, an algorithmic toolbox is provided and a case study is introduced to verify the implemented method. The results of the application examples can be easily reproduced by following the corresponding toolbox. This chapter will also discuss the advantages of using such simple graphical methods in process design as well as the apparent disadvantages.

Keywords: Distillation, process design, short cut methods, reactive distillation, heat integrated distillation, cyclic distillation, intensified processes

Nima Nazemzadeh, Department of Chemistry, Materials and Chemical Engineering “Giulio Natta”, Milano, Italy
Isuru A. Udugama, Rasmus Fjordbak Nielsen, Kristian Meyer, Jakob Kjøbsted Huusom, Jens Abildskov, Seyed Soheil Mansouri, Process and Systems Engineering Center, Department of Chemical and Biochemical Engineering, Technical University of Denmark, Lyngby, Denmark
Eduardo S. Perez-Cisneros, Departamento de Ingeniería de Procesos e Hidráulica, Universidad Autónoma Metropolitana-Iztapalapa, Ciudad de México, Mexico
Mauricio Sales-Cruz, Departamento de Procesos y Tecnología, Universidad Autónoma Metropolitana – Cuajimalpa, Ciudad de México, Mexico

https://doi.org/10.1515/9783110596120-006

6.1 Introduction

The relentless tide of ever tightening economic and environmental needs have led to significant improvements in the general design of chemical processes [1]. These improvements have generally focused on improving a plants energy consumption, raw material depletion, and environmental impact. As such, there is an increasing need for efficiently identifying, designing, and developing appropriate processes and unit operations to achieve these improvements in a cost- and time-efficient manner, which enables industrial chemical plants to remain competitive in today's globally competitive environment.

One key unit operation that is found in many industrial chemical processes is distillation where on an average more than 40% of a plant's energy is consumed [2]. As such, the development of high energy efficient separation technologies can play a major role in reducing the energy consumption of plants and improve its overall efficiency. Distillation is a well-proven and robust technology and has for decades been the industrial work horse for large volume fluid separations. One way to improve the performance of distillation as well as an entire chemical process in general is to use process intensification (PI) options where the overall cost is reduced and sustainability is improved. As defined by Stankiewicz [3], an intensified process is "any chemical engineering development that leads to a substantially smaller, cleaner, and more energy efficient technology." With over 40,000 distillation columns in operation worldwide [4], even small improvements in the technology of distillation will be able to drastically reduce both production and maintenance costs of distillation columns.

One way to achieve such improvements would be to utilize the difference in boiling points between the distillate and bottom products. Energy needs to be supplied for vaporization at a higher temperature that can be recovered by the condensation of the overhead vapor. Hence, energy is degraded along the column, and typical numbers of second law efficiencies for industrial distillation operations are in the range of 10%–20%. Furthermore, separation efficiencies in tray columns are less than optimal due to liquid back mixing and a flow pattern which is not the ideal counter current flow. On the other hand, PI of distillation unit operations can also be used, where distillation is combined with another unit operation to create an application of intensified and multifunctional processes in chemical industry [5]. Several applications of PI principles are realized so far on an industrial scale including reactive distillation, microreactors, rotating packed bed systems, etc. However, reactive distillation with already over 150 industrial applications is one of the most successful intensified processes on an industrial scale [6]. In case of reactive distillation, chemical transformation and separation based on volatility is combined in one equipment. The advantage being the continuous product removal pushing the chemical reaction in a favorable direction and a complicated separation task may be easier if compounds, which are difficult to separate may simply be depleted by the chemical transformation. Some of the applications of reactive distillation in industry are, e.g., production of methyl-tert-butyl-ether

(MTBE) [7], ethyl-tert-butyl-ether (ETBE) [8], and methyl acetate [9]. Figure 6.1(a) represents the schematic configuration of reactive distillation with one reaction zone in the middle, rectifying and stripping sections on top and bottom of the column, respectively. There also exists a catalyst supply stream for the reaction in reactive part.

Besides reactive distillation, which is widely implemented in the chemical industry [10], there are other interesting distillation combinations such as heat integrated distillation (HIDiC) as well as membrane distillation dividing wall columns, and cyclic distillation [11]. There have always been attempts to construct a separation unit that operates with a higher thermodynamic efficiency than the classical column where the stripper and rectification sections are separated and operated at different pressure through the use of vapor compression and a throttle valve. The advance is that by elevating the pressure in the rectifier compared to the stripper, it is possible to obtain a temperature profile in the rectifier which is higher than in the stripper. Therefore, internal heat integration between the two column sections becomes possible leading to partial condensation throughout the rectification section and partial boil-up through the stripper. Hence, the external utility requirement for the reboiler and condenser can be reduced, and in some cases these even become redundant. However, these gains in reboiler duty can be partially offset by the steam/electricity consumption of the compressor. Overall, it can be shown that the HIDiC column achieves the separation at a higher thermodynamic efficiency compared to the classical column. Crucial to success is that the mixture being separated is close boiling such that the pressure elevation can be minimized. A schematic figure of this intensified distillation column is presented in Figure 6.1(b). Cyclic (or periodic) distillation is another attempt to enhance the efficiency of separation in the column. Here the liquid is kept at each tray for a fixed period of time while vapor is flowing through the column as presented in Figure 6.1(c-1). This means that each tray experiences a batch like distillation over this vapor period. At the end of the period a tray draining procedure is performed such that the liquid volume on each tray is transferred to the tray below and reflux liquid and the feed is added on the top tray and on the feed tray, respectively, as illustrated in Figure 6.1(c-2). A liquid amount as distillate and bottom product is also taken out during this period. It is important that the liquid from different trays is not mixed during this so-called liquid flow period. Repeating the vapor period causes the separation to occur as shown in Figure 6.1(c-3). The advantage of the cyclic operation is that higher separation efficiencies on each tray can be achieved, which means that the same separation can be carried out with less trays. Furthermore, there is no need for liquid down comers, which means that the whole tray area is available for the separation.

From a process design point of view, these intensified distillation column designs create inherent complexities and limitations. This is mainly due to the loss in degrees of freedom because of integration of unit operations, functions, or phenomena. From an industrial point of view, these intensified distillation units can be perceived as too hard to design and handle without the use of complex process simulations. As such

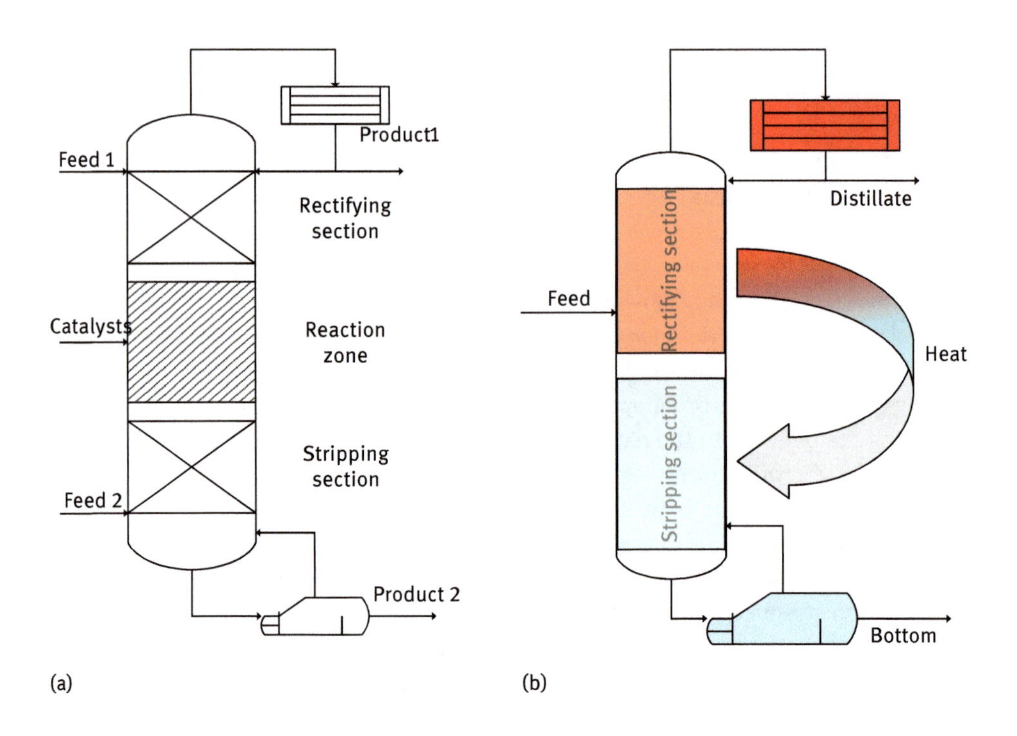

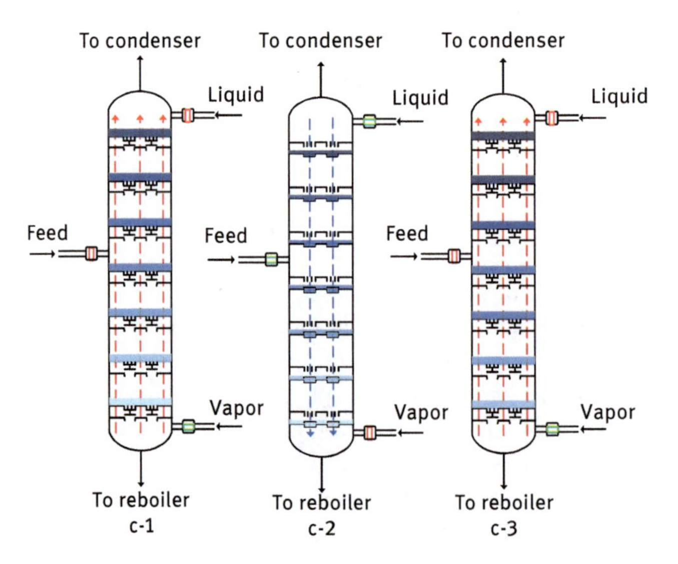

Figure 6.1: Schematic figures of reactive distillation (a), heat integrated distillation (b), cyclic distillation (c). For cyclic distillation, one complete cycle has been presented.

the industrial acceptance of these distillation column designs can be hampered by the lack of simple tools and equations that can be employed to perform first pass calculations. In this regard, simple shortcut methods and graphical representations which are well known and used for classical distillation processes in the engineering community can be used as a basis for developing graphical and shortcut tools that can be used to support the development. Illustration of these intensified distillation columns would be highly valuable, especially in the early stages of process design and synthesis. From an industrial point of view, the development of such shortcut methods for these intensified distillation columns would allow process engineers to quickly and effectively screen through multiple applicable complex distillation columns and understand if these methods would result in significant improvement in process operations/energy efficiency. As such the development of these shortcut methods will facilitate their transition from concepts into implementable solutions. It is also important to note that development of shortcut design methods does not replace the need for rigorous process simulation for final evaluation. Neither can design be separated from the process development and operational issues. Rather the objective of these types of graphical shortcut tools would be to provide a quick assessment for a process engineer that is looking into different types of complex distillation columns.

6.2 Shortcut methods for distillation column design

The synthesis of separation configurations and sequences is a key area of the overall process synthesis as separation is one of the most important elements in any chemical process. As such, a large effort has been placed on determining ways of finding the best separation configurations, especially with regards to distillation columns. Classes of methods for synthesis of separations are methods based on heuristics, algorithmic methods, and graphical methods.

Heuristic methods: Sometimes termed as rule-based methods, these are some of the oldest methods in the field of process synthesis. Many of the heuristic methods that are implemented and found in literature are based on methods from before the 1980s. These methods have been reviewed thoroughly in a prior publication [12]. The consistent use of these methods is the result of their ease of use and quick nature. However, despite this utility, there are some drawbacks to this type of method; on occasion, these rules are conflicting or contradictory in nature that must be reconciled based on personal judgment [13, 14]. Heuristic methods still have their function, especially as preliminary screening methods, to reduce the complexity of a process synthesis problem. As mentioned, many heuristic methods were developed as some of the oldest methods and these still are used or are the foundation for more recent heuristic developments [15–18].

Algorithmic methods: The purpose of developing and using algorithmic methods is to synthesize the optimal separation sequence and/configuration by using optimization methods and tools. The problem-solving ability lies in circumventing the overwhelming combinatorial problems encountered during the trial synthesis of equipment interconnections. A general feature of these methods is that they are principally rigorous and, thereby, computational time and effort is a requirement. Furthermore, many algorithmic methods take advantage of heuristics to reduce the size of search space. It has been shown, as also discussed above, that heuristic methods can be used effectively to guide equipment matching in certain relatively well-defined synthesis problems. Although some of algorithmic methods can also be classified as heuristics, they differ significantly from the heuristic methods previously described because they maintain a certain analytic and algorithmic content and, therefore, they are considered under parametric methods.

Algorithmic methods can be divided into two subcategories: Parametric methods, methods based on thermodynamic insights, and algorithmic methods including mathematical programming methods. Parametric studies try to identify a set of critical parameters and analyse their effect on the optimum separation configuration. Since there are various factors that affect a separation configuration, these parametric methods have limited applications on specific process schemes. Their principal results consist of a set of heuristic guidelines for a quick and simple estimate. Methods based on thermodynamic insights use the pure component properties of the compounds involved in the separation sequence. These methods employ the relationships between physicochemical properties and separation principles; the physical feasibility limits for a specified separation problem are analyzed and the corresponding separation process sequences are determined. Mathematical programming methods, on the other hand, attempt to synthesize the best separation configuration by employing various mathematical-based optimization techniques. The main advantage of these methods is their rigor and the fact that only these methods can guarantee an optimum solution. On the other hand, the main drawback of these methods is their applicability to only simplified problems as they require extensive computational power if they are to be used on rigorous and large problems.

After the review paper of Nishida *et al.* (1981), Westerberg [19] highlights the developments in both algorithmic and heuristic-based tree search algorithms to find the better systems of non-heat integrated distillation columns. In this work, he also provides insight to help with designing energy efficient systems for designing non-sharp separations. Nonetheless, Malone *et al.* [20] comparatively point out some conflicts and inaccuracies of heuristic-based approaches to the selection of distillation sequences and how these can be eliminated by developing analytical criteria. They demonstrate that an appropriate functional form can be derived based on approximate solutions to the design equations and an order of magnitude analysis of the cost equations. Their results highlight that analytical approach gives better predictions than the qualitative heuristics and show that most of generally accepted heuristics are actually incorrect. Gadkari and Govind [21] introduce a rank order function based on total inter-stage flow in an ideal distillation column. Their derived function is superior to several rank order functions proposed in earlier studies [16, 20] and also easier to compute. A more significant advantage of this function lies in the analysis and synthesis of complex and thermally coupled sequences because of its ease of computation. Table 6.1 presents an overview of design methods for distillation column design and their associated pros and cons.

Table 6.1: An overview of design methods for designing distillation columns.

Classification	Method	Pros	Cons
Graphical	– McCabe-Thiele [22] – Ponchon-Savarit [23] – Extended Ponchon-Savarit [24]	– Intuitive – Balance equations are solved graphically – Computationally inexpensive	– Limited (pseudo) binary mixtures (because only two-dimensional representation is possible) – Decision variables are implicitly incorporated into mass or energy balances
Optimization-based	– Superstructure optimization [25, 26] – Stochastic optimization [27]	– Usually generic and flexible economic cost function optimization – Multicomponent mixtures – Model parameter uncertainty issues can be taken into account	– Complete mathematical representation of the model is required – Solution is not trivial – Requires sophisticated solutions strategies depending on the model complexity – Computationally expensive

6.3 Why graphical methods as shortcut design tools for distillation processes?

Graphical tools are well-known methods for carrying out first pass calculations on classical distillation units where these can be used in the determination of optimum column configuration during initial design. This simple tool therefore can facilitate the principal objectives of chemical process design of obtaining economically profitable, safe, and environmentally sustainable process to convert specific feed stream(s) into specific product(s). The information generated from these methods together with mass and energy balances can then be used to find optimal equipment sizes and configuration together with operating conditions. However, with introduction of more intensified distillation designs the application of these graphical tools has become difficult as there are additional requirements that need to be taken into consideration.

While many methods are algorithmic, the use of graphs to visualize thermodynamic behavior is often useful and simpler; hence, it is extensively employed in early stage process synthesis methods. Methods that employ chemical engineering principles, by the way of thermodynamics, try to include a justifiable reason for the steps that are followed for synthesis methods. This utilizes thermodynamic laws and properties to calculate aspects of the separation. Pure component and mixture properties are determined and these values are implemented to solve aspects of the separation sequence through a series of equations or graphs. A simple graphical method that employs ternary information is used by Stichlmair and Herguijuela [28] to design zeotropic and azeotropic distillations. Rather than implementing complex mathematics based on thermodynamics, this method allows for the simple graphical determination of the possible top and bottom fractions in a ternary distillation. This is useful in preliminary design of distillation of zeotropic and azeotropic ternary mixtures as it enables the screening and elimination of impossible separations. The major contribution is that it allows for the design of complete separation of binary azeotropic mixtures through the use of an entrainer. Through implementation of distillation lines, the use of ternary diagrams is a useful method for the screening of possible separation possibilities of ternary mixtures.

Similarly, graphical methods have been employed in representing a driving-force-based approach that is based on thermodynamic information [29]. This method exploits the differences in chemical and or physical properties for the separation system. With this information, graphs are created detailing the difference in the driving force; here the driving force is the difference in the value of a certain property for the separation mixture that is relevant to the task performed. Often the driving force exploits a difference in concentration, or partial pressure for gases. When using the graphical method, it is quickly evident where the driving force is maximized. Then, by employing algorithms this knowledge can be implemented to determining information about the various types of separation sequences. It is shown to be a useful method to determine the optimal, or near optimal, design for separation sequences based solely on phase composition data. This method requires little mathematical effort and visually relates the physical thermodynamic elements to the process design. Another large area of development within graphical methods for separation synthesis is the use of residue curve maps (RCMs), as reviewed by Fien and Liu [30]. These are frequently used for azeotropic and reactive distillation. RCMs are types of ternary diagrams that are used for vapor/liquid separations; they are formed by residue curves, which are found by determining the residual liquid remaining when heated in a single stage. The primary function of these curves generally is limited to entrainer feasibility screening. While applications show that these are valuable tools especially since they provide thermodynamic insight into the situation. Doherty and Caldarola [31] developed a residue curve map analysis for the sequencing of ternary azeotropic distillation columns so feasible and infeasible sequences can be distinguished simply. Based on this analysis, a generalized definition of extractive distillation for separating homogeneous azeotropic mixtures is provided. They conclude that the only requirement of an extractive entrainer is that it does not divide the two components to be separated into different distillation regions. If this is met, the extractive entrainer may introduce new azeotropes into the system without penalty. Distillation column sequences can be easily developed for each entrainer falling in the catalog of feasible residue curve maps. Additionally, Barbosa and Doherty [32] developed residue cure maps for ideal and nonideal distillation of homogenous reactive mixtures. A set of differential equations were used to compute the residue curve maps. The maps show that by allowing the components of a mixture to react, distillation boundaries are either created or eliminated and some reactive azeotropes can be avoided as products of distillation. This result provides fundamental knowledge for the design and synthesis of sequence of reactive distillation columns. Bernot *et al.* [33] developed methods to obtain initial sequence of batch distillation columns during early-stage design. They show that the sequence of separation steps required to achieve a given set of pure component (and possibly azeotropic) cuts depends on which batch distillation region the feed stream lies in. For many systems, there is a preferred region that generates a separation system that is much simpler than the alternatives generated from other regions. This phenomenon suggests a potential to adjust the feed composition, whenever possible, by adjusting upstream variables such as reactor conversion. Wasylkiewicz *et al.* [34] proposed a new algorithm for the design of heterogeneous azeotropic distillation column with top decanter. They apply the method on two design case studies of column sequences with recycle streams and top decanters. Process configuration for difficult separation problems is quickly found by examining the entire component space. Using this method, distillation boundaries can be effectively crossed by using distillation columns with top decanters placed in two-liquid phase regions close to heterogeneous azeotropes.

More recently, Sobočan and Glavič formulated a simple, thermodynamic method for distillation sequences [35]. They employ an extended grand composite curve (EGCC) on a graph of temperature versus enthalpy flow rate difference. This diagram enables the user to design the distillation sequences maximizing the use of heat integration. To measure this, the minimum integrality criterion is introduced to indicate the amount of heat integration that is possible. Various case studies show that this is an effective and quick screening method that often produces the optimum sequence when rigorously simulated and compared. As with all methods, however, this method is limited in that it cannot determine the operating pressure or the absolute temperature of the columns. With the availability of both traditional and novel graphical tools for distillation column design in this chapter, we have focused our extension work on the traditional McCabe-Thiele method and Ponchon-Savarit graphical design methods as well as focusing on the more advanced driving force method. We have then selected case studies from reactive distillation, heat integrated distillation, and cyclic distillation to illustrate how these modified methods would work in practice. Section 6.3 will give a short description of the three mentioned graphical tools, while Section 6.4 will discuss the modification/extension of the graphical distillation column design methods with example case studies.

6.3.1 Driving force approach

The driving force approach is a method to design distillation operations (reactive or nonreactive), which was first proposed by Gani and Bek-Pedersen [36]. Like the McCabe-Thiele method [22] it is based on the graphical representation of vapor-liquid data. However, in this approach, driving force (*DF*), which is a function of vapor and liquid composition is plotted against liquid (or vapor) composition. It is defined as the difference between two coexisting phases (vapor and liquid) and can only represent binary interaction between compounds (for nonreactive systems) or elements (for reactive systems) in two coexisting phases. Furthermore, Sanchez-Daza et al. [37] extended the application of the driving force approach to design reactive distillation columns. A generic driving force diagram is given in Figure 6.2.

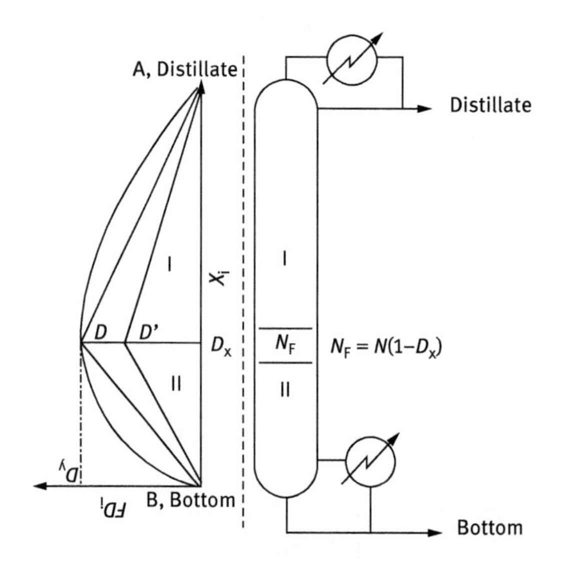

Figure 6.2: Driving force based design of distillation columns – on the left is the driving force diagram and on the right is the corresponding design of the reactive distillation column (adapted from [38]).

In comparison to the other two methods, the driving force approach is much more complex and requires complex calculations and executive algorithms. However, the computational cost of this approach is much less than the optimization based methods and the outcome configuration of the column is optimum in terms of heat duty required for the column itself.

To date, driving force approach has been applied in numerous process synthesis [39, 40], design [29, 36, 37, 41], and process control applications. This approach is very well established as a powerful and simple method for design of separation operations, with or without reactions, that results in optimal/near optimal separation designs both in terms of energy consumption, control, and operation when the process is designed at the maximum driving force.

The driving force is defined as the difference in composition of a light key compound (element [42] or equivalent element [43]) between two coexisting phases. Note however, although the driving-force diagram is plotted for a binary pair of elements or compounds, since all separation tasks are performed for specific binary pairs of compounds (or elements or equivalent elements), this concept can be applied also to multi-compound mixtures as well. Also, the separation of a mixture of *NC* compounds would need *NC*-1 separation tasks and, therefore, *NC*-1 binary pairs of driving forces are involved for each separation task [36]. Note that the element-based reactive driving-force diagram fully considers the extent of reaction on an element basis, and in this work it is applied in the design of reactive distillation columns for chemical equilibrium or kinetically controlled reactions [44]. The supplementary material for this approach is provided in the Appendix.

6.3.2 McCabe-Thiele method

This method uses equilibrium concentration of two coexisting phases to determine the design parameters of a distillation column. Like many other methods, McCabe-Thiele splits a distillation column into rectifying and stripping sections based on the feed location. The key assumption of this method, which is not applicable to other existing methods, is the constant vapor and liquid molar flow rates in each section. Based on the assumption and derivation of material balances of each section, the operating lines of the column can be easily determined. This approach is the most well-known and easiest graphical tool for designing distillation columns as described in an early study in 1925 [22].

6.3.3 Ponchon-Savarit method

In this method, the design of the column is based on the equilibrium data of enthalpy concentration of two coexisting phases. This method is an improvement over the McCabe-Thiele, as in addition to the material balances of each section of the column, energy balances are also taken into account and this method does not assume a constant vapor/liquid flow rate in each distillation section. In so doing, this method is more rigorous and the design parameters are closer to reality in comparison to the McCabe-Thiele method [23].

6.4 Intensified distillation columns: Graphical design

The main focus of the section is on the examination and application of the approaches introduced in Sections 6.3.1 to 6.3.3 on three different categories of intensified distillation units, namely internally heat integrated distillation column (HIDiC), cyclic distillation, and reactive distillation. To this end, the McCabe-Thiele and Ponchon-Savarit are implemented to design the HIDiC example and driving force approach is applied to cyclic and reactive case studies. For each case study, the corresponding toolbox, which explains the design methodology, is provided. Results of the case studies are generated by following the provided algorithms.

In the following, the operation of the three promising case studies will be briefly discussed and the related toolboxes will be provided.

6.4.1 Heat Integrated Distillation

Implementation of HIDiCs in industry has been on a rise since the past decades, because of the capability to reduce the energy of the process significantly. In this configuration, the energy of the sections with higher temperature is exchanged with the sections with lower temperature in order to improve the system energy efficiency and reduce the use of external utility. McCabe-Thiele and Ponchon-Savarit are applied as a graphical tool to design this type of column that are in the following.

6.4.1.1 Graphical design methods

6.4.1.1.1 McCabe-Thiele method for HIDiC

One key assumption that is not applicable in the HIDiC columns but can be applied to conventional distillation columns is the constant molar flow assumption. This is because the heat is integrated in the column. Therefore, the slopes of operating lines change (steps) from stage-to-stage where heat is integrated (the slope is constant for the nonintegrated part). In the proposed algorithm, the heat transfer area and heat transfer coefficient is assumed constant. However, the heat transfer rate varies along the column to satisfy energy balances in HIDiC. McCabe-Thiele diagram of the HIDiC is constructed by an iterative process based on a McCabe-Thiele diagram of the conventional column. The pressure in the rectifying column is initially specified higher than the desired pressure to ensure enough driving force for the initial HIDiC design to be constructed (This way the desired pressure can be specified lower than if one does not start with a high initial pressure). The corresponding toolbox is provided below.

Design toolbox: McCabe-Thiele for HIDiC

Step 1. Set the pressure of the rectifying section a value higher than the desired one.

Step 2. Extract the VLE data of XY diagram for the light key component of the separation both for rectifying and stripping pressures.

Step 3. Draw XY diagram of the system accordingly and draw the diagonal line $y = x$.

Step 4. Considering the purity needed in distillate and bottom for the light key component (x^D, x^B), determine the location of distillate and bottom specifications on the XY diagram regarding the coordinates (x^D, x^D) and (x^B, x^B), respectively.

Step 5. Locate the feed composition on the x-axis and draw a vertical line from that point up to the point it intersects the diagonal line.

Step 6. Draw a line from the intersection point with a slope, which expression is provided below regarding the liquid fraction of the feed (q). Extend the line to intersect the equilibrium curve at rectifying pressure.

$$Slope = -\frac{q}{1-q}$$

Step 7. If the reflux ratio is specified go to step 8, considering the given reflux ratio, else draw a line from (x^D, x^D) to the intersection between feed line and the point derived in step 6 and continue the line to cross the vertical axis. Regarding the expression provided below, find the minimum reflux ratio of the column.

$$Slopeline = \frac{RR_{min}}{RR_{min}+1}$$

Step 8. Relying on the equation provided below, find the operating reflux ratio of the column and calculate the slope of the operating line of rectifier.

$$RR = [1.2,\ 1.5\ ] \times RR_{min} \Rightarrow Slope\ (rectifier) = \frac{RR}{RR+1}$$

Step 9. Draw a line from (x^D, x^D) with the given slope in step 8 and continue the line to intersect the feed line (The rectifier operating line).

Step 10. Draw a line from (x^B, x^B) to cross the intersection point of rectifying and feed lines (The stripper operating line).

Step 11. From (x^D, x^D) draw a horizontal line to intersect the equilibrium curve of rectifying pressure and from that point draw a vertical line to intersect the corresponding operating line. Continue the same procedure to the point that the horizontal line intersects the equilibrium curve to the left of the intersection between two operating lines (The number of horizontal lines are the number of stages in rectifying section).

Step 12. Draw a vertical line from the last point in step 11 to intersect the stripping operating line and draw a horizontal line from that point to intersect the equilibrium curve at stripping pressure. Continue the same procedure to the point it reaches (x^B, x^B) or crosses this coordinate (The number of horizontal lines are the number of stages in stripping section).

Step 13. Count the number of trays determined in steps 11 and 12 to find the total number of stages.

Step 14. Calculate the temperature profile in the column according to the compositions determined from McCabe-Thiele diagram of conventional column by solving bubble point equation; the expression is provided below (For each VLE the equation has different expressions).

$$\sum_{i=1}^{NC} y_i(T,P,x_i) - 1 = 0$$

Step 15. Select the scheme of the heat integration between sections (constant area of heat transfer or constant heat transfer rate).

Step 16. Regarding the mass and energy balances provided below, calculate the profiles of vapor and liquid flow rates in the system for each stage.

$$\text{Rectifier: } V_{n+1}H^V_{n+1} = L_nH^L_n + DH^V_D + \sum_{j=1}^{n} Q_j V_{n+1} = L_n + D$$

$$Q_j = UA\left(T_{\tilde{j}} - T_j\right)\tilde{j} : \textit{the stage which is paired with stage } j$$

$$\text{Stripper: } L_mH^L_m = V_{m+1}H^V_{m+1} + BH^L_B - \sum_{j=m+1}^{n} Q_j\ L_m = V_{m+1} + B$$

$$Q_j = UA\left(T_j - T_{\tilde{j}}\right)\tilde{j} : \textit{the stage which is paired with stage } j$$

Step 17. From the two coordinates (x^D, x^D) and (x^B, x^B) draw lines with slopes of the expressions below.

$$\text{Rectifier: } Slope = \frac{L_n}{V_{n+1}} \text{ Stripper: } Slope = \frac{L_m}{V_{m+1}}$$

Step 18. Consider a point, which is the border of the stripping and rectifying sections (it can be defined as a point near the q line of the feed).

Step 19. From (x^D, x^D) draw a horizontal line to intersect the equilibrium curve of rectifying pressure and from that point draw a vertical line to intersect the corresponding operating line drawn in step 17. Continue the same procedure for all of the operating lines in rectifying section.

Step 20. If the last horizontal line reaches the point defined in step 18, go to step 21, else continue the procedure with the same slope of the last operating line to the point that reaches the aforementioned point.

Step 21. Draw a vertical line from (x^B, x^B) to intersect the equilibrium curve and draw a horizontal line from that point to intersect the corresponding operating line drawn in step 17. Continue the same procedure for all of the operating lines in stripping section.

Step 22. If the last horizontal line reaches the point defined in step 18, go to step 23, else continue the procedure with the same slope of the last operating line to the point that reaches the aforementioned point.

Step 23. If the two McCabe-Thiele diagrams are the same, go to step 24, else take the compositions of each stage from the new diagram as an initial guess and return to step 14.

Step 24. Count the number of stages and design is complete.

Based on this design, the temperatures and flow rates are calculated on each stage for the HIDiC (from mass balances and equilibrium relationships). The initial pressure in the rectifying column is calculated to 3,28 bar, which ensures a temperature difference of at least 20 K between all paired trays. Then the operating lines are plotted for each stage, and the McCabe-Thiele diagram is constructed. This procedure continues until a constant design (two equivalent McCabe-Thiele diagrams) is obtained. Then the pressure is dropped to the desired value, and the method is continued until a constant design is obtained.

6.4.1.1.2 Ponchon-Savarit method for HIDiC

The Ponchon-Savarit method, which includes the energy balance, is considered more rigorous than the McCabe-Thiele method. Since, the McCabe-Thiele extension is based on an iterative procedure on a conventional design of the column;

however, in Ponchon-Savarit method the design of the column will be determined with a straightforward procedure.

In the following, the HIDiC design found with the extended McCabe-Thiele method in Section 6.3.1 is further validated. The Ponchon-Savarit method is not limited to validating other designs, but can be used to design new columns. Here it is appropriate for validating, because the design methods then can be compared and discussed. In the proposed method, a constant heat transfer rate between paired steps is used, so that the heat transfer area variates along the column. This is opposite from the energy model used with the extended McCabe-Thiele method, but is convenient for this algorithm because the design is then solved in a single iteration. Because different energy models are used, the designs, strictly speaking, are not comparable. A mean of the heat transfer rate between paired trays found for the final McCabe-Thiele diagram in Figure 6.3 is used, so the two methods are more comparable. In the following, the operating lines for the rectifying and stripping columns are derived.

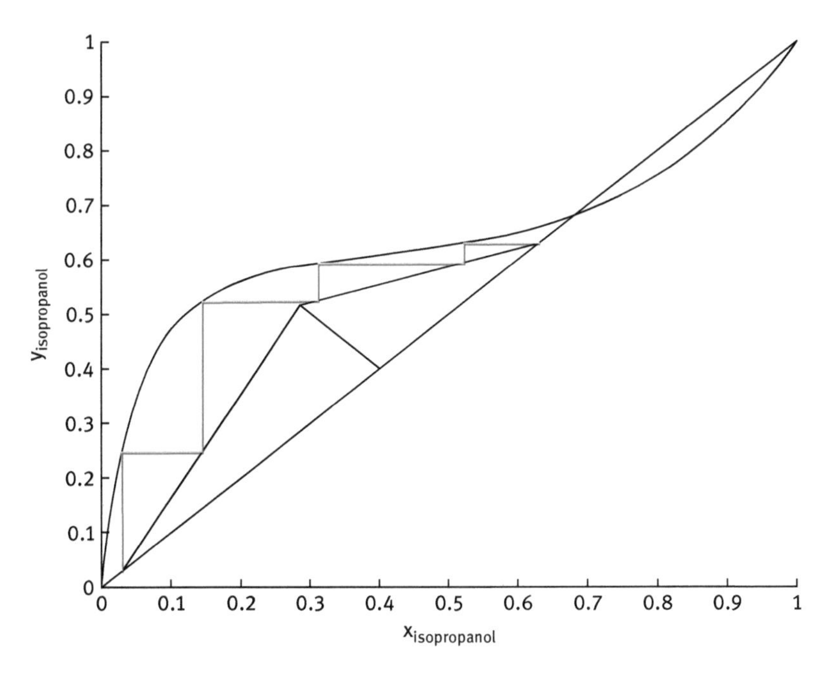

Figure 6.3: McCabe-Thiele diagram of a conventional column. P_{str} = 1.013 bar, RR = 0.5.

Rectifying section. For a HIDiC only the energy balance is different from the conventional column, because heat is integrated in the column. This is expressed as follows:

$$V_{n+1}H_{n+1}^{V} = L_n H_n^L + DH_D^V + \sum_{j=1}^{n} Q_j \tag{6.1}$$

Here the term $\sum_{j=1}^{n} Q_j$ represents the overall internal heat-transfer rate, and is described by Eq. (6.1). Also, a new variable, Q'_D, is defined as follows:

$$Q'_D \equiv \left(H_D^V + \frac{\sum_{j=1}^{n} Q_j}{D} \right) \tag{6.2}$$

Thus, Eq. (6.1) is simplified and expressed as:

$$V_{n+1} H_{n+1}^V = L_n H_n^L + D Q'_D \tag{6.3}$$

The form of Eq. (6.2) is identical to Eq. (6.5), and because the material balances are identical, an operating line in the same form as an operating line by Ponchon-Savarit method for a conventional distillation is found for the HIDiC. Here, the point (x^D, Q'_D) is no longer a stationary point because $\sum_{j=1}^{n} Q_j$ changes every time a paired step is passed.

Stripping section. As for the rectifying section, only the energy balance is different from the conventional column. This is expressed as follows:

$$L_m H_m^L = V_{m+1} H_{m+1}^V + B H_B^L - \sum_{j=m+1}^{n} Q_j \tag{6.4}$$

Similar to the rectifying section, a variable Q'_B, is defined as follows:

$$Q'_B \equiv \left(H_B^L + \frac{\sum_{j=m+1}^{n} Q_j}{B} \right) \tag{6.5}$$

Therefore, Eq. (6.4) can be simplified and expressed as follows:

$$L_m H_m^L = V_{m+1} H_{m+1}^V + B Q'_B \tag{6.6}$$

Here, the point (x^B, Q'_B) is no longer a stationary point because $\sum_{j=m+1}^{n} Q_j$ changes every time a paired step is passed. The forms of Eq. (6.5) and Eq. (6.6) is formally identical to that of being used in Ponchon-Savarit method for conventional distillation columns.

Optimal feed stage. For an HIDiC, the feed line is not a good criteria for where the optimal feed stage is located because it is dependent of the moving points Q'_D and Q'_B. Instead, the feed tie line is used as the criteria when the feed tie line is passed a changeover from one section to the other.

In the following, the corresponding toolbox is provided.

Design toolbox: Ponchon-Savarit method for HIDiC

Step 1. Extract the VLE data (H^V, H^L) v.s. x for key components regarding the pressures of rectifying and stripping sections.

Step 2. Draw the H-X diagram for the light key component for both rectifying and stripping section pressures.

Step 3. Draw two vertical lines from the compositions of distillate and bottom products to cross the corresponding equilibrium curves and name the intersection of the line at distillate composition with upper branch of the curve as V_1. Name the intersection of the line at bottom composition with lower branch of equilibrium curve as L_N.
Step 4. From the feed composition on x-axis, draw a vertical line to cross the equilibrium curves.
Step 5. Draw a line, which is crossing the line drawn in step 3. One of its end points is on the equilibrium curve of rectifier and the other on stripper equilibrium curve in a way that the ratio of the two generated segments is equal to the ratio of liquid and vapor flow rates of feed stream (feed tie line).
Step 6. Based on the components and overall material balances over the entire unit find the flow rate of distillate and bottom product.
Step 7. Define the control volume around the condenser and rectifying section.
Step 8. Regarding y_i and using the VLE equation, find the corresponding liquid composition of first stage (x_i) on the lower branch of equilibrium curve for rectifier.
Step 9. Connect the two liquid and vapor compositions. The line shows the tray number i.
Step 10. Based on Eq. 6.2 find the location of Q'_D on the vertical line at distillate composition for the first stage (considering the paired stages inside the control volume).
Step 11. Connect Q'_D of step 10 and liquid composition determined in step 8. The intersection of the line with upper branch of equilibrium curve will be the vapor composition of the next stage(y_{i+1}).
Step 12. Find the location of Q'_D corresponding to next stage by moving upward from the Q'_D found in step 11 by $\frac{Q_2}{D}$.
Step 13. Draw a line from x_{i+1} to the new Q'_D found in step 12 (operating line of next stage).
Step 14. If the operating line of the stage crosses the feed tie line go to step 15, else return to step 8 and follow the procedure.
Step 15. Define the control volume around reboiler in stripping section and write the energy balance of the control volume as Eq. 6.5.
Step 16. Find the corresponding vapor composition (y_i) in equilibrium with x_i by using VLE equation and connect the two points as the line of the stage.
Step 17. Move vertically down by $\sum_{i=1}^{NS} \frac{Q_i}{B}$ from H_B and locate the point $Q'_B$for that stage (NS is the number of stages exist in the control volume around reboiler in the stripping section).
Step 18. Draw a line from Q'_B found in step 17 to the point found in step 16 (operating line of the stage). The intersection of the line with the lower branch of equilibrium curve is the liquid composition of the next stage(x_{i-1}).
Step 19. If the operating line of that stage is crossing the feed tie line or is located at the right side of feed tie line, go to step 20, else return to step 16.
Step 20. Count the number of stages of the column with respect to the number of stage lines.

6.4.1.2 Graphical design of HIDiC

In this subsection, an internally heat integrated distillation unit is designed regarding the proposed McCabe-Thiele and Ponchon-Savarit algorithms in Sections 6.4.1.2.1 and 6.4.1.2.2, respectively. The corresponding graphs are provided for each method. The distillation unit is designed for refining a water-isopropanol mixture and the detailed operating conditions are as follows: feed flow rate, 100 kmol/h; feed composition (isopropanol), 0.4; liquid feed fraction, 0.5; initial pressure of the rectifying section, 3.28 bar; desired pressure of rectifying section, 2.6 bar; pressure of stripping section, 1.013 bar; heat transfer rate, 9803 W/K; reflux ratio, 0.5; composition in distillate (isopropanol), 0.63; composition in bottoms product (isopropanol), 0.03.

6.4.1.2.1 HIDiC design: McCabe-Thiele application

Figure 6.3 illustrates the McCabe-Thiele diagram that can be constructed for conventional HIDiC column. The graph is drawn by following step 1 to step 13 of McCabe-Thiele toolbox.

After four iterations, the final HIDiC design is found, and can be seen in Figure 6.4. Here, the dotted line represents the equilibrium curve at the rectifying pressure, and the full line represents the equilibrium curve in the stripping section. Two trays are needed in the rectifying column, and three trays are needed in the stripping column to achieve this separation. The feed is introduced on stage three. An asymmetric column is achieved, where heat is exchanged between the paired trays 1–4 and 2–5.

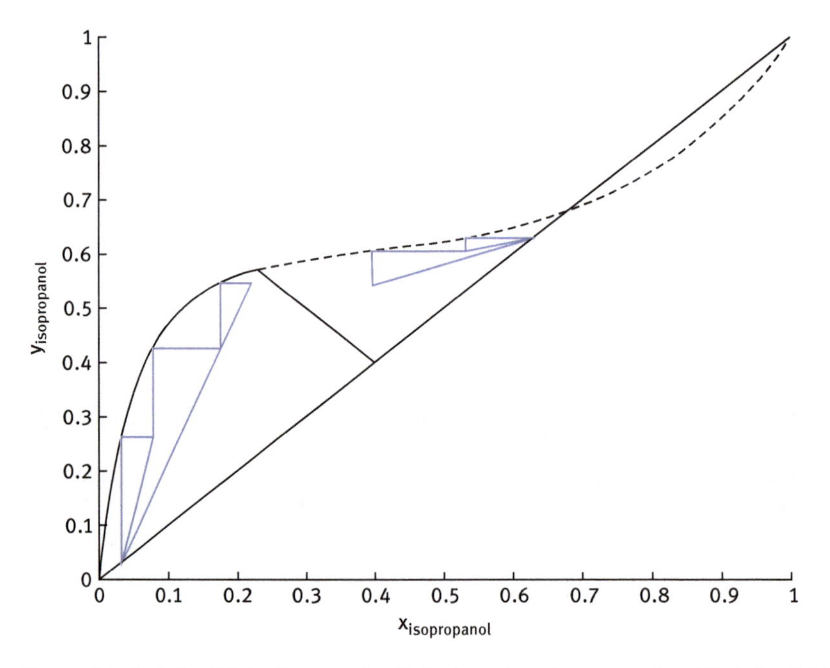

Figure 6.4: McCabe-Thiele diagram of a HIDiC, based on the conventional McCabe-Thiele diagram in Figure 6.3. P_{str} = 1.013 bar, P_{rec} = 2.6 bar.

6.4.1.2.2 HIDiC design: Ponchon-Savarit application

The proposed method is used for designing a HIDiC with the same operating conditions described in Section 6.4.1.2 The heat transfer rate between paired stages is specified to 163 MJ/s, and is the mean of the heat transfer rate used for the McCabe-Thiele design in Figure 6.4. The final Ponchon-Savarit diagram is given in Figure 6.5 by following the Ponchon-Savarit algorithm in Section 6.4.1.1.2. As for the McCabe-Thiele diagram, two trays are needed in the rectifying column and three trays are needed in the stripping column. Again the feed is introduced on stage three. An asymmetric configuration is found, where heat is exchanged between trays 1–4 and 2–5.

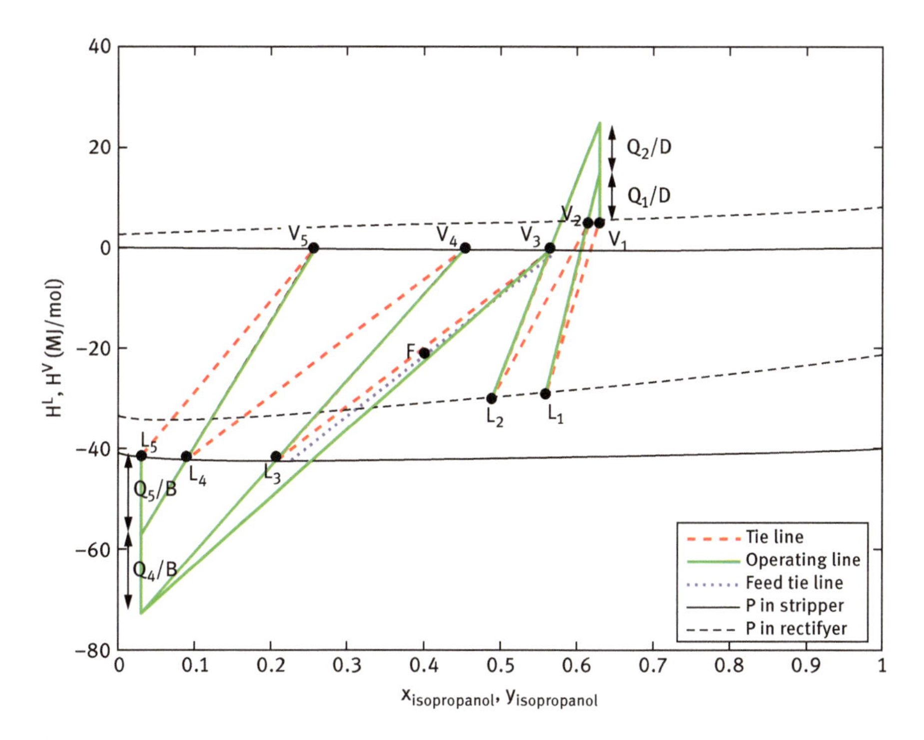

Figure 6.5: Ponchon-Savarit diagram of the HIDiC introduced in section 6.4.1.2.

6.4.2 Cyclic distillation design

Cyclic distillation is another intensification process alternative in distillation unit operations, where the separation efficiency of a distillation unit is improved by using a periodic operation mode by creating vapor flow period and liquid flow period, which together they constitute repeating operating cycles. During the vapor flow period, each tray has a liquid holdup. As a result, there is no requirement for a tray down comer as such the trays can be designed with a higher active surface area [47].

6.4.2.1 Method

For the cyclic distillation, due to the changing compositions in both liquid and vapor phase during a cycle, the design task becomes less trivial in comparison to the original McCabe-Thiele method. One can however use a similar design approach, with the use of operating lines, and corresponding McCabe-Thiele steps.

By plotting the time-averaged vapor composition that enters tray n ($\overline{y_{n+1}}$) against the liquid composition at the tray at the end of the vapor flow period ($x_n^{(V)}$), the operating lines for the cyclic distillation can be drawn. These will be identical to the ones in the original McCabe-Thiele diagram, but with the internal and external flows in terms of per cycle, instead of continuous flows.

The corresponding McCabe-Thiele steps in the cyclic distillation will however be different than the ones used in the classical McCabe-Thiele design procedure, as the tray efficiency of a cyclic tray, where the tray efficiency is defined as:

$$E_T = \frac{\overline{y_n} - \overline{y_{n+1}}}{y_n^{(V)} - \overline{y_{n+1}}}$$

will be approximately 2.

To calculate the tray efficiency enhancement due to the cyclic operation, a backward integration method, like the one suggested by Toftegård and Jørgensen [45] and extended by Pătruţ et al. [46], can be utilized. The design algorithm takes in a specified bottom composition and all the internal and external flows for the column. The algorithm integrates hereafter the mass-balances, for each stage, backwards in time, stage-by-stage. During this procedure, an approximate feed location is found together with the number of needed stages for obtaining the specified separation. This design algorithm was however limited to only model saturated liquid feeds, and which restricted the possibilities of operation.

With an expanded mass balance model, as suggested by Nielsen et al. [47], the design algorithm can be used for mixed phase feeds ($0<q<1$). This makes it possible to obtain a driving force design for the cyclic distillation. The algorithm for this utilizes the method of Pătruţ et al. [46], but includes the parts of the driving force procedure as shown by Gani and Bek-Pedersen [36].

The procedure is the following:

Design toolbox: Driving force approach for cyclic distillation.
Step 1. Find the driving force composition (D_x) for the mixture.
Step 2. Specify product and feed compositions, all external flows, and the number of stages (NP).
Step 3. Specify the internal vapor-flowrate ($V \cdot t_{vap}$) and calculate the rest of the internal flowrates.
Step 4. Adjust q, so the operating lines intersect at $x = D_x$, and calculate the corresponding x_F and y_F.

Step 5. Run the design algorithm for NP stages and place NF where $x_{NF} \approx D_x$.
Step 6. If the composition at start of the vapor flow period, for the upper tray ($x_1^{(L)}$) matches the specified distillate composition, the design has been obtained. Else, go back to step 3 and adjust $V \cdot t_{vap}$ accordingly.

6.4.2.2 Case study

For an ethanol-water separation, with the feed and product compositions as shown in Table 6.2, the driving force design for a cyclic distillation column is now presented. The number of stages used, (including reboiler, excluding condenser) is set to be 12.

Table 6.2: The feed and product compositions.

Component	z_F [−]	x_B [−]	x_D [−]
Ethanol	0.1500	0.0001	0.8300
Water	0.8500	0.9999	0.2700

The operating parameters for the separation can be found in Table 6.3, including the operating parameters for the design obtained using the Pătruţ design method, where $q = 1$. In both designs, the optimal feed location was found to be introduced at stage 2 (the tray above the reboiler).

Table 6.3: The operating parameters for the driving force design compared to the design obtained using the method by Pătruţ.

	$V \cdot t_{vap}/F$ [−]	q [−]	x_F [−]	y_F [−]	RR [−]	RB [−]	$RR + RB$ [−]
Driving force design	0.778	0.837	0.0933	0.4405	4.212	0.950	5.162
Pătruţ design	0.943	1.000	0.1500	–	4.221	1.151	5.372

The driving force design can be seen to reduce the total utility exchange in the column with 3.9%, in comparison to the design obtained using the Pătruţ design method. In Figure 6.6, the driving force diagram for the separation can be seen. The corresponding operating lines and analogous McCabe-Thiele steps can be seen in Figure 6.7.

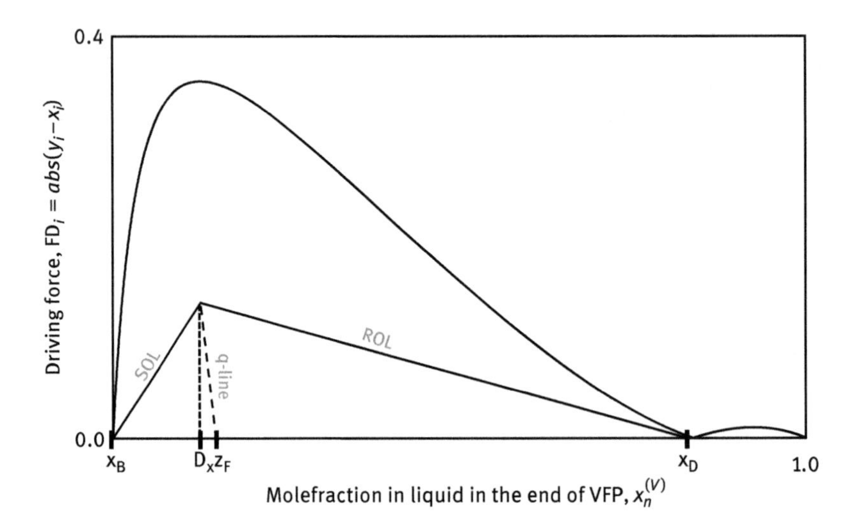

Figure 6.6: Driving force diagram for the separation.

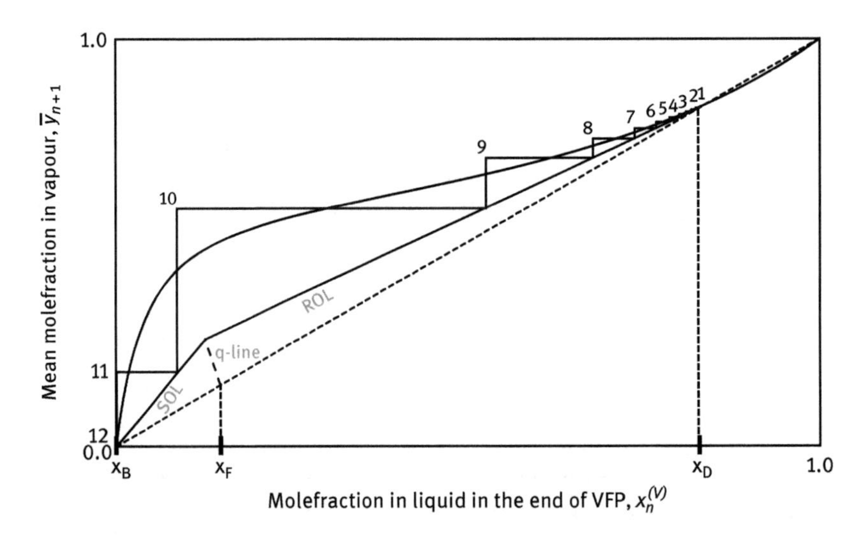

Figure 6.7: Analogous McCabe-Thiele diagram for the cyclic distillation (stage 12 is the reboiler).

In Figure 6.6, the operating lines can be seen intersecting at the driving force composition D_X.

In Figure 6.7, it can clearly be seen that separation with cyclic distillation requires far less trays (approximately 50% less) in comparison to a conventional distillation column with the same internal flows, due to the enhanced tray efficiency.

6.4.3 Reactive Distillation Design: Driving Force Approach

Reactive distillation is an intensification unit operation where the operations of a reactor and a distillation unit are combined to a single unit operation to achieve synergistic process efficiency improvements. The reactive distillation units are generally implemented a higher reaction rate and an equilibrium can be achieved by continuous removal of the reaction product or inhibitors from the system [42].

6.4.3.1 Method

The methodology of designing this type of distillation column is provided in the toolbox below.

Design Toolbox: Driving Force Approach for Reactive Distillation.

Step 1. Specify the number of compounds NC and number of reactions NR.

Step 2. Calculate the number of elements, using the equations.

$$NE = NC - NR$$

Step 3. Define the NE elements as molecules or parts of molecules present in the reaction.

Step 4. Generate a formula matrix, providing the occurrence of each element in each compound.

Step 5. Identify light key, heavy key, and non-key elements, using the following rules.

a. The element(s) that are contained in the lightest component should not be specified as the heavy key element.
b. The element(s) that are contained in the heaviest component should not be specified as the light key element.
c. Both key elements should be contained in both distillate and bottom products.

Step 6. Generate a phase diagram and a driving force diagram for the reactive system based on the light and heavy key elements, using equivalent compositions, defined as the provided equations.

$$W_{eq} = \frac{W_{LK}}{W_{LK} + W_{HK}}$$

$$DF_{eq} = \left| W_{LK}^{l} - W_{LK}^{v} \right|$$

Step 7. Identify area of operation in the driving force diagram and rescale the area of operation between 0 and 1 in the composition domain for both the driving force diagram and generate the corresponding McCabe-Thiele diagram.

Step 8. Identify the feed composition(s) W_{eq}^{F} and desired bottom and distillate compositions (W_{eq}^{B}, W_{eq}^{D}) in the driving force diagram and the McCabe-Thiele diagram.

Step 9. Identify the maximum driving force composition (D_x, D_y) in the driving force diagram.

Note: The mixture of the two feeds must be consistent with the composition of the maximum driving force diagram. If it is not, adjust the vapor fraction of the two feeds to comply with the condition.

Step 10. Draw the stripping and rectifying operating lines from the driving force composition to the bottom and distillate compositions (W_{eq}^{B}, W_{eq}^{D}) in the driving force diagram.

Step 11. Calculate the minimal reflux ratio RR_{min} using the equation provided below:

$$RR_{min}=\frac{W_D-W_g}{D_y}-1$$

Step 12. Draw operating lines with a reflux ratio of $RR=1.2\cdot RR_{min}$ in the McCabe-Thiele diagram, where the operating lines still intersect at the driving force composition D_x, and draw steps starting from the desired distillate composition W_{eq}^D until stepping over the desired bottoms composition W_{eq}^B.
Step 13. The number of triangles drawn indicate the number of needed reactive stages.

6.4.3.2 Case study

The design objective is to obtain the reactive distillation column design operating at the maximum driving force to produce methyl acetate with a purity of 99% ± 0.5% on a molar basis. The reaction between methanol (MeOH) and acetic acid (HOAc) yields methyl acetate (MeOAc) and water (H_2O). The reaction takes place in liquid phase over a catalyst. It is exothermic with a heat of reaction pf −5.42 kJ/mol and is given as follows:

$$MeOH(CH_4O)+HOAc(C_2H_4O_2)\leftrightarrow MeOAc(C_3H_6O_2)+Water(H_2O)$$

The design targets for a reactive distillation column with only reactive section is obtained from Jantharasuk et al. (2011) and is given as follows:

Table 6.4: Design targets and product specifications [43].

Component	Structure	Feed (1)	Feed (2)	Distillate	Bottom
Methanol	C_4H_8	1	0	0.0694	0
Acetic Acid	CH_4O	0	1	0.0089	0.3345
Methyl Acetate	C_4H_8	0	0	0.7612	0
Water	$C_5H_{12}O$	0	0	0.1606	0.6651

Feed (1): 230.28 kmol/h methanol; Feed temperature and pressure: 328 K and 1 atm.
Feed (2): 230.28 kmol/h acetic acid; Feed temperature and pressure: 328 K and 1 atm.

Note however, the design targets specified in Table 6.4 will be used in course of distillation column design. As it is mentioned previously, the final distillation column design target (including both reactive and nonreactive sections) is to obtain 99% pure methyl acetate on a molar basis.

A discussion by Pöpken et al. (2001) specified that any side reaction is completely suppressed by using near-stoichiometric feeds (1:1 ratio in this case – see Table 6.4). Therefore, in this case there are four compounds and one reaction, thereby the reaction mixture is represented by three elements and the formula matrix is given in Table 6.5.

Table 6.5: Elements representing the system and formula matrix.

Methanol (CH_4O) + Acetic Acid ($C_2H_4O_2$) ↔ Methyl Acetate ($C_3H_6O_2$) + Water (H_2O) Element definition: $A = CH_4O$ $B = C_2H_2O$ $C = H_2O$ Element reaction: $A + BC \leftrightarrow AB + C$				
Formula Matrix				
	Methanol	Acetic Acid	Methyl Acetate	Water
A	1	0	1	0
B	0	1	1	0
C	0	1	0	1

In order to identify the key elements, the rules of key element selection are applied. Therefore, element C is selected as the non-key element, element A is the light key element (LK), and element B is the heavy key element (HK). In this step, the SRK equation of state is selected to calculate the vapor phase fugacity coefficients and Wilson model is used to calculate liquid phase activity coefficients. The phase diagram for this reactive system based on the equivalent binary elements is presented in Figure 6.8.

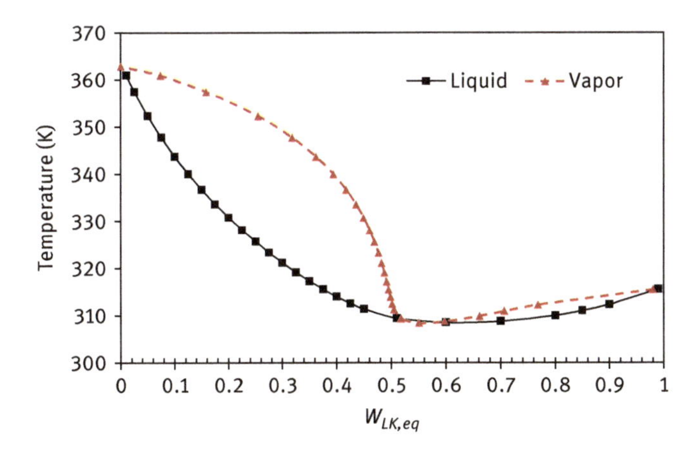

Figure 6.8: Phase diagram for methyl-acetate multi-element system at 1 atm.

The driving force diagram is constructed and the area of operation for the reactive distillation column without nonreactive stages is identified as depicted in Figure 6.9.

The area of operation is rescaled between 0 and 1 in the composition domain on the *x*-axis of the driving force diagram and the corresponding McCabe-Thiele diagram is constructed. Note that in this particular case, the composition of the feeds (W_k,W_h) and the design targets in the distillate and bottom compositions (W^D,W^B).

That is, the light key equivalent element composition is in pure state in one feed ($W_k = 1$) and does not exist in the other feed ($W_h = 0$). Figure 6.10 shows the result to design methyl acetate multi-element reactive distillation column with two feeds.

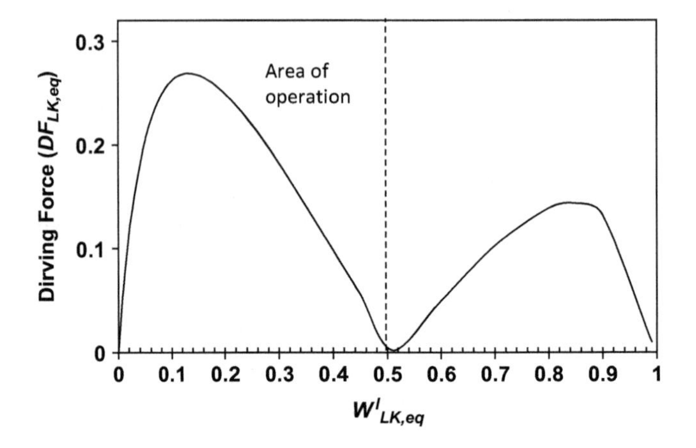

Figure 6.9: Reactive driving force diagram for methyl-acetate multi-element system at 1 atm.

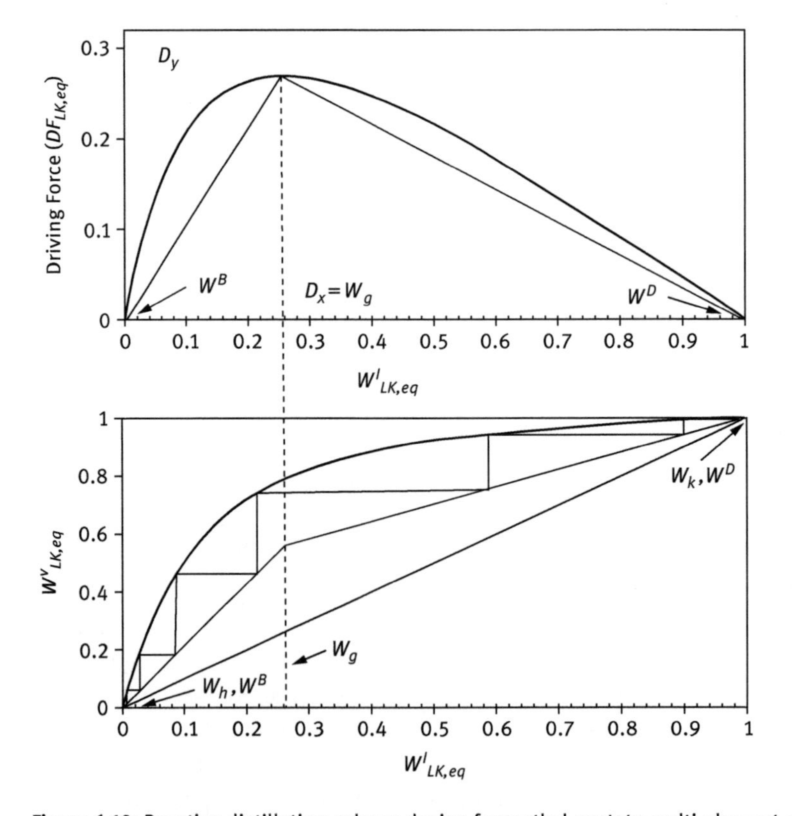

Figure 6.10: Reactive distillation column design for methyl-acetate multi-element system at 1 atm.

Therefore, the reactive distillation column without nonreactive stages and two feeds has six reactive stages plus nonreactive reboiler and condenser. Therefore, the reactive distillation column has eight stages. The feed that contains light key element (methanol) is introduced at the first reactive stage and the other feed (which does not contain the light key element – acetic acid) is introduced at the last reactive stage. The reflux ratio is determined to be 2.2 according to the driving force method.

6.5 Discussion and perspective

The use of graphical design tools in the first pass design and estimation of distillation processes has been a tried and tested practice in process engineering. However, with the widespread industrial use of intensified distillation columns such as reactive distillation and the introduction of new intensified distillation columns such as HIDiC and cyclic distillation the traditional graphical tools need to be updated/enhanced to be used in the design of these intensified distillation units. In future, the use of intensified distillation units are only likely to increase. This is because even a marginal decrease in energy usage an intensified distillation column provides over a classical distillation column can be both commercially and environmental significant when considering over 40% of energy used in the petrochemical industry goes toward distillation unit operations [48].

From a design perspective, intensified distillation units can present a formidable challenge as there are many variables and parameters that can influence the design and operation process. The use of first principal process models based on thermodynamics such as industrial process simulators is one way to design these units. However, even with the availability of industrial process simulation, this can be a tedious task requiring expert understanding of the process model. In comparison a classical distillation column model can be easily set up and solved on an industrial process simulator. As such, extension and application of practical, robust, and well-known graphical tools is a necessity both in the current and future context.

From a practical point of view, all three of the intensified distillation columns are significantly more complicated to design, manufacture, and operate than standard distillation columns. As such, they should only be commercialized under certain techno-economic conditions. To this end, heat integrated distillation columns are highly recommended when energy efficiency is a prime economic driver. This column configuration in particular has a complicated operation procedure as such controllability of these columns can be complicated. Cyclic distillation is one of the best intensified option for tight boiling point mixtures where traditional distillation maybe unable to perform sufficient separation efficiency. In comparison to heat integrated distillation, the operation of these columns can be easier to operate, but requires complex column internals. Reactive distillation column integrates reaction and distillation in one unit operation, and should be used where continual removal of products/inhibitors from a reaction provides a notable economic benefit. However, this unit operation needs much.

The extension of McCabe-Thiele graphical tool to be applicable for intensified distillation columns is one such example. As illustrated in Section 6.4.1, the case study of a heat integrated distillation column (HIDiC) has been designed using the McCabe-Thiele method then further validated using the Ponchon-Savarit method. However, the accuracy of these graphical approaches, in particular McCabe-Thiele approach, can lack that of an industrial process simulation, but, for a first pass estimate when multiple distillation models are being evaluated this type of analysis allows for a quick and efficient results and is much more economical to develop compared to the first principal process model. In Section 6.3.1 the authors have discussed driving force approach and illustrated its use in cyclic and reactive distillation in sections 6.4.2 and 6.4.3. Again in comparison with the development of the first principals process models, this graphical method provides quick and economical results but might lack in accuracy.

The modified McCabe-Thiele method for designing HIDiC is not so easy to follow. As illustrated in Section 6.4.1.1.1, determining a good "first guess" for the rectifying pressure to make sure a sufficient driving force heat exchange is not very straightforward. This procedure, which has been asserted as a trial and error procedure, sometimes takes long to reach a satisfactory convergence. On the other hand, Modified Ponchon-Savarit is considered as the easiest method for HIDiC design. This method uses a very straightforward procedure similar to conventional Ponchon-Savarit to design the unit operation. In contrast, Driving force method for reactive distillation columns contains complex mathematical calculations but is the only way to design this intensified unit operation.

From a techno-economic perspective, the extension of established graphical tools to the "realm" of intensified distillation processes allows a quick first pass calculation of the operating expenses, initial estimates of product rates, utilities consumption to proposed distillation unit. In industry, this economic calculation is of importance as the decision to retrofit or choose a column will always be based on the economic performance of the column. As such the graphical tools discussed in above sections as well as other methods enable these types of calculations to be performed quickly and efficiently. Similarly, the first pass estimation of factors such as number of trays, operating pressure, and heat duty can also be incorporated into capital expense calculation, which would be the next step in developing the economic evaluation process. As such, this type of graphical tools despite being simple, provide sufficient information to make initial stop/go decisions. These types of graphical tools are also an easy way to carry out diagnosis analysis on as-built columns.

From an educational perspective, the extension of these graphical methods allows students to get a glimpse into the actual operation of these types of columns and allows them to understand the fundamental operations of intensified distillation units, which are often hard to understand from a mathematical point of view due to the complex derivations and mathematical operations required for them to be solved. As such, the graphical tools approach provides them with a platform where they can carry out practical exercises to understanding how each operating condition change will affect the overall performance of these intensified distillation columns. As such, these graphical tools can serve as a learning tool especially at an undergraduate level where the students can develop an intuitive understanding of how process conditions and column geometry and other conditions will affect the operations and how each of them might interact with each other.

6.6 Concluding remarks

This chapter has discussed briefly the necessity of implementation of intensified distillation units in process industries. Intensification of distillation units can reduce the overall energy demand of the process significantly, which is beneficial from commercial and environmental point of view. Accordingly, there have been great efforts to develop new methodologies for design of these types of distillation units. Optimization based methods and graphical tools are the two options for design of these unit operations. Although the first method category is much more precise and can cover the majority of these units, complex mathematical calculations make it extremely difficult to use as a first pass design tool. In contrast, graphical tools can intuitively design the columns and minimize the need for complex mathematical procedure. To this end, these tools are much more convenient than the optimization-based methods, especially in first pass design. However, these tools are not generally applicable to all unit types and for each type; the standard methodology used to design simple distillation units needs to be extended. The main focus of this chapter has been on application of graphical tools on some of the intensified distillation units that are more common and important in industry; including internally heat integrated distillation columns (HIDiC), cyclic distillation, and reactive distillation.

Extensions of McCabe-Thiele and Ponchon-Savarit are introduced for designing the HIDiCs. These two methods are well known for designing of ordinary distillation units, but for designing a heat integrated systems a few modifications are needed that have been explained in detail in the corresponding toolboxes. The main difference of these methods compared to the ordinary methodology is that the energy balances play a pivotal role in the design procedure. By so doing, Ponchon-Savarit still can be used as a straightforward method. However McCabe-Thiele becomes an iterative procedure that needs a very basic design of the conventional column with the same specifications. Finally, a case study is analyzed with these two methods and the final design outcomes have been compared. The procedure has shown that

Ponchon-Savarit is more convenient to be implemented in designing of these column types.

Afterward, a novel driving force-based method approach for design of columns has been discussed. Although driving force needs more mathematical calculation, it is considered as one of the most reliable methods when carried out in conjunction with McCabe-Thiele diagram, and graphically shows the design configuration. The advantage of using this method is that the design configuration is one of the optimum designs possible for the system. The application of driving force is analyzed on two important case studies of cyclic distillation and reactive distillation. For each case study a toolbox has been provided and the design procedures are explained step by step. The corresponding results can be reproduced by following the algorithms.

References

[1] A. Carvalho, R. Gani, H. Matos, Design of sustainable chemical processes: Systematic retrofit analysis generation and evaluation of alternatives, Process Saf. Environ. Prot. 86 (2008) 328–346. doi:10.1016/j.psep.2007.11.003.

[2] M.A. Kraller, I.A. Udugama, R. Kirkpatrick, W. Yu, B.R. Young, Side draw optimisation of a high-purity, multi-component distillation column, Asia-Pacific J. Chem. Eng. 11 (2016) 958–972. doi:10.1002/apj.2030.

[3] A.I. Stankiewicz, J.A. Moulijn, Process Intensification: Transforming Chemical Engineering, Chem. Eng. Prog. (2000) 22–34.

[4] A.A. Kiss, Distillation technology – still young and full of breakthrough opportunities, J. Chem. Technol. Biotechnol. 89 (2014) 479–498. doi:10.1002/jctb.4262.

[5] N.M. Nikačević, A.E.M. Huesman, P.M.J. Van den Hof, A.I. Stankiewicz, Opportunities and challenges for process control in process intensification, Chem. Eng. Process. Process Intensif. 52 (2012) 1–15. doi:10.1016/j.cep.2011.11.006.

[6] G.J. Harmsen, Reactive distillation: The front-runner of industrial process intensification, Chem. Eng. Process. Process Intensif. 46 (2007) 774–780. doi:10.1016/j.cep.2007.06.005.

[7] D. Panda, A. Kannan, Equilibrium and Rate Based Simulation of MTBE Reactive Distillation Column, 8 (2014) 998–1004.

[8] M.G. Sneesby, M.O. Tade, T.N. Smith, Two-point control of a reactive distillation column for composition and conversion, J. Process Control. 9 (1999) 19–31. doi:10.1016/S0959-1524(98)00007-9.

[9] T. Pöpken, S. Steinigeweg, J. Gmehling, Synthesis and Hydrolysis of Methyl Acetate by Reactive Distillation Using Structured Catalytic Packings: Experiments and Simulation, Ind. Eng. Chem. Res. 40 (2001) 1566–1574. doi:10.1021/ie0007419.

[10] H.G. Schoenmakers, B. Bessling, Reactive and catalytic distillation from an industrial perspective, Chem. Eng. Process. Process Intensif. 42 (2003) 145–155. doi:10.1016/S0255-2701(02)00085-5.

[11] V. Aneesh, R. Antony, G. Paramasivan, N. Selvaraju, Distillation technology and need of simultaneous design and control: A review, Chem. Eng. Process. Process Intensif. 104 (2016) 219–242. doi:10.1016/j.cep.2016.03.016.

[12] N. Nishida, G. Stephanopoulos, a. W. Westerberg, A review of process synthesis, AIChE J. 27 (1981) 321–351. doi:10.1002/aic.690270302.
[13] J.J. Siirola, G.J. Powers, D.F. Rudd, Synthesis of system designs: III. Toward a process concept generator, AIChE J. 17 (1971) 677–682. doi:10.1002/aic.690170334.
[14] R. Nath, R.L. Motard, Evolutionary synthesis of separation processes, AIChE J. 27 (1981) 578–587. doi:10.1002/aic.690270407.
[15] M.D. Lu, R.L. Motard, Computer-Aided Total Flowsheet Synthesis, Comput. Chem. Eng. 9 (1985) 431–445.
[16] V.M. Nadgir, Y.A. Liu, Studies in Chemical Process Design and Synthesis : Part V: A Simple Heuristic Method for Systematic Synthesis of Initial Sequences for Multicomponent Separations, AIChE J. 29 (1983) 926–934. doi:10.1002/aic.690290609.
[17] S.-H. Cheng, Y.A. Liu, Studies in chemical process design and synthesis. 8. A simple heuristic method for the synthesis of initial sequences for sloppy multicomponent separations, Ind. Eng. Chem. Res. 27 (1988) 2304–2322. doi:10.1021/ie00084a016.
[18] Y.A. Liu, T.E. Quantrille, S.H. Cheng, Studies in chemical process design and synthesis. 9. A unifying method for the synthesis of multicomponent separation sequences with sloppy product streams, Ind. Eng. Chem. Res. 29 (1990) 2227–2241. doi:10.1021/ie00107a007.
[19] A.W. Westerberg, The synthesis of distillation-based separation systems, Comput. Chem. Eng. 9 (1985) 421–429. doi:10.1016/0098-1354(85)80020-X.
[20] M.F. Malone, K. Glinos, F.E. Marquez, J.M. Douglas, Simple, analytical criteria for the sequencing of distillation columns, AIChE J. 31 (1985) 683–689. doi:10.1002/aic.690310419.
[21] P.B. Gadkari, R. Govind, Analytical screening criterion for sequencing of distillation columns, Comput. Chem. Eng. 12 (1988) 1199–1213. doi:10.1016/0098-1354(88)85071-3.
[22] W.L. McCabe, E. Thiele, Graphical design of fractionating columns, Ind. Eng. Chem. 17 (1925) 605–611. doi:10.1021/ie50186a023.
[23] J.M. Ledanois, C. Olivera-Fuentes, Modified Ponchon-Savarit and McCabe-Thiele methods for distillation of two-phase feeds, Ind. Eng. Chem. Process Des. Dev. 23 (1984) 1–6. doi:10.1021/i200024a001.
[24] T.J. Ho, C.T. Huang, L.S. Lee, C.T. Chen, Extended ponchon-savarit method for graphically analyzing and designing internally heat-integrated distillation columns, Ind. Eng. Chem. Res. 49 (2010) 350–358. doi:10.1021/ie9005468.
[25] H. Yeomans, I.E. Grossmann, A systematic modeling framework of superstructure optimization in process synthesis, Comput. Chem. Eng. 23 (1999) 709–731. doi:10.1016/S0098-1354(99)00003-4.
[26] V. Bansal, J.D. Perkins, E.N. Pistikopoulos, A Case Study in Simultaneous Design and Control Using Rigorous, Mixed-Integer Dynamic Optimization Models, Ind. Eng. Chem. Res. 41 (2002) 760–778. doi:10.1021/ie010156n.
[27] S.P. Ramanathan, S. Mukherjee, R.K. Dahule, S. Ghosh, I. Rahman, S.S. Tambe, D.D. Ravetkar, B.D. Kulkarni, Optimization of Continuous Distillation Columns Using Stochastic Optimization Approaches, Chem. Eng. Res. Des. 79 (2001) 310–322. doi:10.1205/026387601750281671.
[28] J.G. Stichlmair, J.-R. Herguijuela, Separation regions and processes of zeotropic and azeotropic ternary distillation, AIChE J. 38 (1992) 1523–1535. doi:10.1002/aic.690381005.
[29] E. Bek-Pedersen, R. Gani, Design and synthesis of distillation systems using a driving-force-based approach, Chem. Eng. Process. Process Intensif. 43 (2004) 251–262. doi:10.1016/S0255-2701(03)00120-X.
[30] G.-J.A.F. Fien, Y.A. Liu, Heuristic Synthesis and Shortcut Design of Separation Processes Using Residue Curve Maps: A Review, Ind. Eng. Chem. Res. 33 (1994) 2505–2522. doi:10.1021/ie00035a001.

[31] M.F. Doherty, G.A. Caldarola, Design and synthesis of homogeneous azeotropic distillations. 3. The sequencing of columns for azeotropic and extractive distillations, Ind. Eng. Chem. Fundam. 24 (1985) 474–485. doi:10.1021/i100020a012.

[32] D. Barbosa, M.F. Doherty, The simple distillation of homogeneous reactive mixtures, Chem. Eng. Sci. 43 (1988) 541–550. doi:10.1016/0009-2509(88)87015-5.

[33] C. Bernot, M.F. Doherty, M.F. Malone, Feasibility and separation sequencing in multicomponent batch distillation, Chem. Eng. Sci. 46 (1991) 1311–1326. doi:10.1016/0009-2509(91)85058-6.

[34] S.K. Wasylkiewicz, L.C. Kobylka, F.J.L. Castillo, Synthesis and design of heterogeneous separation systems with recycle streams, Chem. Eng. J. 92 (2003) 201–208.

[35] G. Sobočan, P. Glavič, A simple method for systematic synthesis of thermally integrated distillation sequences, Chem. Eng. J. 89 (2002) 155–172. doi:10.1016/S1385-8947(02)00012-8.

[36] R. Gani, E. Bek-Pedersen, Simple new algorithm for distillation column design, AIChE J. 46 (2000) 1271–1274. doi:10.1002/aic.690460619.

[37] O. Sánchez-Daza, E.S. Pérez-Cisneros, E. Bek-Pedersen, R. Gani, Graphical and Stage-to-Stage Methods for Reactive Distillation Column Design, AIChE J. 49 (2003) 2822–2841. doi:10.1002/aic.690491115.

[38] D.K. Babi, R. Gani, Hybrid Distillation Schemes: Design, Analysis, and Application, in: A. Gorak, E. Sorensen (Eds.), Distill. Fundam. Princ., Elsevier, London, 2014: pp. 357–381. doi:10.1016/B978-0-12-386547-2.00009-0.

[39] D.K. Babi, P. Lutze, J.M. Woodley, R. Gani, A process synthesis-intensification framework for the development of sustainable membrane-based operations, Chem. Eng. Process. Process Intensif. 86 (2014) 173–195. doi:10.1016/j.cep.2014.07.001.

[40] A.K. Tula, M.R. Eden, R. Gani, Process synthesis, design and analysis using a process-group contribution method, Comput. Chem. Eng. 81 (2015) 245–259. doi:10.1016/j.compchemeng.2015.04.019.

[41] E. Bek-Pedersen, R. Gani, O. Levaux, Determination of optimal energy efficient separation schemes based on driving forces, Comput. Chem. Eng. 24 (2000) 253–259. doi:10.1016/S0098-1354(00)00474-9.

[42] O. Sánchez-Daza, E. Pérez-Cisneros, E. Bek-Pedersen, M. Hostrup, Tools for reactive distillation column design: Graphical and stage-to-stage computation methods, Comput. Aided Chem. Eng. 9 (2001) 517–522. doi:10.1016/S1570-7946(01)80081-X.

[43] A. Jantharasuk, R. Gani, A. Górak, S. Assabumrungrat, Methodology for design and analysis of reactive distillation involving multielement systems, Chem. Eng. Res. Des. 89 (2011) 1295–1307. doi:10.1016/j.cherd.2011.04.016.

[44] M.L. Michelsen, Calculation of multiphase equilibrium, Comput. Chem. Eng. 18 (1994) 545–550. doi:10.1016/0098-1354(93)E0017-4.

[45] B. Toftegård, S.B. Jørgensen, Design Algorithm for Periodic Cycled Binary Distillation Columns, Ind. Eng. Chem. Res. 26 (1987) 1041–1043.

[46] C. Pătruţ, C.S. Bîldea, I. Liţă, A.A. Kiss, Cyclic distillation – Design, control and applications, Sep. Purif. Technol. 125 (2014) 326–336. doi:10.1016/j.seppur.2014.02.006.

[47] R.F. Nielsen, J.K. Huusom, J. Abildskov, Driving Force based design of Cyclic Distillation, Ind. Eng. Chem. Res. 56 (2017) 10833–10844. doi: 10.1021/acs.iecr.7b01116

[48] D.C. White, Optimize Energy Use in Distillation, Chem. Eng. Prog. (2012) 35–41.

Appendix

This approach provides the basis for the determination of important reactive distillation column design variables in terms of two parameters, the location and the size of the maximum driving force, D_x and D_y, respectively. The feed stage location (N_F) and the minimum reflux ratio, *RR* (and/or the reboil ratio, *RB*) are determined from these two parameters for a given feed and product specification. A driving force diagram together with the distillation design parameters is given in Figure A.1.

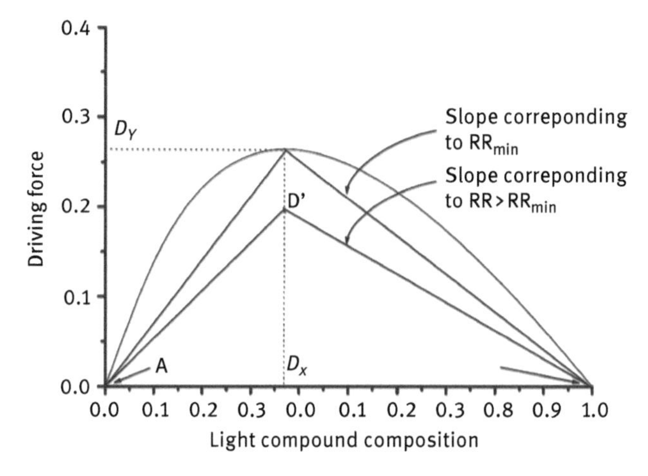

Figure A.1: A generic driving force diagram with the important distillation design parameters [29].

The driving force is defined by Gani and Bek-Pedersen [41] as follows:

$$F_{D_i} = y_i - x_i \tag{A.1}$$

This method is based on investigating the vapor-liquid equilibrium to find the liquid composition D_x of the light key component *i* (based on the boiling points of the individual components in a binary mixture) where the driving force (F_{Di}) is the largest. The optimal or near optimal feed stage is identified based on the location of maximum driving force for a given (calculated) total number of plates N_P. The reasoning behind this operation method lies in the operating lines in the conventional continuous distillation as they are also used in McCabe-Thiele method [22]. For a binary mixture, the following equations describe the operating lines. The operating line for the rectifying and stripping sections in a continuous distillation column are expressed as follows:

$$y_i = x^D \frac{D}{D+L} + x_i \frac{L}{D+L} \tag{A.2}$$

$$y = -x^B \frac{B}{V} + x \frac{V+B}{V} \tag{A.3}$$

By defining the external reflux ratio of the column as $RR = L/D$, and external reboil ratio of the column as $RB = V/B$; Eqs. (A.2) and (A.3) can be simplified as follows:

$$y_i = x^D \frac{1}{RR+1} + x_i \frac{RR}{RR+1} \tag{A.4}$$

$$y_i = -x^B \frac{1}{RB} + x_i \frac{RB+1}{RB} \tag{A.5}$$

Hence, subtracting the liquid composition from both sides of Eqs. (A.4) and (A.5) and rearranging them with respect to x^D and x^B gives the operating line for rectifying and stripping sections based on driving force as follows:

$$x^D = (RR+1)F_{D_i} + x_i \tag{A.6}$$

$$x^B = x_i - RB \cdot F_{D_i} \tag{A.7}$$

Energy is used in both the reboiler and condenser to vaporize liquid and cool down cooling-liquid respectively. To obtain the separation in the column with minimum energy consumption, the sum of reboiler and condenser duties must be minimized. This is by assuming the liquid is at its boiling point in the reboiler and the vapor is at the dew point in the condenser. Furthermore, it is also assumed that the distillate product is liquid at its boiling point when it exits the condenser. The following equations are given to account for the

$$|Q_{reoboiler}| = \overline{\Delta H}_{vap,\,reobiler} \cdot V_s \tag{A.8}$$

$$|Q_{condenser}| = \overline{\Delta H}_{vap,\,condenser} \cdot V_r \tag{A.9}$$

where $V_r$and V_s are the vapor flow in the rectifying and stripping sections, respectively. Therefore, the sum of RB and RR can be written as follows:

$$RB + RR = \frac{V_r}{D} + \frac{V_s}{B} - 1 \tag{A.10}$$

Isolating $V_r$and V_s in Eqs. (A.8) and (A.9) and substituting in Eq. (A.10) gives the following:

$$RB + RR = \frac{|Q_{reoboiler}|}{B \cdot \overline{\Delta H}_{vap,\,reobiler}} + \frac{|Q_{condenser}|}{D \cdot \overline{\Delta H}_{vap,\,condenser}} - 1 \tag{A.11}$$

Thus, if the sum of the two ratios is minimized, the amount of energy transferred in the column will be minimized. The operation where the sum of the ratios is minimized, can be found by investigating a driving-force diagram, where the driving force is plotted against x_i, where the component i is the most volatile component. An example of a driving-force diagram can be seen in Figure A.2. Therefore, the slopes of the two lines (rectifying section and stripping section) are also readily identified from Eqs. (A.6) and (A.7) in terms of driving force. Therefore, given the slopes if they intersect at the point corresponding to the maximum driving force ($x = D_x$) for a given product specification then the sum of boilup and reflux ratios will be minimum. This corresponds to minimum energy required for the given separation task.

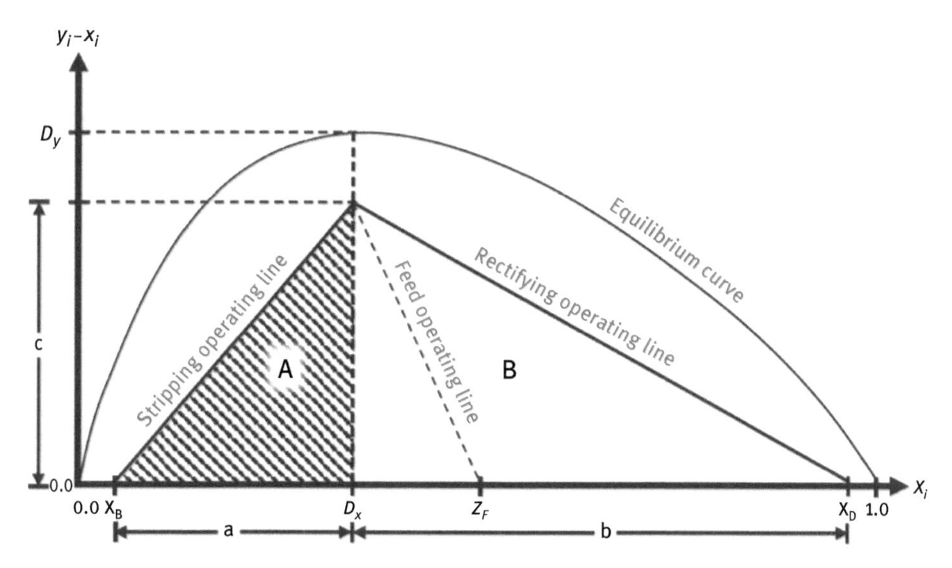

Figure A.2: A driving-force diagram with the operating lines intersecting at x = D_x. The measurements a, b and c are indicated on the figure, for better understanding of the reasoning behind using driving force to obtain the lowest column energy transfer.

Consider the area under the operating lines to consist of two right-angled triangles A and B (see Figure A.2). The slope of the hypotenuse on triangle A (the stripping operating line) is, $\frac{c}{a} = \frac{1}{RB}$ and similarly for triangle B is, $\frac{-c}{b} = -\frac{1}{RR+1}$. Moreover, then it can be also written such that the sum of lengths a and b is equal to the difference of top and bottom product compositions on the x-axis. That is:

$$a + b = x^D - x^B \tag{A.12}$$

Therefore, isolating the lengths a and b in the slopes of the hypotenuse for triangles A and B and substituting into Eq. (A.11), the following expression is obtained:

$$c \cdot RB + c \cdot RR + c = x^D - x^B \tag{A.13}$$

Isolating the sum of ratios in Eq. (A.13) will give the following expression:

$$RB + RR = \frac{x^D - x^B}{c} - 1 \quad (A.14)$$

Thus, if the minimum sum of ratios is sought, for a fixed amount of stages N, and given product specifications, the operating lines must intersect at the highest possible $F_{Di} = c = (y_i - x_i)$ value, that results in the need of exact this number of trays. If the operating lines intersect where the driving force is the largest, it is possible for the operating lines to intersect at the highest driving force value, and still only needing the number of stages N. This is the reason for the driving force approach being able to minimize the sum of ratios for a distillation column. To obtain a distillation process that will operate under these conditions, the distillation parameters must be adjusted, so the operating lines intersects at $x = D_x$. The parameter needed to be adjusted, so this happens, is the fraction of liquid in the feed q. If $q = 1$, the feed will be pure saturated liquid, if $q = 0$ the feed will be pure saturated vapor and if $q > 1$ the feed will be subcooled liquid, and if $q < 0$ the feed will be superheated vapor. The intersection of the two operating lines can be found by solving for x as follows:

$$x = \frac{(RR + 1) \cdot z_F + (q - 1) \cdot x^D}{q + RR} \quad (A.15)$$

To obtain the driving force based operation, the necessary fraction q can be found, as when the intersection happens at $x = D_x$, given a reflux ratio RR obtained from Eq. (A.6):

$$q = RR \cdot \frac{z_F - D_x}{D_x - x^D} + \frac{z_F - x^D}{D_x - x^D} \quad (A.16)$$

Note that the reflux ratio also depends on the fraction q, so the expression above is not an explicit way of calculating q. It is observed that to obtain the driving force based operation where $D_x \neq z_F$, requires a $q \neq 1$. This may be problematic when adapting the driving force based operation to other types of distillation processes. Therefore, further considerations must be taken into account.

J. Rafael Alcántara-Avila

7 Optimization methodologies for intensified distillation processes with flexible heat integration networks

Abstract: The innovation of chemical processes strongly depends on finding a new technology or combining two or more existing technologies in new ways. Nevertheless, at early design stages, many of the degrees of freedom are restricted to known structures or operating conditions. In this chapter, an optimization methodology is proposed to generate energy-efficient separation processes in which preestablished structures do not restrict the optimal structure and operating conditions.

Keywords: Heat integration, Process synthesis, MILP optimization

7.1 Introduction

The design of chemical processes is often restricted to predefined structures based on experience or mature process modeling techniques. However, intensified processes combine two or more features from typical processes into single equipment or step. Therefore, the resulting innovative process or equipment rises more degrees of freedom to be determined. The degrees of freedom can be structural or operational. Nevertheless, initial designs of intensified structures are very restricted to predefined feasible structures at the early design stage. Therefore, new optimization methodologies that do not depend on predefined structures at early design stages are necessary for the innovation of intensified processes.

Takase and Hasebe proposed a synthesis method for the separation of a ternary mixture, and problem was formulated as a linear programming (LP) problem in which the composition space was discretized [1]. The ternary mixture consisted of benzene (A), toluene (B), and o-xylene (C). There are two possible sequences of conventional columns to separate a ternary mixture: the direct sequence (withdraw A first, then B, and C) and the indirect sequence (withdraw C first, then B, and A) as shown at the left side of Figure7.1. However, the results showed that a process structure close to the Petlyuk column (Figure 7.1 right side), which can separate A, B, and C simultaneously, was derived as the optimal solution when the operation cost was minimized. The derived solution shows the importance of proposing optimization methodologies that are not restricted to predefined structures.

J. Rafael Alcántara-Avila, Department of Chemical Engineering, Kyoto University, Katsura Campus Nishikyo-ku, Kyoto, Japan

https://doi.org/10.1515/9783110596120-007

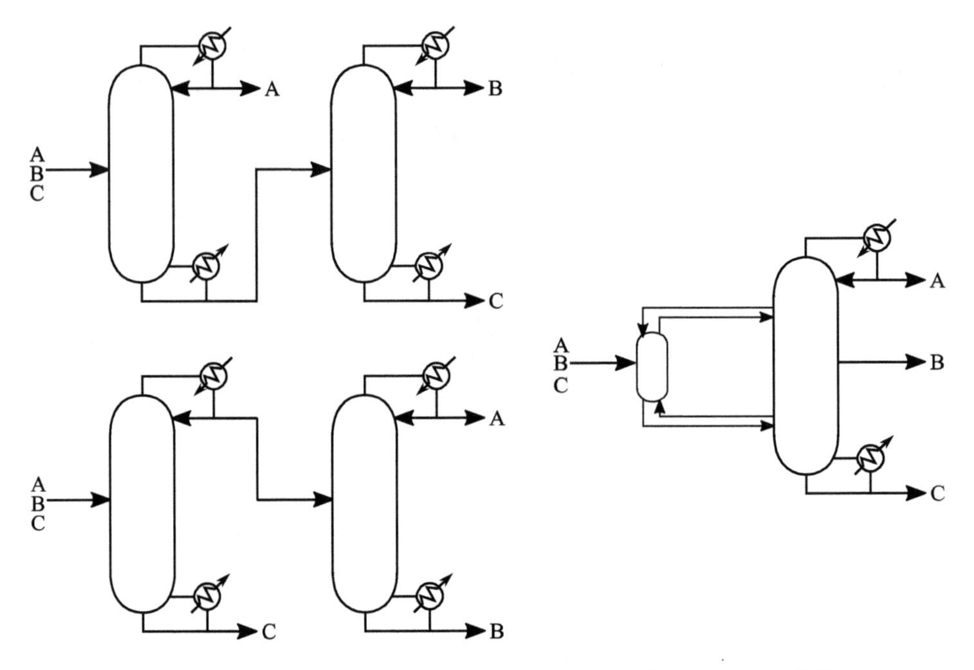

Figure 7.1: Distillation structures to separate a ternary mixture: typical sequences (left), and Petlyuk column (right).

This chapter presents a methodology to design intensified separation processes with heat integration combinations that are not restricted or limited to predefined patterns or structural constraints.

7.2 Structures with heat integration

Distillation is the most used technology to separate liquid mixtures in the Chemical Industry, and it spends between 10% to 15% of the world's energy consumption [2]. It uses the difference in boiling point between components to make the separation. Therefore, distillation repetitively boils liquid streams and condenses vapor streams from bottom to top resulting in a technology that consumes enormous amounts of energy. The chemical industry in Japan consumed 2,373 PJ of energy, which corresponded to 26.9% of the energy consumed by the manufacturing industry in 2016 [3]. The U.S. Department of Energy estimates that there are more than 40,000 distillation columns and they consume 5,064 PJ of energy, which is the 40% of the processing energy used in refining and continuous chemical processes [4]. Another report on energy consumption in the U.S. shows that distillation itself constitutes 90%–95% of all separations in the chemicals and petroleum refining industries, and it accounts for 49% of the energy consumption [5]. Therefore, energy consumption reduction in distillation is necessary.

An alternative to reduce the energy consumption in chemical processes is through heat integration. Heat integration uses the available energy in process streams and uses it instead of using energy from utilities. The basic principle to enforce heat integration is that one stream at high temperature (i.e., heat source) will supply heat to a stream at low temperature (i.e., heat sink). Thus, typical heat integration in distillation columns has been done between the vapor leaving the top of a distillation column at a high temperature and supplying that heat to some of the liquid leaving the bottom of a distillation column at a low temperature. Since there is a phase change in the vapor stream, all its latent heat is used in the heat integration.

7.2.1 Heat integration

Typical heat integration is done between the heat source that has the lowest temperature in the rectifying section (i.e., top vapor stream), and the heat sink that has the highest temperature in the stripping section (i.e., bottom liquid). Thus, heat integration occurs at the worst thermal condition.

Previous research has exploited the idea of realizing heat integration between heat sources at temperatures higher than that in the condenser and heat sinks at temperatures lower than that in the reboiler. Heat integration between stages in rectifying sections and stripping sections of different columns has been proposed for the separation of binary [6] and ternary mixtures [7] in which one column operated at high pressure and one at low pressure. Kim performed simulations that combined heat integration between sections and multi-effect distillation [8].

A compressor is used to generate the necessary pressure difference, thus temperature difference to realize heat integration between a rectifying section and a stripping section within the same distillation column. This kind of heat integration is called Heat-Integrated Distillation Column (HIDiC). Typical HIDiC structures consider heat integration either along all or almost all the rectifying or stripping section. Cabrera-Ruiz et al. performed simulations for several ternary mixtures to assess their potential for saving energy, and they found that HIDiC is not the best choice in all cases [9]. Matsuda et al. performed simulations that considered heat and mass transfer between sections through a rate-based model, and they found good agreement between experimental results and simulation results that predicted the overall heat transfer coefficient with the mass transfer correlation [10]. Bisgaard et al. performed steady-state and dynamic simulations of HIDiC structures to design control configurations that allow such structures to operate stably [11].

On the other hand, HIDiC structures with few heat integrations have also been proposed, and it has been demonstrated that they outperform typical HIDiC.

Wakabayashi and Hasebe explicitly evaluated the relationship between heat integration at stages in HIDiC and the reduction of reboiler duty for the separation of a benzene/toluene mixture, and they found that few heat integrations were better than typical HIDiC [12]. Then, Wakabayashi and Hasebe extended their design method to the separation of a multicomponent mixture by adopting the idea of a quasi-binary mixture, and the optimal solution showed that only four heat integrations realized the lowest cost with energy savings above 50% [13]. Wakabayashi et al. used their graphical design method to design a discretely heat-integrated distillation column for commercial purpose as the world first commercial application of HIDiC technology [14]. The proposed structure was able to attain 56% energy savings in comparison with the conventional distillation column. Heat exchange took place at four discrete stage locations.

Heat integration between stages of different columns has been extended in complex distillation processes involving chemical reaction [15] and thermal coupling [16] to find innovative ways to enhance current intensified processes with energy savings up to 22% and 47% higher than the conventional reactive distillation sequence, respectively.

7.3 Optimization of heat-integrated structures

The U.S. Department of Energy estimates that built-in models in process simulation software have the potential to optimize distillation column operations and to save approximately 56 PJ per year by 2020 [4]. Regarding heat integration between stages, there are three challenging issues, which can be summarized as follows:

1. **Complex combinations**. If all the M stages in a high-temperature rectifying section are regarded as heat sources, and all the N stages in a low-temperature stripping section are regarded as heat sinks, there will be $M \times N$ possible combinations to realize only one heat integration. In addition, if the number of heat integrations increases, the combinations of pairs will also increase.
2. **Changes in the internal flows and composition**. If heat is removed from a heat source, it means that part of the vapor will condense. Thus, the ascending vapor flow will decrease. Similarly, if heat is supplied to a heat sink, part of the liquid will vaporize. Thus, the descending liquid flow will decrease. Also, the composition of internal flows will change. When part of the vapor is condensed, the components with a low boiling point will condense at a rate lower than those with a high boiling point. Therefore, the temperature will decrease in the stages subject to heat integration in the rectifying section. Oppositely, when part of the liquid is vaporized, the components with a high boiling point will vaporize at a rate lower than those with a low boiling point. Therefore, the temperature will increase in the stages subject to heat integration in the stripping section.

3. **Energy balance**. For the typical heat integration between a condenser and a reboiler, the energy supplied by the condenser is equal to the energy removed at the reboiler. Nevertheless, in case of heat integration between stages, this equality is not valid [7, 12]. There are complex and implicit relations of the energy in HIDiC because it is difficult to estimate the combinations of heat integration and the effective reduction of the condenser or reboiler duty. To transform this inequality into equality, the concept of '*compensation terms*' has been introduced as a linear approximation to estimate the reboiler and condenser duty resulting from heat integration [7]. However, the previous approach has still some limitations because it does not consider the interaction between heat integrated stages, but only the independent relationship between heat exchanged at a stage and the reduction of the condenser or reboiler duty. Figure 7.2 summarizes the three issues in heat integration between stages.

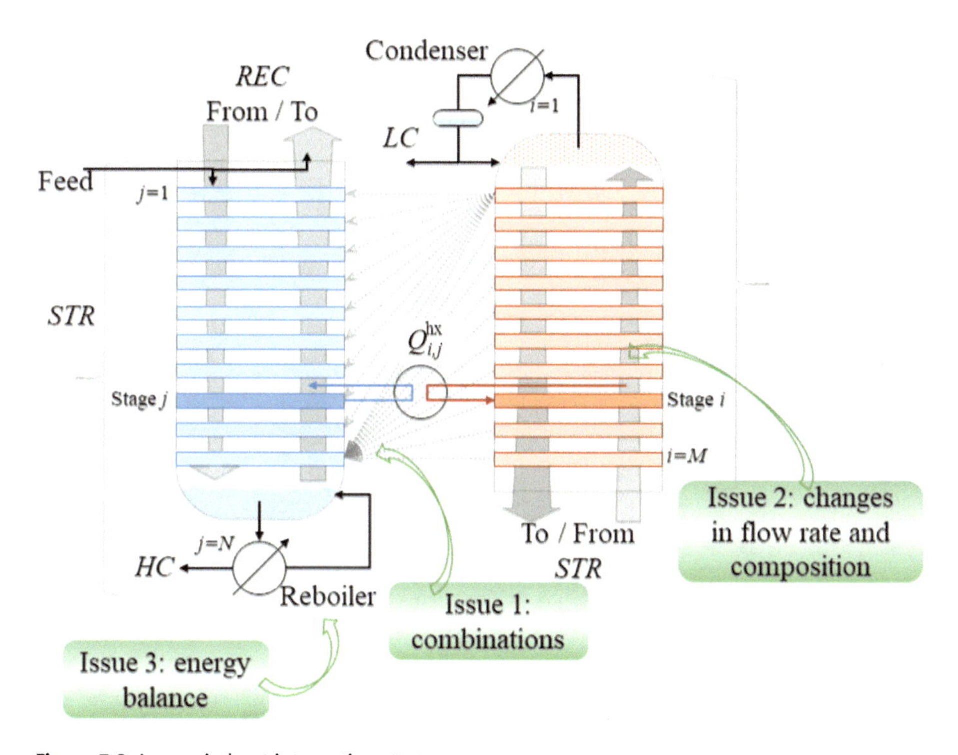

Figure 7.2: Issues in heat integration at stages.

In the figure, the section at the right side is the rectifying section, and that at the left side is the stripping section. *LC* (*HC*) denote the stream with low (high) boiling point components. *REC* (*STR*) is the set of i (j) stages in the rectifying (stripping) section. $Q^{hx}_{i,j}$ is the amount of heat transferred from stage i to j. The thick grey arrows at the back of each section represent the changes of liquid and vapor flows. The dotted grey lines represent the combinations of possible heat integration between stages. These combinations can be extended to include the condenser, reboiler, and other stages but for the sake of simplicity, they are not drawn.

This work aims to propose an optimization methodology to find optimal heat integration networks in HIDiC by simultaneously considering structural variables (i.e., number of stages in each section, locations of heat-integrated stages) and operational variables (e.g., heat transfer amount, reflux, stages pressure, temperature, and component flows). A superstructure that comprises all heat integration possibilities is proposed and formulated as a mixed integer linear programming (MILP) problem. The data input into the optimization problem is taken from rigorous simulations of each stage and updated through an interface that combines optimization and simulation successively. The reader is encouraged to refer to [7, 17] for further details.

Figure 7.3 shows the superstructure that comprises all the possible heat integrations between a rectifying section and a stripping section.

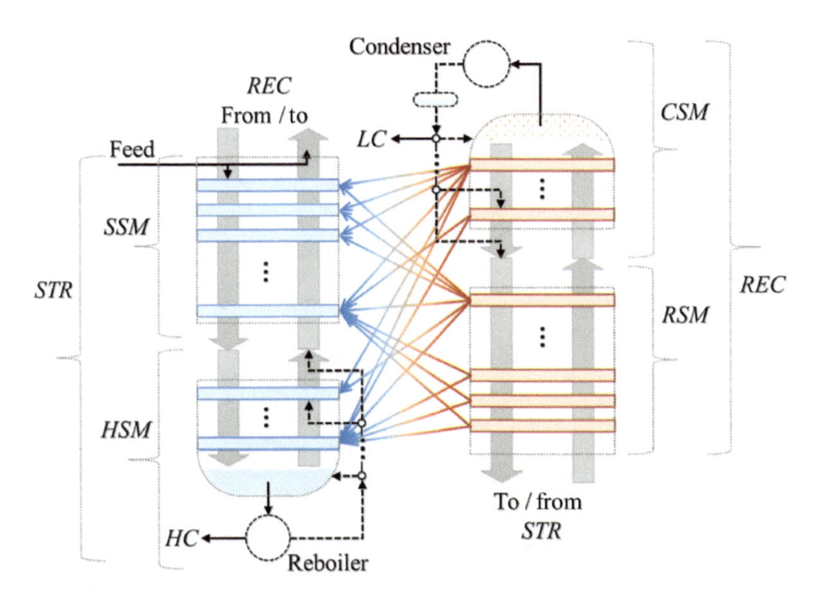

Figure 7.3: Superstructure representation for heat integration between stages.

In the figure, each section is further divided into modules. *RSM* and *SSM* are the rectifying and stripping stages module, respectively, and these modules have a fixed number of stages. Contrarily, *CSM* and *HSM* are the cooling and heating stages module, respectively, and these modules have a variable number of stages, which means that the vapor and liquid streams can bypass stages. Therefore, the stages in each section and the combinations for heat integration vary simultaneously in the optimization stage. It can also be seen that any stage in *CSM* or *RSM* can supply heat to any stage in *STR* and that any stage in *SSM* or *HSM* can accept heat from any stage in *REC*. The liquid stream leaving the condenser and reflux drum can return to any stage in *CSM*, which means that the stages above the liquid reflux can be omitted in the final solution. Similarly, the vapor stream leaving the reboiler can return to any stage in *HSM*, which means that the stages below the vapor recirculation can be omitted in the final solution. Previous mathematical formulations have considered bypassing liquid, or vapor steams in stages of distillation columns [18] and HIDiC structures [19].

Optimization of HIDiC structures has been based on mathematical programming [19, 20], stochastic methods [21, 22], and graphical methods [12–14]. Mathematical programming approaches have used a superstructure representation and then formulated it as a nonlinear programming problem [19] or as a linear programming problem [20]. Stochastic methods have been addressed mainly to HIDiC structures in all stages that enforce heat integration; however, the HIDiC structure was not the best in terms of total annual cost in most cases. Finally, the graphical method based on the Ponchon–Savarit diagram is available to represent the enthalpy conditions in each stage subject to heat integration; however, it is limited to binary or quasi-binary mixtures.

7.3.1 Mathematical formulation

To find intensified solutions with heat-integrated stages, an optimization methodology that combines process simulation and optimization has been developed [17]. In the process simulation step, the stage-by-stage MESH equations are solved while in the optimization step, part of the energy balance is decoupled from the MESH equations and linearized at each stage. These linearizations explicitly approximate the relationship between the heat exchanged at stage j (Q_j) and the condenser duty (Q^{cn}) or reboiler duty (Q^{rb}). The advantage of linearizing these relations is that they can deal with the issues of complex combinations and the energy balance in heat-integrated stages. The remaining issue is addressed by iteratively combining simulation and optimization. The simulation results will update the composition and temperature profiles in the distillation columns, and by setting a convergence criterion [17], a solution can be derived.

The superstructure in Figure 7.3 can be formulated as a MILP problem if the pressure between sections is treated as a discrete variable, the energy balance and combination constraints are denoted by linear equations, and the objective function is approximated as a linear equation.

Figure 7.4 shows the general representation of a stage k and the MESH equations that apply for that stage. Complex thermodynamics to calculate VLE (i.e., $k_{i,k}$),

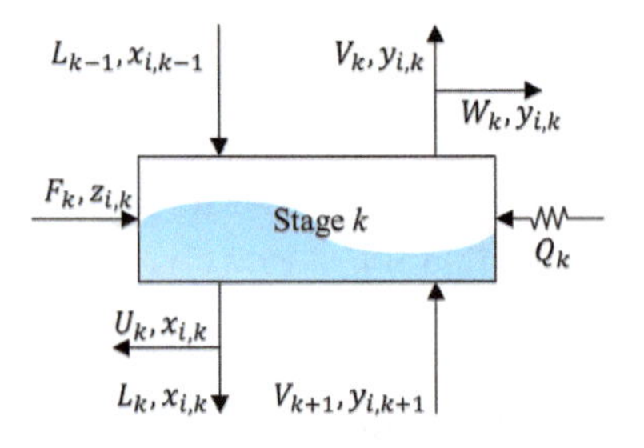

Mass: $L_{k-1}x_{i,k-1} + V_{k+1}y_{i,k+1} + F_k z_{i,k} - (L_k + U_k)x_{i,k} - (V_k + W_k)y_{i,k} = 0$

Energy: $L_{k-1}h_{L_{k-1}} + V_{k+1}h_{V_{k+1}} + F_k h_{F_k} - (L_k + U_k)h_{L_k} - (V_k + W_k)h_{V_k} + Q_k = 0$

Equilibrium: $y_{i,k} - k_{i,k}x_{i,k} = 0$

Summation: $\sum_{i \in C} y_{i,k} - 1 = 0 \quad \sum_{i \in C} x_{i,k} - 1 = 0$

Figure 7.4: Stage representation and its ruling equations.

enthalpy values (e.g., h_{L_k}), and bilinear terms (e.g., $L_{k-1}x_{i,k-1}$) are implicitly calculated in the simulation software. However, the heat transfer (i.e., Q_k) is explicitly calculated in the optimization software. In the figure, Q_k, Q^{cn}, and Q^{rb} are known in advance. Their relationships can be treated as linear functions in such a way that Q_k can be decoupled from the heat balance.

Energy Minimization: Equations 7.1 through 7.22 are used to find the solution with the minimum energy consumption. The adopted objective function minimizes the energy consumption of a distillation process

$$\min Energy = \sum_{\substack{i \in HU \\ j \in HSM}} Q_{i,j}^{hx} + \eta W \tag{7.1}$$

where W is the compressor work duty, and η is a factor that converts the consumed electricity by the compressor in equivalent primary energy with an empirical power generation efficiency for Japan of 1/0.366 [23].

Energy balance constraints: Equations 7.2 through 7.8 are the energy balance constraints for heat integration at stages. Changes in the condenser and reboiler duties due to heat integration at stages can be predicted in advance from steady-state simulations of a column without any heat integration. Since heat is supplied to a stripping section, its vapor flow rate toward the compressor and condenser will increase. If a compressor is used, the compressor work duty will increase for the same given compressor outlet pressure, and the condenser duty will increase to condense that vapor. Similarly, since heat is removed from a rectifying section, the liquid flow rate toward the reboiler will increase. Thus, the reboiler duty will increase to vaporize that liquid.

When realizing heat integration, only some of the heat will usefully reduce energy consumption in distillation. The unuseful heat can be regarded as the difference between the heat integrated and the heat usefully reduced. In this study, the amount of unuseful heat can be estimated by using an approximation based on the concept of "*compensation terms*" [7].

Equations 7.2 and 7.3 show linear approximations to the relationship between heat integration at stage i and the condenser ($i = 1$) and stage j and the reboiler ($j = N$)

$$\Delta Q_i = Q_i - \left(Q_0^{cn} - Q^{cn}\right) \; i \in REC \tag{7.2}$$

$$\Delta Q_j = Q_j - \left(Q_0^{rb} - Q^{rb}\right) \; j \in STR \tag{7.3}$$

where Q_0^{cn} and Q_0^{rb} are the condenser duty and reboiler duty before any heat integration. Q^{cn} and Q^{rb} are the condenser duty and reboiler duty resulted from heat integration at stages. ΔQ_i and ΔQ_j are the compensation terms for the stages subject to heat integration in the rectifying section(s), stripping(s). Since $Q_i \geq \left(Q_0^{cn} - Q^{cn}\right)$ and $Q_j \geq \left(Q_0^{rb} - Q^{rb}\right)$, ΔQ_i and ΔQ_j are positive values that denote the inefficiency in the condenser and reboiler duty reduction resulted from heat integration. Therefore, large values must be avoided.

Equations 7.4 to 7.6 show the energy balance when heat integration is realized between the modules in Figure 7.3

$$Q^{cn} = \sum_{\substack{i \in CSM \\ j \in STR}} Q_{i,j}^{hx} = Q_0^{cn} - \sum_{\substack{i \in RSM \\ j \in STR}} Q_{i,j}^{hx} + \sum_{i \in REC} \Delta Q_i + \sum_{j \in STR} \Delta Q_j \tag{7.4}$$

$$Q^{rb} = \sum_{\substack{i \in REC \\ j \in HSM}} Q_{i,j}^{hx} = Q_0^{rb} - \sum_{\substack{i \in REC \\ j \in SSM}} Q_{i,j}^{hx} + \sum_{j \in STR} \Delta Q_j + \sum_{i \in REC} \Delta Q_i \tag{7.5}$$

$$W = W_0 + \sum_{i \in REC} \Delta W_i + \sum_{j \in STR} \Delta W_j \tag{7.6}$$

where W_0 is the work duty before any heat integration. W is the work duty resulted from heat integration at stages. ΔW_i and ΔW_j are the compensation terms for the compressor resulting from heat integrations between rectifying and stripping sections.

The solid line in Figure 7.5 shows the relation between ΔQ_j and Q_j. It was calculated from sensitivity analyses and steady-state simulations. It can be seen that the unuseful heat steadily increases as the amount of heat integration increases. Moreover, it holds a nonlinear relationship. In this study, for the sake of keeping a linear model, a piecewise linear function was used (dotted lines in Figure 7.5) to approximate the relationship between Q_i and ΔQ_i in *REC* as well as Q_j, and ΔQ_j in *STR*.

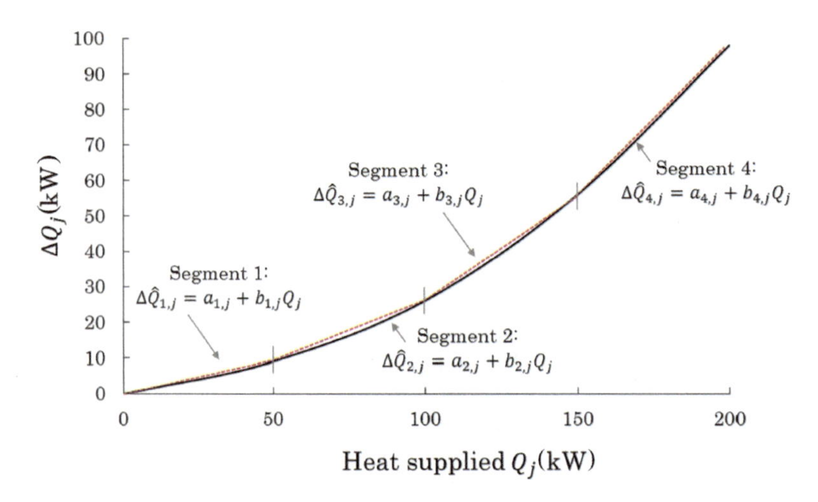

Figure 7.5: Linear approximation to ΔQ_j for heat supplied at stage j in a striping section.

Equations 7.7 to 7.10 show the explicit piecewise linear function to approximate the values of ΔQ_i, ΔQ_j, ΔW_i, and ΔW_j, respectively.

$$\Delta Q_i \approx \Delta\hat{Q}_{k,i} = \max_{k \in PW}\{a_{k,i}\delta_i + b_{k,i}Q_i\} \; i \in REC \tag{7.7}$$

$$\Delta Q_j \approx \Delta\hat{Q}_{k,j} = \max_{k \in PW}\{a_{k,j}\delta_j + b_{k,j}Q_j\} \; j \in STR \tag{7.8}$$

$$\Delta W_i \approx \Delta\hat{W}_{k,i} = \max_{k \in PW}\{c_{k,i}\delta_i + d_{k,i}Q_i\} \; i \in REC \tag{7.9}$$

$$\Delta W_j \approx \Delta\hat{W}_{k,j} = \max_{k \in PW}\{c_{k,j}\delta_j + d_{k,j}Q_j\} \; j \in STR \tag{7.10}$$

where $\Delta\hat{Q}_{k,i}$, $\Delta\hat{Q}_{k,j}$, $\Delta\hat{W}_{k,i}$, and $\Delta\hat{W}_{k,j}$ are piecewise linear functions. PW is the set of segments k in the piecewise linear function. δ_i and δ_j are dummy variables, which are zero when Q_i and Q_j are zero. $a_{k,i}$, $b_{k,i}$, $c_{k,i}$, $d_{k,i}$, $a_{k,j}$, $b_{k,j}$, $c_{k,j}$, and $d_{k,j}$ are the intercept and slope parameters in the linear functions. A thorough discussion of these equations can be found in [7].

Temperature constraints: The temperature difference between stages i and j will create the driving force to realize heat integration. Such a driving force depends directly on the pressure in each section. As the pressure difference between sections increases, temperature difference also increases. In this study, the log mean temperature difference (ΔT_{LM}) is used to determine the temperature driving force and feasibility for heat transfer. Figure 7.6 shows the heat integration at a stage in the rectifying section and that at a stage in the stripping section, respectively.

Equation 7.11 represents the ΔT_{LM} between heat sources and sinks while Eq. 7.12 shows the feasibility criterion based on the "*Big-M*" formulation, which is a valid representation for linear constraints [18].

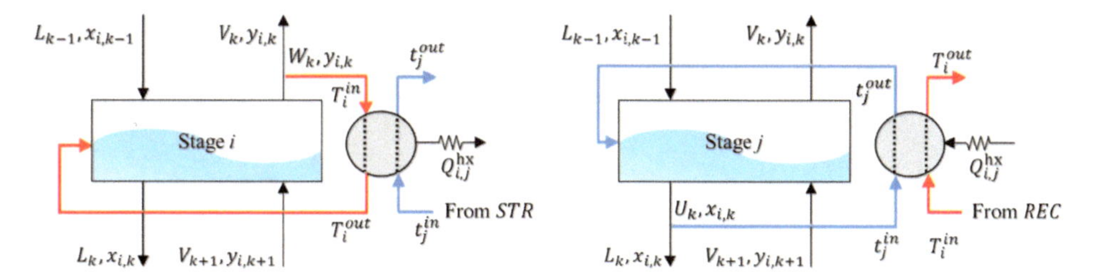

Figure 7.6: Heat integration in a rectifying section (left) and a stripping section (right).

$$\Delta T_{LM_{i,j}} = \begin{cases} \dfrac{\left(T_i^{\text{in}} - t_j^{\text{out}}\right) - \left(T_i^{\text{out}} - t_j^{\text{in}}\right)}{\ln\left[\left(T_i^{\text{in}} - t_j^{\text{out}}\right) / \left(T_i^{\text{out}} - t_j^{\text{in}}\right)\right]} & \text{if } T_i^{\text{in}} - t_j^{\text{out}} \geq \Delta T_{\text{min}} \text{ and } T_i^{\text{out}} - t_j^{\text{in}} \geq \Delta T_{\text{min}} i \in HS, j \in CS \\ 0 & \text{otherwise} \end{cases} \tag{7.11}$$

$$Q_{i,j}^{\text{hx}} - M\Delta T_{LM_{i,j}} \leq 0 \; i \in HS, \; j \in CS \tag{7.12}$$

In Eq. 7.11, T_i^{in} and T_i^{out} are the inlet and outlet temperature of the heat source i while t_j^{in} and t_j^{out} are the inlet and outlet temperature of the heat sink j. HS and CS are the sets of heating sources (i.e., $HS = REC \cup HU$) and cooling sources (i.e., $CS = STR \cup CU$), respectively. CU is the set of cooling utilities while HU is the set of heating utilities. ΔT_{min} is the minimum temperature difference allowed, and M is a positive value. Equation 7.12 eliminates all infeasible heat exchanges because $Q_{i,j}^{\text{hx}}$ is zero when $\Delta T_{LM_{i,j}}$ is zero.

Equation 7.13 shows the calculation of the heat transfer are $A_{i,j}$ between heat sources and sinks

$$A_{i,j} = \frac{Q_{i,j}^{\text{hx}}}{U\Delta T_{LM_{i,j}} + \varepsilon} \tag{7.13}$$

where U is the overall heat transfer coefficient, and it has a different value for condensers, reboilers, and process-to-process heat integration because they use fluids with different properties. ε is a small value (i.e., 0.0001) to avoid division by zero.

Connectivity constraints: The combination of stages subject to heat integration can be represented by binary variables (i.e., variables that can take zero or one values), which are defined by Eq. 7.14

$$Y_{i,j}^{\text{hx}} \in \{0, 1\} \; i \in HS, \; j \in CS \tag{7.14}$$

$Y_{i,j}^{\text{hx}}$ is one if heat integration takes place between a stage i in the rectifying section and a stage j the stripping section, and zero otherwise.

Equation 7.15 is a constraint that enforces a determined number of heat exchangers in the rectifying and stripping sections

$$\sum_{\substack{i \in HS \\ j \in CS}} Y_{i,j}^{\mathrm{hx}} \leq N^{\mathrm{hx}} \tag{7.15}$$

where N^{hx} is a parameter which limits the maximum number of heat exchangers to be installed.

Equation 7.16 is an additional equation that is necessary to allocate heat in feasible heat integrations

$$Q_{i,j}^{\mathrm{hx}} - Y_{i,j}^{\mathrm{hx}} Q_{i,j}^{\max} \leq 0 \quad i \in HS, \ j \in CS \tag{7.16}$$

where $Q_{i,j}^{\max}$ is the upper bound for heat integration between heat sources and sinks.

Variation in the number of stages: As mentioned in Sect. 3, the rectifying section can be divided into two modules: *CSM* and *RSM*. Similarly, the stripping section can be divided into two modules: *SSM* and *HSM*. Heat integration is possible between stages in these modules. Equations 7.17 and 7.18 show the calculation for the number of stages in each section

$$NSR = \mathrm{NS_R} + NS_k \tag{7.17}$$

$$NSS = \mathrm{NS_S} + NS_n \tag{7.18}$$

where $\mathrm{NS_R}$ and $\mathrm{NS_S}$ are the number of stages in *RSM* and *SSM*. NS_k and NS_n denote the stage number where the condenser and reboiler are located. These variables are are integer. *NSR* and *NSS* are the number of total stages in the rectifying and stripping sections.

As the number of stages in *CSM* and *HSM* increases, the condenser and reboiler duty will decrease. Thus, less energy consumption can be attained. Contrary, as that number of stages decreases, more energy must be used to perform the separation.

Figure 7.7 shows the relationship between energy consumption and NS_k and NS_n for the separation of the quaternary mixture in Sect. 4. In this case, heat integration at stages is not included. Thus, the effect of adding stages is evaluated individually. The simulations were done in the process simulation software Aspen Plus V.8.8. The horizontal axes represent the number of stages in *CSM* and *HSM* while the vertical axes represent the condenser and reboiler duty, respectively.

The relationship between the number of stages in *CSM* and *HSM* can be useful to determine the cost of the distillation process. The relationships in Figure 7.7 are represented as linear functions and added to the optimization problem.

Equations 7.19 and 7.20 show the condenser duty ($\hat{Q}_0^{\mathrm{cn}}$) and reboiler duty ($\hat{Q}_0^{\mathrm{rb}}$) as a function of the number of stages in *CSM* and *HSM*, respectively.

$$\hat{Q}_0^{\mathrm{cn}} = \hat{Q}_{0,\min}^{\mathrm{cn}} + \alpha NS_k \tag{7.19}$$

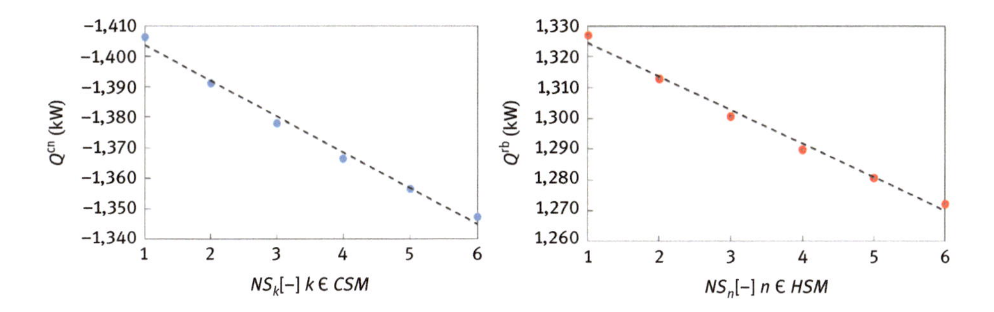

Figure 7.7: Relation between the number of stages and the energy consumption for a condenser (left) and that for a reboiler (right).

$$\hat{Q}_0^{rb} = \hat{Q}_{0,\max}^{rb} - \beta NS_n \tag{7.20}$$

where α and β are the slopes of the functions. $\hat{Q}_{0,\min}^{rb}$ and $\hat{Q}_{0,\max}^{rb}$ are the condenser and reboiler duty values when only one stage exists in *CSM* and *HSM* ($NS_k = NS_n = 1.0$).

If there is only one stage in *CSM* and *HSM*, Eqs. 7.4 and 7.5 can use the values of $Q_0^{cn} = \hat{Q}_{0,\min}^{cn}$ and $Q_0^{rb} = \hat{Q}_{0,\max}^{rb}$, respectively. However, if the number of stages in *CSM* and *HSM* is more than one, Eqs. 7.21 and 7.22 are used in the optimization problem

$$Q^{cn} = \sum_{\substack{i \in CSM \\ j \in STR}} Q_{i,j}^{hx} = \hat{Q}_0^{cn} - \sum_{\substack{i \in RSM \\ j \in STR}} Q_{i,j}^{hx} + \sum_{i \in REC} \Delta Q_i + \sum_{j \in STR} \Delta Q_j \tag{7.21}$$

$$Q^{rb} = \sum_{\substack{i \in REC \\ j \in HSM}} Q_{i,j}^{hx} = \hat{Q}_0^{rb} - \sum_{\substack{i \in REC \\ j \in SSM}} Q_{i,j}^{hx} + \sum_{j \in STR} \Delta Q_j + \sum_{i \in REC} \Delta Q_i \tag{7.22}$$

when energy is minimized, there is always the possibility of the obvious answer: as the number of stages increases, indeed, the energy consumption decreases (it is at least true for conventional distillation process without chemical reaction or heat integration), but the equipment cost increases. Therefore an economic criterion can better represent a good balance between energy and cost.

Total Annual Cost (TAC) minimization: Equations 7.2 through 7.34 are used to find the solution with the minimum cost

$$\min TAC = OC + \frac{CC}{\text{payback time[yr]}} \tag{7.23}$$

where *OC* is the operating cost resulted from using cooling water, steam, electricity or other utilities. *CC* is the capital cost, which comprises the cost of the distillation column, column internals such as trays, heat exchangers, and compressor. The payback time is the period in which the capital cost is recovered.

Economic constraints: Equation 7.24 shows in detail the calculation of the operating cost

$$OC = \mathrm{OH}\left[\sum_{\substack{i\in CSM\\ j\in CU}} Q_{i,j}^{\mathrm{hx}} C_j^{\mathrm{cool}} + \sum_{\substack{i\in HU\\ j\in HSM}} C_i^{\mathrm{heat}} Q_{i,j}^{\mathrm{hx}} + WC^{\mathrm{elec}}\right] \tag{7.24}$$

where OH is the annual operation hours, C_j^{cool} is the cost of j cooling utilities, C_i^{heat} is the cost of i heating utilities, and C^{elec} is the cost of electricity.

Equation 7.25 shows the capital cost calculation

$$CC[\$] = \frac{CEPCI\ in\ 2015}{CEPCI\ in\ 2006}\left[C_{\mathrm{shell}} + C_{\mathrm{tray}} + C_{\mathrm{comp}} + \sum_{\substack{i\in HS\\ j\in CS}} C_{hex_{i,j}}\right] \tag{7.25}$$

where *CEPCI* is the Chemical Engineering Plant Cost Index. The reference value is the one corresponding to 2006, and it is 500 [24]. In this study, the used *CEPCI* value corresponds to the average one of 2015 (556.8) [25]. C_{shell}, C_{tray}, C_{comp}, and C_{hex} are the cost of the column shell, trays, compressor and heat exchangers including condensers and reboilers [24].

Equation 7.26 shows the calculation of the column shell

$$C_{\mathrm{shell}} = \mathrm{F_M} C_v + 300.9 \cdot D^{0.63316} L^{0.80161} \tag{7.26}$$

where $\mathrm{F_M}$ is the material factor. In this study, all of the equipment material is assumed to be carbon steel, therefore, $\mathrm{F_M} = 1.0$. D is the column diameter, and L is the column height. C_v is calculated from the column weight according to Eqs. 7.27 and 7.28

$$G = 612.5\pi(D + 1.25)(L + 0.8\,D) \tag{7.27}$$

$$C_v = \exp\left\{7.2756 + 0.18255 \ln G + 0.02297(\ln G)^2\right\} \tag{7.28}$$

The tray cost is calculated according to Eq. 7.29

$$C_{\mathrm{tray}} = N_T F_{NT} \mathrm{F_{TT}} \mathrm{F_{TM}} C_{BT} \tag{7.29}$$

Where N_T is the number of trays, $\mathrm{F_{TT}}$ is the factor associated with the type of tray, $\mathrm{F_{TM}}$ is the tray material factor. In this study, $\mathrm{F_{TT}} = \mathrm{F_{TM}} = 1.0$ because carbon steel sieve trays were assumed to be used. F_{NT} is a factor associated with the number of trays in the column, and it is calculated from Eq. 7.30.

$$F_{NT} = \begin{Bmatrix} 1 & if \quad N_T \geq 20 \\ \frac{2.25}{1.0414^{N_T}} & if \quad N_T < 20 \end{Bmatrix} \tag{7.30}$$

C_{BT} is calculated from the tray diameter according to Eq. 7.31,

$$C_{BT} = 468 \ \exp(0.1739\,\mathrm{D}) \tag{7.31}$$

and the compressor cost is calculated according to Eq. 7.32

$$C_{\mathrm{comp}} = \mathrm{F_D F_M} C_B \tag{7.32}$$

where $\mathrm{F_D}$ and $\mathrm{F_M}$ are the factors associated with the type of compressor and construction material, respectively. In this study, the assumed values of these factors are $\mathrm{F_D} = 1.25$ and $\mathrm{F_M} = 1.0$. C_B is calculated from the needed work duty to run the compressor. In this study, the compressor duty is in the range between 10 and 750 Hp. Therefore, C_B is calculated from Eq. 7.33

$$C_B = \exp(8.1238 + 0.7243 \ln W) \tag{7.33}$$

where W is the compressor work duty in horsepower.

Equation 7.34 shows the calculation of the heat exchangers. It includes condensers and reboilers

$$C_{hex_{ij}} = \begin{Bmatrix} 0 & if \quad A_{i,j} \leq 0.2 \\ 10^{\left(3.0238 + 0.0603 \log A_{i,j}\right)} \times (0.74 + 1.21) \times 1.18 & if \quad 0.2 \leq A_{i,j} \leq 4 \\ 10^{\left(3.2138 + 0.2688 \cdot \log A_{i,j} + 0.07961 \log A_{i,j})^2\right)} \times (1.8 + 1.5) \times 1.18 & if \quad 4 \leq A_{i,j} \leq 10 \\ 10^{\left(3.4338 + 0.1445 \cdot \log A_{i,j} + 0.1079 \log A_{i,j})^2\right)} \times (1.8 + 1.5) \times 1.18 & otherwise \end{Bmatrix} \tag{7.34}$$

Equipment cost linearization: The equations to calculate the equipment cost are nonlinear. Therefore, C_{shell}, C_{tray}, C_{comp}, and C_{hex} must be approximated to linear relationships with the optimization variables. Equation 7.23, can be rewritten in a linearized form as shown in Eq. 7.35

$$\min \widehat{TAC} = OC + \frac{\widehat{CC}}{\text{payback time [yr]}} \tag{7.35}$$

where the term OC is linear itself, thus any linearization is not needed.

Once the pressure difference between section is set, the column height is a function of the number of stages and the column diameter, which depends on the internal vapor flow rate. Since the compressor work duty and the total heat transfer area will not change much from the reference case design (i.e., structure without heat integration), the linearization of these equipment costs is possible near the reference case design. Therefore, Eqs. 7.36 to 7.40 show the linearized equipment cost.

$$C_{\text{shell}} \approx \hat{C}_{\text{shell}} = \sum_{i \in REC} (o_i + p_i L_i) + \sum_{j \in STR} (o_j + p_j L_j) \tag{7.36}$$

$$C_{\text{tray}} \approx \hat{C}_{\text{tray}} = \sum_{i \in REC} (q_i + r_i NS_i) + \sum_{j \in STR} (q_j + r_j NS_j) \tag{7.37}$$

$$C_{\text{comp}} \approx \hat{C}_{\text{comp}} = s + tW \tag{7.38}$$

$$C_{hex_{i,j}} \approx \hat{C}_{hex_{i,j}} = u + vA_{i,j} \tag{7.39}$$

$$\widehat{CC} = \frac{CEPCI\ in\ 2015}{CEPCI\ in\ 2006} \cdot \left[\hat{C}_{\text{shell}} + \hat{C}_{\text{tray}} + \hat{C}_{\text{comp}} + \sum_{\substack{i \in HS \\ j \in CS}} \hat{C}_{hex_{i,j}} \right] \tag{7.40}$$

where $\hat{C}_{\text{shell}}$, $\hat{C}_{\text{tray}}$, $\hat{C}_{\text{comp}}$, and $\hat{C}_{hex_{i,j}}$ are the equipment cost values calculated from linear relationships. o, p, q, r, s, t, u, and v are parameters used to approximate the values of the equipment cost.

7.3.2 Optimization methodology

The proposed optimization methodologies combine rigorous simulation and optimization through an iterative procedure. Process simulation is used to readily know in advance the pressure and temperature in each stage, the condenser duty in the rectifying section and the reboiler duty in the stripping section. It is also used to know the relation between heat transferred at stages and heat reduced in the condenser and reboiler. Then, the obtained values in the simulation are input into the optimization as parameters. Once the optimization problem is solved, the selected stages subject to heat integration, the heat transferred between stages, and the number of stages in each section are known. Finally, since heat integration causes changes in the temperature of heat-integrated stages, the simulation is updated with the optimization values, then the optimization problem is updated with the results from the simulation, and solved again. Optimizations and simulations are repeated until a termination criterion is met.

The optimization methodology in Figure 7.8 minimizes energy consumption (Eq. 7.1) in HIDiC structures while in a post-optimization step the total annual cost is calculated (Eq. 7.23).

In Figure 7.8, ΔP is the pressure difference between each *REC* and *STR* section. The maximum pressure difference ($\Delta P_{\max}$) is the one that creates a temperature difference higher than $\Delta T_{\min}$ to realize heat integration between the condenser and the reboiler. The parameters $a_{k,i}$, $a_{k,j}$, $b_{k,i}$, $b_{k,j}$, $c_{k,i}$, $c_{k,j}$, $d_{k,i}$, and $d_{k,j}$ are used in Eqs. 7.7

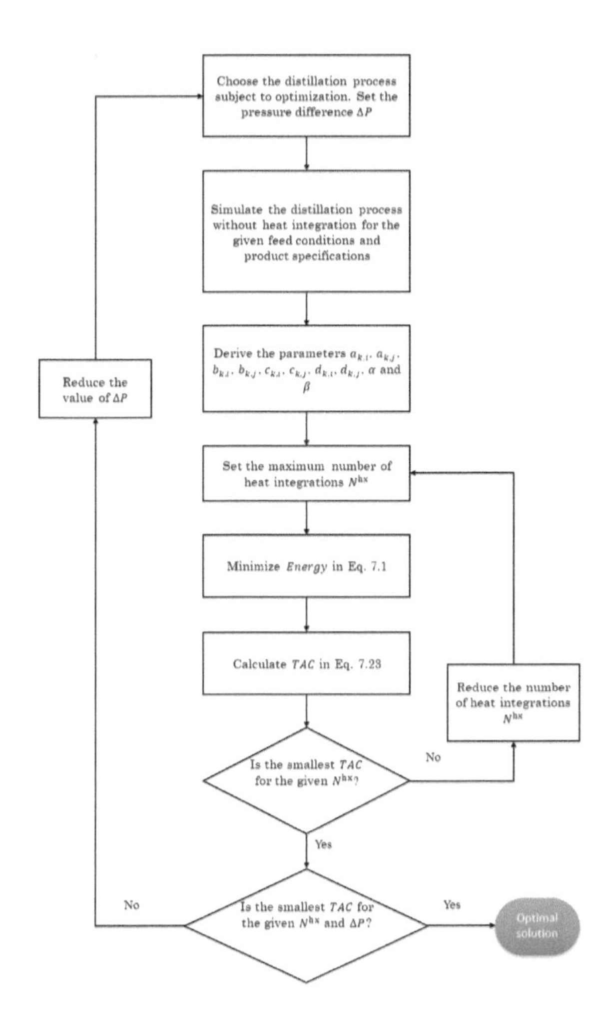

Figure 7.8: Flowchart for the proposed optimization methodology that minimizes energy consumption.

to 7.10. The minimization of Eq. 7.1 is subject to constraints in Eqs. 7.2 to 7.23. However, if the number of stages in each section is considered as optimization variable, Eqs. 7.19 to 7.22 replace Eqs. 7.4 and 7.5.

The optimization methodology in Figure 7.9 minimizes the linearized total annual cost (Eq. 7.35) in HIDiC structures while in a post-optimization step the total annual cost is calculated (Eq. 7.23) to determine the deviation between $\widehat{TAC}$ and TAC.

In Figure 7.9, the minimization of $\widehat{TAC}$ is subject to constraints in Eqs. 7.2 to 7.23 and Eqs. 7.36 to 7.40. However, if the number of stages in each section is considered as optimization variable, Eqs. 7.19 to 7.22 replace Eqs. 7.4 and 7.5.

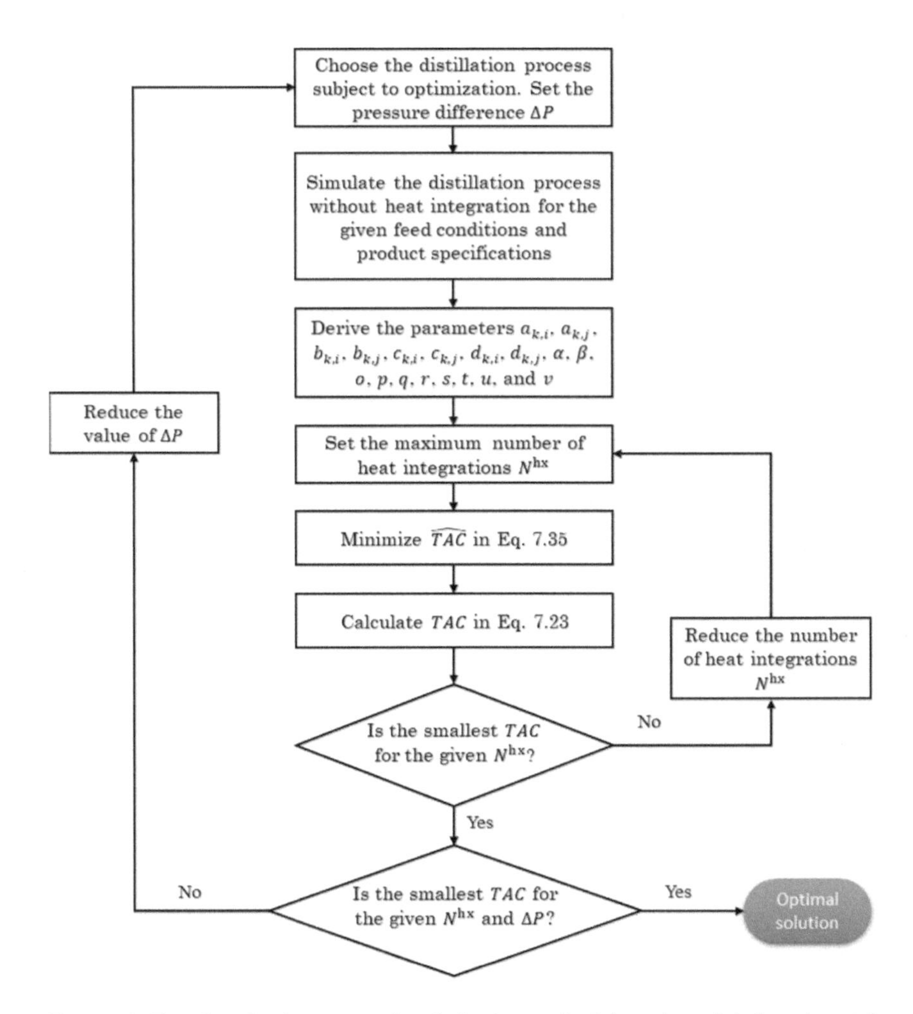

Figure 7.9: Flowchart for the proposed optimization methodology that minimizes the total annual cost.

7.3.3 Iterative optimization procedure

The optimal solution depends on the parameters calculated from simulations, which at the same time it depends on the previous optimization solution. Therefore, successive iterations are necessary to find a solution in which the result converges to the optimal solution. Because the simulation software and the optimization software (IBM ILOG CPLEX Optimization Studio 12.5) cannot be linked directly, an interface in Microsoft Excel VBA was programmed. Figure 7.10 explains the simulation-based optimization procedure to find optimal solutions.

In Figure 7.10, the initialization values of heat transfer and temperature in each stage are the ones when heat integration does not take place ($Q^{hx}_{i,j,s} = 0$ for $s = 0$). After

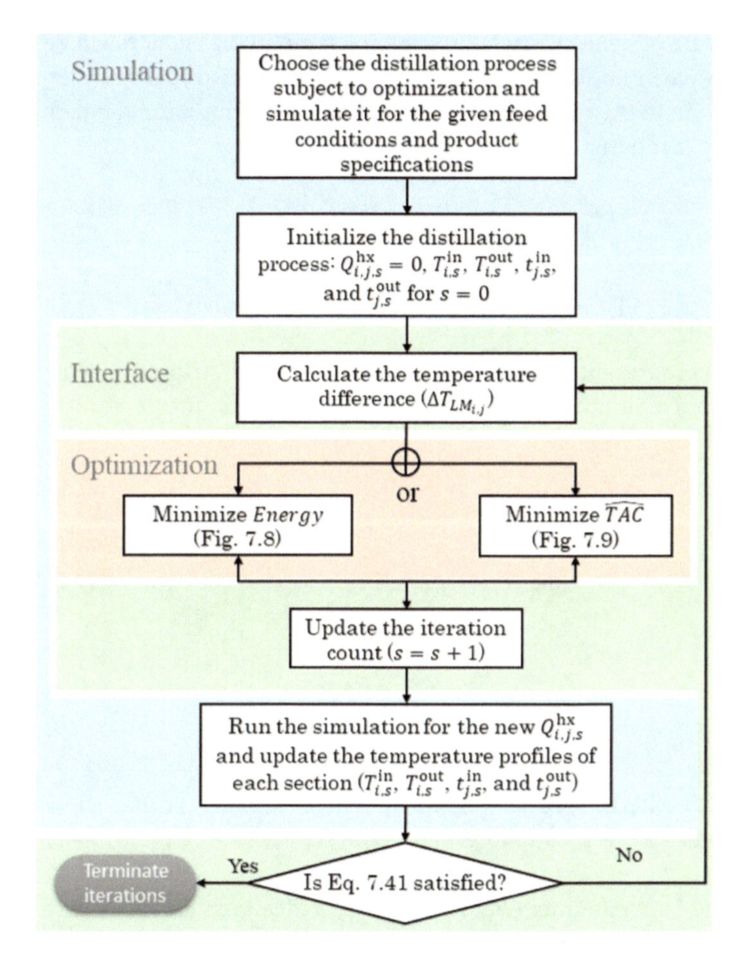

Figure 7.10: Flowchart of the iterative procedure to link the simulations with the optimization problems.

executing an optimization, the iteration counter is updated ($s = s + 1$), and the optimization results are exported to the interface. Then, these values are imported from the interface to run the simulation ($Q^{hx}_{i,j,s} \geq 0$ for $s > 0$). After running the simulation, the new temperature values ($T^{in}_{i,s}$, $T^{out}_{i,s}$, $t^{in}_{j,s}$, and $t^{out}_{j,s}$) are exported to the interface, and the optimization step is repeated until it meets the termination criterion in Eq. 7.41.

$$\sqrt{\sum_{i \in REC} (t_{i,\,s} - t_{i,\,s-1})^2 + \sum_{j \in STR} (T_{j,\,s} - T_{j,\,s-1})^2} \leq N\varphi \tag{7.41}$$

where N is the number of all the stages subject to changes in its temperature ($N = NSR + NSS$), and φ is the tolerance of the simulation, which typically is a small value (e.g., 0.0001).

Since the temperature of heat-integrated stages in the rectifying section will decrease, and those in the stripping section will increase, Eqs. 7.42 and 7.43 represent the temperature reduction in the rectifying section by taking the minimum temperature value at each stage through all the iterations

$$T_i^{\text{in}} \leq \min_{s \in S} \{ T_{i,s}^{\text{in}} \} i \in REC \tag{7.42}$$

$$T_i^{\text{out}} \leq \min_{s \in S} \{ T_{i,s}^{\text{out}} \} i \in REC \tag{7.43}$$

while Eqs. 7.44 and 7.45 represent the temperature increase in the stripping section by taking the maximum temperature value at each stage through all the iterations

$$t_j^{\text{in}} \leq \max_{s \in S} \left\{ t_{j,s}^{\text{in}} \right\} j \in STR \tag{7.44}$$

$$t_j^{\text{out}} \leq \max_{s \in S} \left\{ t_{j,s}^{\text{out}} \right\} j \in STR \tag{7.45}$$

7.4 Case study

The first case study deals with the separation of a toluene/benzene mixture because it has been widely used as a case study to evaluate the energy saving performance of HIDiC [12, 19–23, 26].

Table 7.1 shows the feed conditions and product specifications for the separation of the mixture. Also, the Peng-Robinson equation of state is used to estimate the mixture thermodynamic properties.

The second case study deals with the separation of a multicomponent mixture containing benzene (A), toluene (B), ethylbenzene (C), and styrene (D) [19, 26]. The mixture containing the four components is separated so as the light components (benzene, toluene, and ethylbenzene) are separated from styrene. Styrene is a monomer that is largely used in the manufacture of polystyrene, styrene-derived polymers, and other important copolymers. It is produced from the ethylbenzene dehydrogenation [27]. Nevertheless, styrene can easily polymerize at temperatures over 100 °C. Therefore sulfur or p-TBC (p-tert-butyl catechol) are added as polymerization inhibitors [28].

Table 7.2 shows the feed conditions and product specifications for the separation of the mixture. The reference case [19] uses the Raoult law and the Antoine equation to calculate the VLE. However, this study uses the ideal model in the process simulation to estimate the mixture thermodynamic properties [27]. To avoid polymerization reactions inside the HIDiC processes, the pressure of the stripping

Table 7.1: Feed conditions and product specifications for the separation of the binary mixture.

Condition	Value
Feed flow rate as saturated liquid [kmol/h]	100
Feed composition [mol%] (Benzene/Toluene)	(50/50)
Product purity [mol%] (Benzene/Toluene)	(99/99)
Stages in *CSM* [-]	1~4
Stages in *RSM* [-]	14
Stages in *SSM* [-]	17
Stages in *HSM* [-]	1~4
Pressure in the rectifying section [kPa]	110~270
Pressure in the stripping section [kPa]	100
Compressor efficiency [%]	80

Table 7.2: Feed conditions and product specifications for the separation of the quaternary mixture.

Condition	Value
Feed flow rate as saturated liquid [kmol/h]	36
Feed composition [mol%] (A/B/C/D)	(1.3/0.6/33.6/64.5)
Styrene purity in the bottom of *STR* [mol%]	99
Styrene purity in the top of *REC* [mol%]	1
Stages in *CSM* [-]	1~6
Stages in *RSM* [-]	35
Stages in *SSM* [-]	37
Stages in *HSM* [-]	1~6
Pressure in the rectifying section [kPa]	34~54
Pressure in the stripping section [kPa]	20
Compressor efficiency [%]	80

section operates at 20 kPa while that of the rectifying section is an optimization variable. The pressure drop inside the sections is neglected.

Table 7.3 shows the additional parameters used in optimization.

Table 7.3: Additional parameters used in the optimization methodologies.

Condition	Value
Cooling cost (C^{cool}) [$/kWh]	0.0013
Heating cost (C^{heat}) [$/kWh]	0.0506
Electricity cost (C^{elec}) [$/kWh]	0.1880
Annual operation hours (OH) [h/yr]	8500
Payback time [yr]	3
Overall heat transfer coefficient [kW/m^{2}K]	
Condenser:	0.7
Reboiler:	1.0
Heat exchanger:	0.7
Big-M parameter (M) [-]	50
Minimun Temperature difference (ΔT_{min}) [K]	8

7.5 Results and discussion

7.5.1 Binary mixture separation

Figure 7.11 shows the *TAC* when $\widehat{TAC}$ is minimized. In this case, the number of stages in *CSM* and *HSM* can change at each iteration through the optimization procedure. However, in all cases, the number of stages are 18 and 21 in *REC* and *STR*, respectively. It means that the number of stages in each section is at the upper bound. Also, the total annual cost in HIDiC is dominated by the operating cost; therefore the solution tends to a configuration with many stages that can operate at low reflux conditions.

The compression ratio is defined by Eq. 7.46 as the ratio between the pressure in the rectifying (P^{rec}) section and that in the stripping section (P^{str})

$$CR = \frac{P^{rec}}{P^{str}} \tag{7.46}$$

The best solution is the one with $CR = 2.5$, and it has three heat integrations between stages. Although the compression ratio is high, for values lower than 2.5, the heat integrations will be realized at locations away from the condenser and reboiler, which

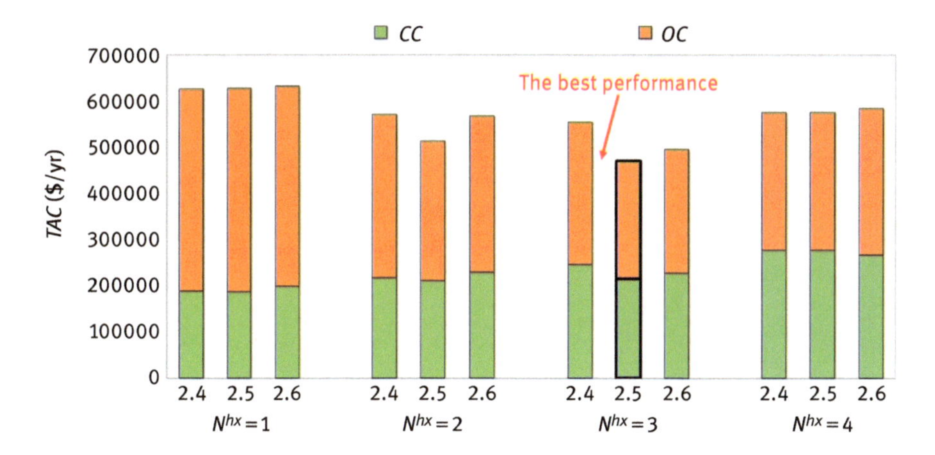

Figure 7.11: Relationship between *CR* and N^{hx} for the benzene/toluene separation.

implies that for the same amount of heat exchanged, the vapor in *REC* and the liquid in *STR* will have high internal flowrates.

Figure 7.12a shows the reference solution [23] in which the energy consumption is 692 kW, which saves 47% of energy in comparison with its conventional counterpart ($Q^{rb}_{conv} = 1305$ kW). This solution has four heat integrations: R1-S1 (top of each section), R8-S4, R14-S17, and R15-S18 (bottom of each section). The total integrated heat between sections is 1,586 kW. Figure 7.12b shows the solution when there is only one stage in *CSM* and *HSM*. The features of this solution are: it does not use external heating sources (no reboiler), it consumes less electricity although it has a compression ratio (*CR*) higher than that in Figure 7.12a, and it integrates less heat than Figure 7.12a between sections (1,304 kW). The combination of these three features results in a better-intensified process because excessive heat integration causes the adverse effect in increasing the internal vapor and liquid flows, which results in more reboiler duty, compressor work, and condenser duty. High values of *CR* can increase the driving force for heat integration. Thus there are more possibilities to have more combinations for heat integrations. The proposed solution in Figure 7.12b avoids heat integration at stages near the rectifying section bottom and stripping section top. Finally, Figure 7.12c shows the solution in which the distillation structure and heat integration combinations can change simultaneously. This solution has three heat integrations: R2-S8, R13-S20, and R14-S21. The energy consumption in the solution is 553 kW, and the total integrated heat between sections is 1,077 kW.

Table 7.4 summarizes the cost and energy consumption of the solutions in Figure 7.12. Solutions with heat integration can attain high energy savings. However, for the reference and energy minimization cases, the *TAC* is higher than the one for the conventional column. When the cost is minimized, energy and economic savings can be attained (58% and 3%, respectively) in comparison with the conventional

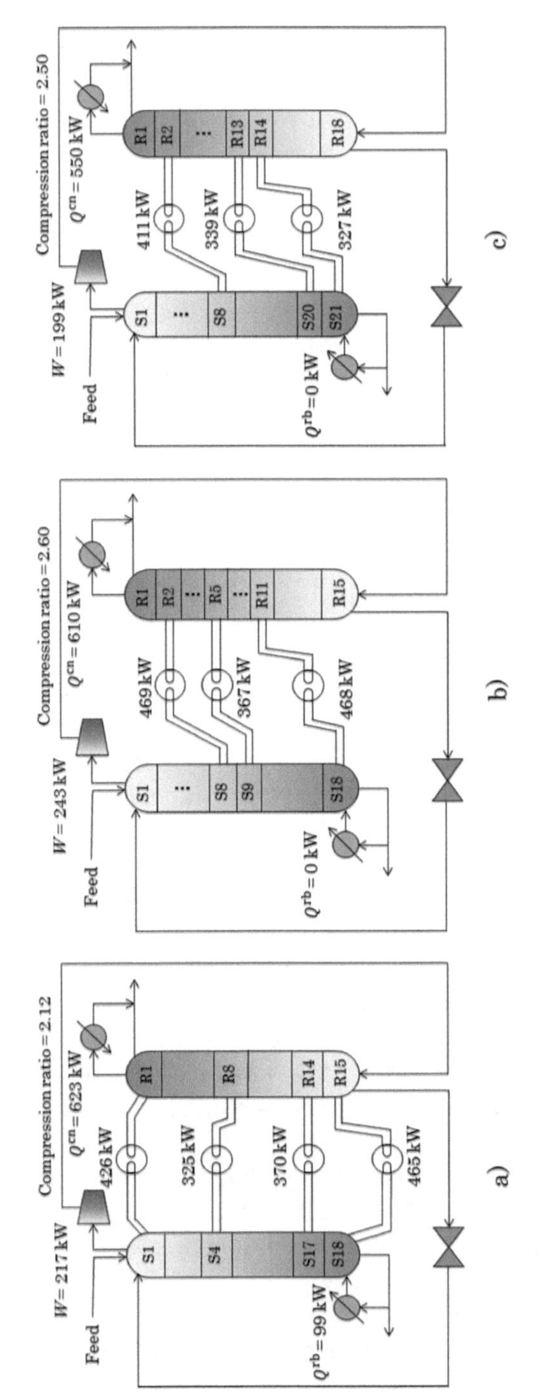

Figure 7.12: Solutions for the separation of the Benzene/Toluene mixture.

Table 7.4: Cost and energy consumption for the binary separation.

	Conventional	Reference [23]*	Minimize *Energy*	Minimize $\widehat{TAC}$
C_{shell} [$]	132,816	135,375	136,128	141,730
C_{tray} [$]	31,776	22,619	22,619	33,478
C_{hex} [$]				
(condenser)	46,732	36,460	31,156	31,336
(reboiler)	48,967	29,633	0	0
(heat exchangers)	0	228,881	238,927	189,420
C_{comp} [$]	0	259,997	311,733	243,439
CC [$]	260,291	712,966	740,563	639,404
Cooling [$/yr]	15,236	6,479	6,743	6,087
Heating [$/yr]	364,120	33,040	0	0
Electricity [$/yr]	0	255,005	285,341	234,071
OC [$/yr]	379,356	294,524	292,084	240,158
TAC [$/y]	466,120	532,179	538,938	453,293
Energy [kW]	1,305	702	674	553
$\sum_{i \in HS, j \in CS} A_{i,j}$ [m²]	74	244	306	206

*Simulated from available data.

column. The best solution (Figure 7.12c) has the smallest heat transfer area, which implies that an excessive heat transfer is counterproductive.

7.5.2 Multicomponent mixture separation

Figure 7.13 shows the *TAC* when $\widehat{TAC}$ is minimized. In this case, the number of stages in *CSM* and *HSM* can change at each iteration through the optimization procedure. However, in all cases, the number stages are 36 and 43 in *REC* and *STR*, respectively. Contrary to the previous case study, the total annual cost in HIDiC is dominated by the equipment cost; therefore the solution tends to a configuration with a few stages in *REC*. Although the consumption of cooling water in *REC* will increase as the number of stages in *REC* will decrease, it is still beneficial to do that in order to reduce the equipment cost because the cooling cost is low. Contrastingly, the cost of

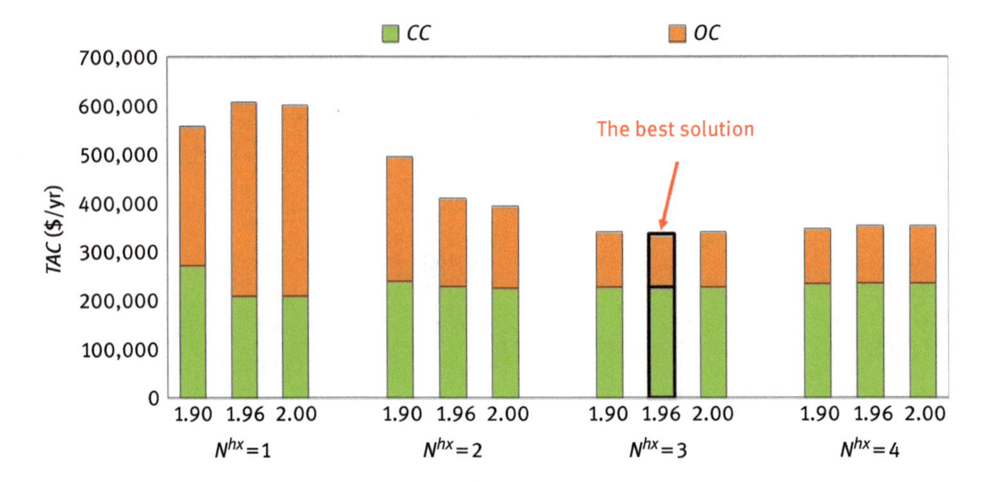

Figure 7.13: Relationship between *CR* and N^{hx} for the separation of the quaternary mixture.

heating is higher than that of cooling. Therefore it is preferred to use more stages in order to reduce the vapor flow that must be boiled in *STR*. Finally, the best solution is the one with $CR = 1.96$, and it has three heat integrations between stages.

Figure 7.14a shows the reference solution [19] in which the energy consumption is 342 kW, which saves 75% energy in comparison with its conventional counterpart ($Q^{rb}_{conv} = 1395$ kW). This solution has three heat integrations: R1-S1 (top of each section), R2-S17, and R23-S38. The total integrated heat between sections is 1,672 kW. Figure 7.14b shows the derived solution when there is only one stage in *CSM* and *HSM*. The features of this solution are: it does not use external heating sources (no reboiler), it consumes less electricity although it has a compression ratio higher than that in Figure 7.14a, and it integrates less heat than Figure 7.14a between sections (1,327 kW). Similar to case study one, the combination of these three features results in a better-intensified process. Figure 7.14b avoids heat integration at stages near the rectifying section bottom and stripping section top. Finally, Figure 7.14c shows the solution in which the distillation structure and heat integration combinations can change simultaneously. This solution has three heat integrations: R1-S41, R2-S42, and R3-S43. The energy consumption in the solution is 253 kW, and the total heat integration between sections is 1,274 kW.

Table 7.5 summarizes the cost and energy consumption of the solutions in Figure 7.14. Solutions with heat integration can attain high energy savings. When the total annual cost is minimized, energy and economic savings can be attained (82% and 53%, respectively) in comparison with the conventional column. The best solution (Figure 7.14c) has the smallest heat transfer area, which implies that an excessive amount of heat transfer between section is counterproductive.

The presented results in the two cases have shown that few heat integrations have the potential to attain energy and economic savings that are higher than typical HIDiC

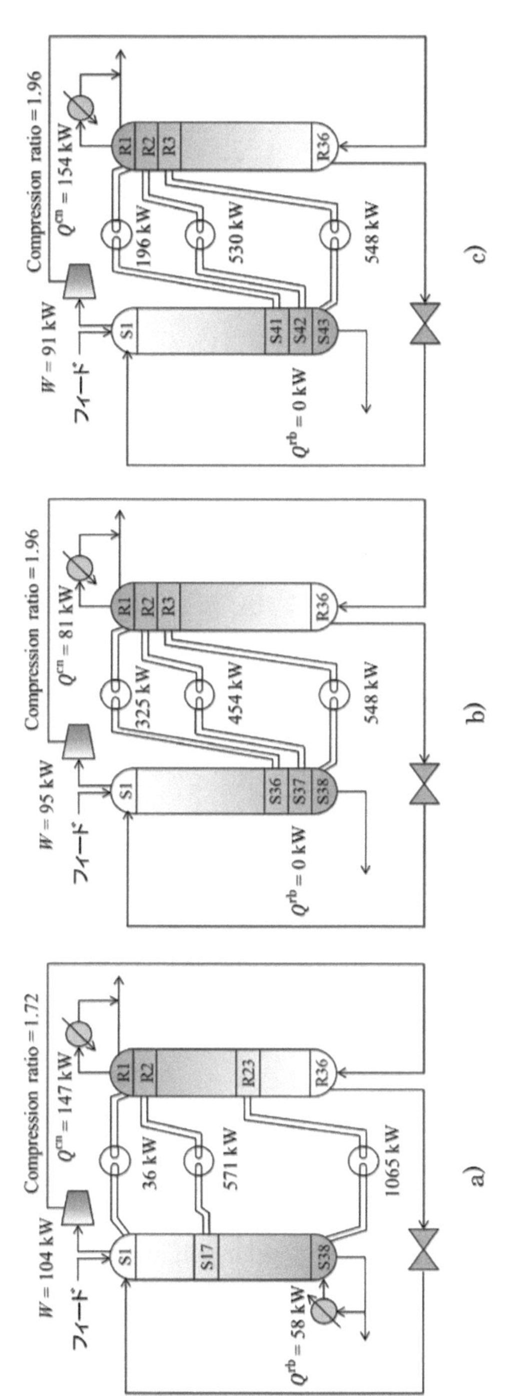

Figure 7.14: Solutions for the separation of the quaternary mixture.

Table 7.5: Cost and energy consumption for the separation of the quaternary mixture.

	Conventional	Reference [19]*	Minimize *Energy*	Minimize $\widehat{TAC}$
C_{shell} [$]	125,920	216,047	212,278	216,003
C_{tray} [$]	86,906	87,211	85,392	93,806
C_{hex} [$]				
(condenser)	47,673	14,486	3,730	3,713
(reboiler)	61,420	3,696	0	0
(heat exchangers)	0	226,348	221,911	213,216
C_{comp} [$]	0	181,422	158,024	153,124
CC [$]	321,919	729,210	681,336	679,863
Cooling [$/yr]	15,406	1,487	889	2,997
Heating [$/yr]	600,033	23,879	0	0
Electricity [$/yr]	0	135,141	111,684	106,932
OC [$/yr]	615,439	160,507	112,573	109,928
TAC [$/y]	722,745	403,577	339,685	336,549
Energy [kW]	1,395	342	261	253
$\sum_{i \in HS, j \in CS} A_{i,j}$ [m²]	98	272	259	241

*Simulated from available data.

structures. It explains why the latter [21, 22] fails to achieve such high savings. As a final remark, the cost values in Tables 7.4 and 7.5 correspond to those calculated by Eq. 7.26 to 7.34 and not the linearized values in Eqs. 7.36 to 7.39.

7.6 Conclusions

This work proposes optimization methodologies to find optimal HIDiC structures. The proposed methodologies use the superstructure approach, which was then reformulated as an MILP problem. Process simulation and optimization were combined iteratively to solve the optimization problem, and to deal with the inherent issues in HIDiC structures.

Two case studies were taken, and the proposed approach showed a performance higher than the current optimization approaches in both cases. Also, a solution with a less number of heat integrations as well as less heat transfer between section was derived in each case study. It means that an excessive amount of heat integration is unnecessary and even adverse from the economic and energy viewpoints.

The presented methodology can be extended to separation systems including chemical reaction and more complex structures with recycle streams.

Acknowledgments: The author wants to thank Prof. Ken-Ichiro Sotowa and Dr. Toshihide Horikawa from the Department of Applied Chemistry at Tokushima University for their valuable comments regards the elaboration of this research. Also to Mr. Hitoshi Yasuhira, Masataka Terasaki, and Morihiro Tanaka, graduate students from the Department of Applied Chemistry in Tokushima University for helping in the elaboration of the simulation, optimization, and interface tools as well as for the execution of the case studies.

References

[1] Takase H, Hasebe S. Optimal structure synthesis of ternary distillation system. Comput Aided Chem Eng 2015, 37, 1097–1102.
[2] Sholl DS, Lively RP. Seven chemical separations to change the world. Nature 2016, 532, 435–437.
[3] Trend of Energy Consumption by Category in the Annual report on energy in Fiscal Year 2016. Tokyo, Japan: Agency for Natural Resources and Energy, 2018. (Accessed on June 7th, 2018, at http://www.enecho.meti.go.jp/about/whitepaper/2017html/2-1-2.html in Japanese)
[4] Distillation column modeling tools, U.S. Department of Energy, 2001. (Accessed on March 29th, 2019, at https://www1.eere.energy.gov/manufacturing/resources/chemicals/pdfs/distillation.pdf)
[5] Materials for separation technologies: energy and emission reduction opportunities, Oak Ridge National Laboratory, 2005. (Accessed on March 29th, 2019, at https://www1.eere.energy.gov/manufacturing/industries_technologies/imf/pdfs/separationsreport.pdf)
[6] Zhang X, Huang K, Chen H, Wang S. Comparing three configurations of the externally heat-integrated double distillation columns (EHIDDiCs). Comput Chem Eng 2011, 35, 2017–2033.
[7] Alcántara Avila JR, Hasebe S, Kano M. New synthesis procedure to find the optimal distillation sequence with internal and external heat integrations. Ind Eng Chem Res 2013, 52(13), 4851–4862.
[8] Kim HY. Energy conservation of a multi-eect distillation column with internal heat integration. J Chem Eng Jpn 2012, 45(10), 840–849.
[9] Cabrera-Ruiz J, Jiménez-Gutiérrez A, Segovia-Hernández JG. Assessment of the implementation of heat-integrated distillation columns for the separation of ternary mixtures. Ind Eng Chem Res 2011, 50, 2176–2181.
[10] Matsuda K, Iwakabe K, Nakaiwa M. Heat and mass transfer of internally heat integrated distillation column (HIDiC). J Jpn Petrol Inst 2015, 58(4), 189–196.
[11] Bisgaard T, Skogestad S, Abildskova J, Huusom JK. Optimal operation and stabilising control of the concentric heat-integrated distillation column (HIDiC). Comput Chem Eng 2017, 96, 196–211.
[12] Wakabayashi T, Hasebe S. Effect of internal heat exchange rate distribution on energy saving in heat integrated distillation column (HIDiC). Kagaku kogaku ronbunshu 2011, 37(6), 499–505. (In Japanese)

[13] Wakabayashi T, Hasebe S. Higher energy saving with new heat integration arrangement in heat-integrated distillation column. AIChE J 2015, 61(10), 3479–3488.
[14] Wakabayashi T, Yoshitani K, Takahashi H, Hasebe S. Verification of energy conservation for discretely heat integrated distillation column through commercial operation. Chem Eng Res Des 2019, 142, 1–12.
[15] Alcántara Avila JR, Terasaki M, Lee H-Y, Chen J-L, Sotowa K-I, Horikawa T. Design and control of reactive distillation sequences with heat-integrated stages to produce diphenyl carbonate. Ind Eng Chem Res 2017, 56, 250–260.
[16] Lee H-Y, Chen C-Y, Chen J-L, Alcántara Avila JR, Terasaki M, Sotowa K-I, Horikawa T. Design and control of diphenyl carbonate reactive distillation process with thermally coupled and heat-integrated stages configuration. Comput Chem Eng 2019, 121, 130–147.
[17] Alcántara Avila JR, Sillas-Delgado AH, Segovia-Hernández JG, Gómez-Castro FI, Cervantes-Jauregui JA. Optimization of a reactive distillation process with intermediate condensers for silane production. Comput Chem Eng 2015, 78, 85–93.
[18] Yeomans H, Grossmann IE. Disjunctive Programming Models for the Optimal Design of Distillation Columns and Separation Sequences. Ind Eng Chem Res 2000, 39, 1637–1648.
[19] Harwardt A, Marquardt W. Heat-integrated distillation columns: Vapor recompression or internal heat integration? AIChE J 2012, 58(12), 3740–3750.
[20] Takase H, Hasebe S. Optimal structure synthesis of internally heat integrated distillation column. J Chem Eng Jpn 2015, 48(3), 222–229.
[21] Shahandeh H, Ivakpour J, Kasiri N. Internal and external HIDiCs (heat-integrated distillation columns) optimization by genetic algorithm. Energy 2014, 64, 875–886.
[22] Gutiérrez -Guerra R, Cortez-González J, Murrieta-Dueñas R, Segovia-Hernández JG, Hernández S, Hernández-Aguirre A. Design and optimization of heat-integrated distillation column schemes through a new robust methodology coupled with a Boltzmann-based estimation of distribution algorithm. Ind Eng Chem Res 2014, 53, 11061–11073.
[23] Wakabayashi T, Hasebe S. Design of heat integrated distillation column by using H-xy and T-xy diagrams. Comput Chem Eng 2013, 56, 174–183.
[24] Seider WD, Seader JD, Lewin DR, Widagdo S. Product and Process Design Principles: Synthesis, Analysis and Design. Wiley, 2010.
[25] Economic Indicators, CHEMICAL ENGINEERING, April issue, 2016.
[26] Tanaka M. Optimum design and performance evaluation of internal heat-integrated distillation columns (HIDiC). M.S. Thesis, Tokushima University, 2018. (In Japanese)
[27] Parra-Santiago JJ, Guerrero-Fajardo CA, Sodré JR. Distillation process optimization for styrene production from a styrene-benzene-toluene system in a Petlyuk column. Chem Eng Process 2015, 98, 106–111.
[28] Ishiyama T. Styrene. Kobunshi 1970, 19, 703–709. (In Japanese)

Oscar Andrés Prado-Rubio, Javier Fontalvo and John M. Woodley

8 Conception, design, and development of intensified hybrid-bioprocesses

Abstract: The exponential growth in demand for energy, food/feed, and commodities driven by the increasing world population imposes tremendous stress on optimization of technology. The need for process development toward a greener industry and circular economy was the motivation to propose the UN sustainable development goals for 2050. Within this trend, bioprocesses offer the exploitation of renewable resources to be transformed into food, feed, chemicals, materials, or energy promising a long-term sustainable industry. Despite huge interest, many bioprocesses have shown techno-economic constraints and even infeasibility due the trade-off between sustainability performance indexes. Therefore, further research is necessary to developing novel technologies for bioprocesses that can overcome current process constraints. However, this is not a straightforward task due to the particular complexity of bioprocesses such as the variance of raw materials characteristics and availability, lack of process understanding, complex systems interactions, bio systems sensitivity, reliability and reproducibility of experiments (uncertain information), and monitoring difficulties. These challenges imply that methodologies to conceive, design, scale-up, and operate intensified bioprocesses are still under development.

This chapter highlights the importance of intensified hybrid bioprocesses and discusses the interdisciplinary approach required to accomplish it. We outline bioprocess hybrid technologies reported in literature and current challenges of bioprocesses intensification at different levels. We present an overview of important ideas addressed within methodologies proposed for designing hybrid bioprocesses. As case studies, hybrid membrane bioreactors for biofuels and organic acids production are used to show how process understanding drives developments in hybrid bioprocesses. Finally, some forthcoming challenges and perspectives are presented for future bioprocess development.

Keywords: Process intensification, hybrid bioprocesses, fermentation-pervaporation, fermentation-dialysis

Oscar Andrés Prado-Rubio, Javier Fontalvo, Department of Chemical Engineering, National University of Colombia, Campus La Nubia, Bloque L 170003, Manizales, Caldas, Colombia
John M. Woodley, Department of Chemical and Biochemical Engineering, Technical University of Denmark, DK-2800 Kgs, Lyngby, Denmark

https://doi.org/10.1515/9783110596120-008

8.1 Introduction

Bioprocesses are characterized by the use of one or more enzymes either in growing or resting microbial cells, or in solution, to carry out the conversion of renewable feedstocks to chemicals and proteins. In the last half century, they have become the focus of increasing attention due to the need to use renewable raw materials as a feedstock for the future. This is an essential requirement if greenhouse gas emissions are to be cut and thereby the effects of global warming and climate change be reduced. Besides, this addresses societal problems such as the availability of water and food, energy and material resources, transport, and health. From other perspectives, bioprocesses also use renewable catalysts and are in general highly selective, meaning still further gains from an environmental perspective. Many synthetic organic chemists in industry and academia, therefore, see bioprocesses as an alternative and/or complementary method for chemical production in the future [1]. Nevertheless, despite this interest, it has become clear in recent years that implementing such processes is fraught with difficulty. A major reason for this is that the economic requirements to ensure a commercially viable process demand the use of conditions far from those where microorganisms or enzymes usually grow in nature. For example, very high product concentrations are required in order to ensure the downstream product recovery operation does not become too expensive. This is particularly important because many bioprocesses operate in aqueous solution, meaning that at some point water must be removed. The removal of water is particularly expensive due to its relatively high vaporization heat. Likewise, effective processes require sufficient productivity (space-time yield) as well as yield on substrate. These demands for high process performance metrics mean that very often the performance of bioprocesses in the laboratory are far from what is required in industry. This mismatch of expectations results in far fewer bioprocesses implemented in industry than would be desirable.

In principle, there are two complementary approaches to solve this problem which require interdisciplinary methods, let us call them Bottom-up and Top-down approaches [2]. The first method is to carry out genetic or protein engineering resulting in improved biocatalyst properties, such as increased reaction rate, improved yields, improved tolerance to high concentrations of product/substrate, use of multiple carbon sources or increasing tolerance for operating conditions (removing inhibitions). While kinetic improvements can be made, as well as increases in stability, the thermodynamics of reactions cannot be altered by such techniques. Biological reactions in nature have been evolved based on the requirements in nature. That does not necessarily mean that all aspects have been optimized, but that all have been improved against a given requirement to such an extent as to allow the next generation to survive. Indeed, a relatively recent article highlights the challenge of nonoptimized enzyme properties found in nature [3]. This also means that improvements of given traits against defined targets are also possible and this has been the subject of huge research in the past two decades both targeted at improving metabolism as well as specific enzyme properties. Today directed evolution, which accelerates the approach used in Nature, can be assisted by computer-aided tools, and genetic engineering driven by genome-scale models. These developments are very exciting from a scientific perspective but also hold the promise of improved biocatalysts. Nevertheless, this is often not adequate for process implementation since screening against multiple traits is difficult and requires excellent analytical procedures. Therefore, while in many cases biocatalyst improvement is a necessary requirement for bioprocess implementation in industry, it not only takes time but is also often insufficient. For example, improving the tolerance of an organism to high product concentrations is one of the hardest of all traits to engineer [4].

Hence the second approach, whereby the process is engineered is very often essential. Unfortunately, it seems likely that in many cases bioprocess intensification is a "last resort" rather than the "go to" solution to provide the techno-economic viability required. Several reasons lie behind this including the fact that engineers need to be involved in process development from a much earlier stage aligned with efforts for biocatalyst modification. This is particularly important given the extra degree of freedom afforded by alterations to the biocatalyst. The second approach is focused on process intensification techniques, of which operation in hybrid processes is one of the most important methods. In this chapter, the possibilities of hybrid technologies will be outlined, illustrated by some examples to show the power of this approach to enable implementation of bioprocesses. Of particular interest is the application of computer-aided modeling tools to help build a library, screen and evaluate a range of options quickly, and efficiently and subsequently focus experimental validation and testing only on the most likely successes.

8.2 Overview of hybrid technologies for bioprocesses intensification

Processes intensification is relevant within bioprocesses engineering as an approach to overcome their natural limitations. The topic is sufficiently broad that it has motivated several recent reviews and critical papers [5–14] and very recently a book [15].

Process intensification as a design philosophy focuses on process bottleneck identification and then proposal process structural changes in order to achieve "substantial" improvements in performance indexes such as footprint, environmental impact, energy efficiency, productivity, and safety [11, 16]. From a processing perspective, this implies a compact and safe equipment design providing high conversion and low by-product formation aiming at a less energy demanding downstream processing. Within biotransformations, it might be unrealistic to expect the promised quantum leaps through novel technology due to the inherent challenges of the bioprocesses such as: dilution (implying high footprint and energy demanding harvesting, collection and processing), complex systems interactions (i.e., nonlinear time variant systems), bio systems sensitivity and stability, unwanted by-products generation, monitoring difficulties (including uncertain data and experiments reproducibility), bioprocesses controllability issues, lack of processes understanding thus reliable models, among others [2]. However, accounting for the need of future sustainable processes, reaching a techno-economic viability in the trade-off of environmental requirements might be enough for the near future instead of tremendous changes in processing performance.

As conventional process intensification, bioprocess intensification can be accomplished from different perspectives integrating/hybridizing various levels of abstraction [17, 18]:

- Integration of known unit operations: hybrid reaction-reaction, reaction-separation or separation-separation systems.
- Integration of functions: incorporate new functionality into a known operation based on the bioprocess limitations.
- Integration of phenomena: identify target key phenomena to accomplish a biotransformation and customize the bioprocess design to put them together.

Using these levels of abstraction within the process intensification taxonomy proposed by Stankiewicz and Moulijn, the hybrid methods used for bioprocesses intensification are summarized in Table 8.1 [14, 16, 19].

Table 8.1: Hybrid methods used for bioprocesses intensification (extended from [14]).

Multifunctional reactors	**Hybrid separations**	**Alternative energy sources**	**Other methods**
Membrane bioreactors	Membrane absorption	Centrifugal fields	Dynamic operation
Bioreactive extraction	Membrane distillation	Ultrasound	Miniaturization (microscale)
Bioreactive crystallization	Adsorptive distillation	Solar energy	
Bioreactive adsorption	Distillation pervaporation	Microwaves	
Bioreactive extrusion	Electrodialysis and bipolar membranes	Electric fields	
Chromatographic bioreactors	Dividing wall distillation	Light	
Reactive comminution			
Fuel cells			
Multienzymatic reactors			

Among the options, is clear that substantial efforts have been made to develop multifunctional bioreactors through the integration of bioreaction and separation units, referred to as *in situ* product removal (ISPR) [8]. Depending on the level of integration, different configurations are proposed as shown in Figure 8.1.

Optimal techno-economical hybrid technology design using integrated units usually leads to a trade-off between designs (a) and (b) in Figure 8.1 [20]. Conventionally,

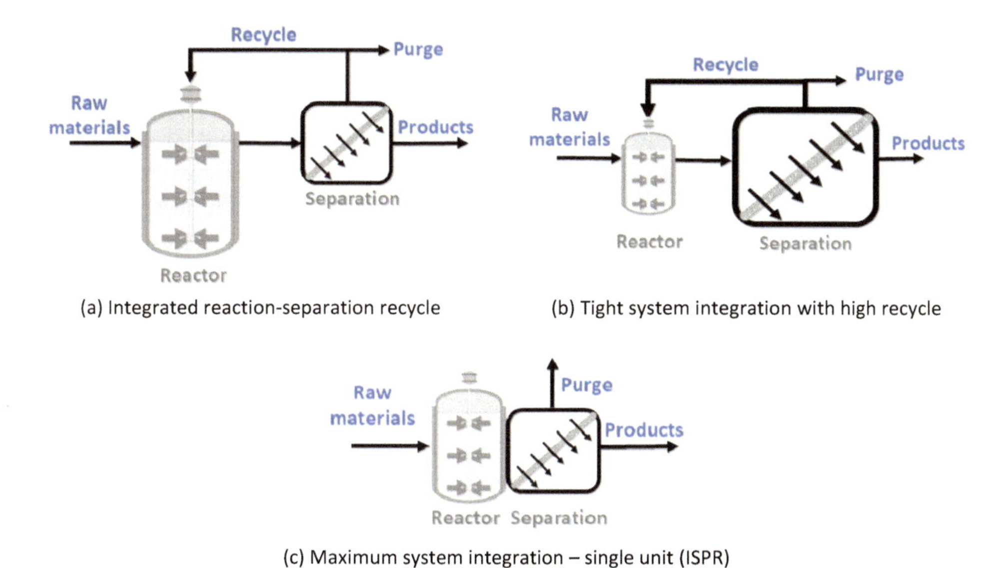

Figure 8.1: Processing intensification from conventional process integration to hybrid technology in a single unit. Adapted from [21].

big reactors with large conversion (constrained by inherent reaction limitations) imply small separation systems and low recycle (design a). On the other hand, a small reactor with low conversion leads to a big separation unit with a large recycle (design b). Therefore, the optimal integrated system design corresponds to a trade-off between CAPEX and OPEX for both units. In particular, it has been shown for a simple case that if the level of interaction is very high, meaning a high recycle, then both units can be considered as a single one from the dynamic point of view. In that case, there is a dramatic drop of CAPEX and OPEX meaning the limit of system integration [21]. Under these circumstances the maximum process intensification is achieved when multi-tasking systems are integrated in a single unit. So, the intensified process is the limit of the maximum process integration [21, 22]. This particular vision might not be general and some technical difficulties arise for a fully integrated hybrid bioprocess design and operation. For instance, if the design is mechanically/hydraulically viable, if there are controllability issues beyond which the intensified bioprocess has a faster response (due to lower holdup) or fewer degrees of freedom, and the appearance of new limiting factors (design or operational) due to the interacting phenomena in a single unit. Those challenges should be addressed during each particular hybrid process development.

It is interesting how broad-spectrum membrane separation processes have been used for hybrid bioprocessing. This is not surprising given the separation capabilities of membrane separation evident from nature, leading research toward biomimetic systems. Efforts have been made in membrane bioreactors (MBRs) using: microfiltration, ultrafiltration, nanofiltration, reverse osmosis, pervaporation, vapor permeation, membrane distillation, dialysis, Donnan dialysis, electrodialysis, electrodialysis with bipolar membranes, membrane chromatography, liquid membranes, membrane contactors, and electrophoresis. Membrane bioreactors are remarkable since in addition to the expected increase in performance indexes due to ISPR and biocatalyst confinement, MBRs enable continuous operation at higher dilution rates, favor the control of the cultivation parameters, facilitate sterile operation, and ease scalability due to the modular design [2]. For these reasons, examples of these processes are further discussed in this chapter.

Besides multifunctional reactors, hybrid separation systems are used in bioprocesses for downstream processing, appropriate since in many biotransformations the recovery and purification stages are economic bottlenecks. In particular, significant research has been done within first and second-generation biorefineries where sustainable production can be achieved when combining biofuels production with top building block chemicals to be transformed into new families of useful molecules [23, 24]. Within biorefineries, the combination of multifunctional bioreactors and intensified downstream processes can provide the trade-off between the multi-objective sustainability function. Interestingly membrane separation processes play an important role in providing hybrid separation-separation technologies for bioprocesses.

Alternative energy sources allow reaction or transfer bottlenecks to be overcome within unit operations. Then, incorporation of functionalities into systems involve structural design changes to create new technologies such as spinning disk bioreactors, ultrasound assisted bioreactors, bioreactor heated by microwave radiation or electrically enhanced membrane separation process such as electro-ultrafiltration or reverse electro-enhanced dialysis. As other methods for bioprocess intensification, microscale phenomena either in reaction or separation processes have also been used. From bioreaction point of view, microbioreactors have played an important role in speeding up bioprocess development through high throughput devices and they have been hybridized with ISPR or immobilization [25, 26]. Besides, microscale phenomena have facilitated increasing heat and mass transfer in order to develop compartmentalized intensified separation systems useful for ISPR. Finally, dynamic operation of systems has been used to increase driving forces within bioreactive or separation systems to achieve intensification. In most of the cases the dynamic operation is accomplished using periodic system operation.

8.3 Methodologies for hybrid bioprocesses design

The expected performance breakthroughs by bioprocesses intensification are not easily accomplishable for the reasons mentioned previously. As a consequence, it is understandable that there is a lack of consolidated methodologies to deal with structural changes to bioprocesses within the conventional paradigm for unit operation design, besides the still underdeveloped computational methods and tools to achieve them. Therefore, many developments in hybrid bioprocesses have been accomplished based on the separate unit, function or phenomena understanding, semi-qualitative insights and tremendous experimental efforts. Consequently, technological milestones toward technology maturity (stepping up the Technology Readiness Level – TRL) are energy, time, and cost-demanding explaining why many intensified bioprocesses are still at laboratory and pilot scale.

Relatively recent efforts have been made to develop model-based frameworks as powerful tools to assist intensified process design and implementation. From a general perspective, there have been proposed frameworks based on integrated systems with recycle asymptotic analysis [21] or using a phenomena based framework as building blocks [17, 18, 22, 27, 28]. However, less work has been carried out into hybrid bioprocess and methodologies where the phenomenological approach [29], a sensitivity analysis including controllability constraints [30, 31] or a conceptual retro-design philosophy [32] have been applied. The available literature highlights the efforts in building model-based frameworks and solving the problem either using optimization or sensitivity analysis.

Due to the diversity of intensified processes and the particular challenges associated with each technology, here a conceptual model-based framework is presented in Figure 8.2 with some relevant topics that have to be handled for hybrid bioprocesses development to increase their technology readiness level (TRL).

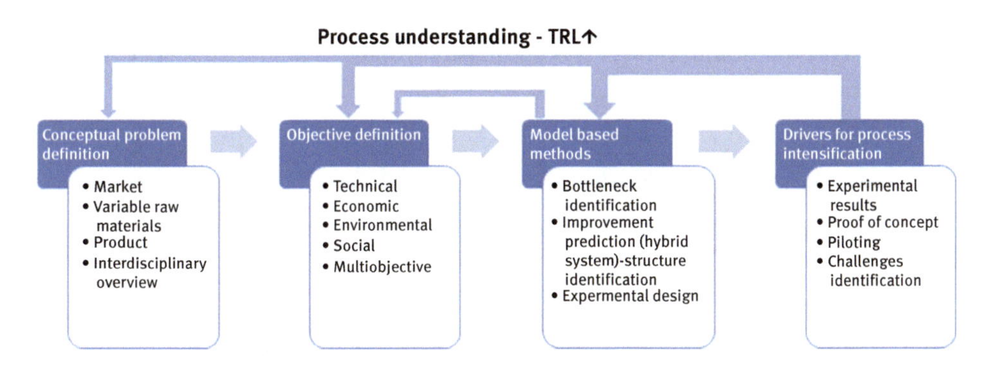

Figure 8.2: Overview for bioprocess intensification development.

As seen in the figure, and analogous to chemical processes, bioprocesses design starts by defining raw materials, desired products and specifications, market perspective and identifying technology knowledge (experimental or model based). Due to the broad scope of this problem definition for bioprocesses, it might be needed a multi-disciplinary overview [2]. Then, a single or multiple objective functions for design are defined depending of the processing possibilities, available technologies, and processes understanding. The particular objective function for the intensified bioprocess design must be aimed to overcome known limitations, this is proposed in terms of performance indexes such as:

- Technical: productivity, yield, selectivity, raw material utilization, product concentrations, footprint
- Economic: capital expenses, operating cost, net present value
- Environmental impact: waste reduction algorithm, Eco-indicator 99, life cycle assessment
- Other: social impact indexes, security, controllability.

Subsequently, here is where a model-based approach complemented by experiments is determinant. This is particularly challenging for model based design of bioprocesses hybrid technology, due to the models complexity, lack of process understanding, and high uncertainty. Sensitivity based (bottleneck identification) or optimization algorithms could be used to determine equipment size and operating conditions from single units to an integrated plant perspective which minimize or maximize the defined objective function/functions. Improvements into the systems performance might be investigated in an earlier stage of process design and the models even can be used to propose new experiments for model validation and upgrade. Particular strategies to setup and solve the problems depend on the available models, process structure, and available computational tools. At this stage of the conceptual bioprocess design, technically achievable solutions might be obtained. However, experience has shown that at least economic and environmental viability become a problem considering the low economic margins of certain markets and the usually found antagonist nature of the objective functions. This particular situation has been seen in second generation biofuels production where in many cases its production alone proved unviable [33]. Then, the objective function definition might be reconsidered or this could be seen as the driver to propose alternative design options as multi-product bioprocesses, to develop intensified bioprocesses and/or optimizing processing configurations.

Finally, process understanding become the driver force for bioprocesses intensification based on encouraging experimental results leading to proof of concept, piloting and scaling up in order to identify new challenges. In general, this approach implies returning to earlier stages of the process development methodology targeting the increase process understanding thus promoting process development toward an increase on technology readiness level. It is expected an iterative procedure, where the feedback is larger toward a closer previous stage. However, the markets dynamics, evolution of needs, new regulations etc., require to reconsider an update the objective function or even the conceptual problem definition.

From a conventional design approach, it is questionable the starting point and strategy for hybrid bioprocess design and which are the appropriate drivers for intensified processes development. For instance, for a defined biotransformation, the enzyme or microorganism operating conditions are identified, the entire process design depends on the fixed degrees of freedom. That might impose downstream constraints resulting in a complex combination of unit operations. However, if the biotransformation could be modified by genetic or protein engineering, or bioprocess intensification, this gives a completely different picture for the downstream process design. As an example, the relevance of the conceptual problem definition and initial overview has been shown for the succinic acid production where the conventional bioprocess design leads to a costly separation and recycle plant whereas changing the pH tolerance for the bioproduction, to avoid undesired byproduct formation giving excessive downstream costs, may prove a better strategy [32]. As mentioned in the introduction, engineers need to be involved in process development from a much earlier stage aligned with, or even driving, efforts for biocatalyst modification thus an interdisciplinary approach favors defining goals for bioprocess intensification. At this point, it is appropriate to define the market in terms of future product and variable raw materials constrained by energy and water supply; aiming for a better processing perspective of opportunities and limitations.

A model-based approach for the hybrid bioprocess design is very convenient in order to facilitate the individual unit operation, functionality, or phenomena integration and assessment. This stage might include also include technology screening based on the bioprocess bottleneck as the potential intensification principle: maximizing the effectiveness of intra- and inter-molecular events, homogenization at the molecular level, maximizing driving forces and transfer areas and, more importantly, for hybrid technology, maximizing the synergistic effects of integration [16]. For instance, simple heuristics can be used for the selection of an appropriate ISPR technique as shown by [8] or embedded into the optimization problem [27, 29].

We believe a particularly important part of designing hybrid technology for bioprocesses corresponds to the integration itself. From the model-based perspective, this step involves several challenges: separate processes/functionality/phenomena models' availability (i.e., validity and predictive power) and level of complexity (i.e., time and scale), understanding of those individual systems, parameters availability and their uncertainty, and numerical problems during the integrated system solution. As shown later in this chapter, gaining process understanding through sensitivity analysis prior to, and during, the integration process is very useful instead of using an optimization strategy as a first attempt. Additionally, from system analysis and understanding, ideas arise for models' upgrade, experimental design, prototyping, or anticipating process control limitations. Then, new experiments can be carried out in order to provide evidence which allows the system development at scale and consequently in technology readiness level. Due to the iterative strategy involved into developing hybrid bioprocesses, it is relevant to update conceptual problem definition and objective functions.

8.4 Hybrid bioprocesses design – examples

Among the possibilities to design hybrid bioprocesses, *in situ* product removal is particularly important due to the difficulties of overcoming product inhibition through metabolic or protein engineering. In this section, two examples of model-based design and operation of membrane bioreactors are presented. The core of the methodology is the individual process understanding and dealing with difficulties of the integration, and this is emphasized in the case studies. A combination of first principles models with Process System Engineering tools allow not only the investigation of design possibilities but also operating scenarios at an early stage of detailed process design.

8.4.1 Hybrid fermentation and pervaporation system

This example shows the effect of product removal on productivity, operation time, and yield for a hybrid fermentation using pervaporation. Thus, a methodology for system design and analysis is proposed. Two issues are addressed. First, when is it convenient to start the product removal after the fermenter starts its operation? Second, how selective should the membrane be? This is a question also important for membrane producers. Alternatively, is a very selective membrane convenient for the hybrid fermentation-pervaporation process? The aim of this example is not just to address these issues but, more importantly, to show some conceptual design aspects for hybrid fermentation-pervaporation systems.

Figure 8.3 shows a flow diagram of a general fed-batch fermenter with in situ product removal. In this particular system, an outlet stream is presented for generalization purposes. Temperature and concentration at the outlet stream are equal to the ones inside the fermenter, which corresponds to a stirred tank reactor (STR).

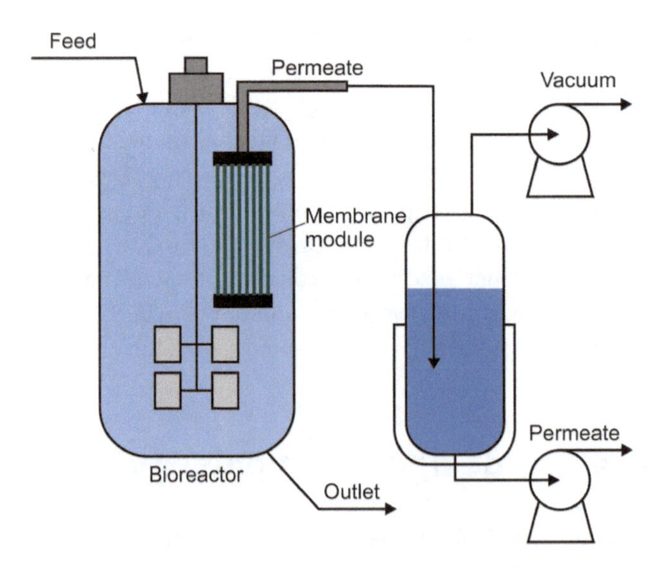

Figure 8.3: Flow diagram of a hybrid bioreactor with removal by submerged pervaporation.

8.4.1.1 Mathematical model

For modeling a hybrid fermentation and pervaporation system, two sets of equations are required, related to each unit operation. According to Figure 8.3, the fermentation model is presented below.

The global mass balance for the hybrid bioreactor is:

$$F_f\rho_f - F_o\rho_o - \sum_{i=1}^{n} R_i = \frac{d(\rho V)}{dt} \tag{8.1}$$

In Eq. 8.1, n corresponds to the total number of components including biomass.

For component i, the mass balance becomes:

$$F_f P_i^f - F_o P_i^o - R_i + r_i V = \frac{d\left(V P_i^o\right)}{dt} \tag{8.2}$$

Equation 8.2 is used for all nc components in the system including biomass (X). Thus, if for instance the fermentation is performed in an aqueous media the water mass balance is described also by Eq. 8.2.

Net reaction rates (r_i) in Eq. 8.2 correspond to the production of component i, based on reaction volume at the concentration and temperature conditions inside the reactor that are the same of the outlet stream. Additionally, within the value of the reaction rate the stoichiometry coefficient is included. For a substrate, its reaction rate (r_i) has a negative net value.

The accumulative mass of removed component i is given by the following equation:

$$J_i A = \frac{d\left(m_{p,i}\right)}{dt} \tag{8.3}$$

where $m_{p,i}$ corresponds to the total amount of component i removed by the membrane to the permeate. It requires a suitable model for J_i according to the removal or membrane technology used.

The fermentation of this example proceeds in aqueous phase with the following net reaction:

$$A \rightarrow B + C + X \tag{8.4}$$

For this example, the kinetic parameters as function of temperature (°C) are shown in Table 8.2 [34].

This model considers the effect of temperature that it is not important for this example, but is critical for final applications because an increase of temperature reduces ethanol yield and productivity. Besides, the model also includes inhibition of substrate uptake, product formation, and biomass growth. In the original work of Atala et al. [34], equations for concentration of viable and dead cells are presented. Furthermore, the model can be used for a high cell density (around 60 kg/m^3). If cell

Table 8.2: Kinetic parameters for reaction given in Eq. (8.4) as function of temperature (°C) [34].

Parameter	Meaning	Units	Expression or value
μ_{max}	Maximum growth rate	h^{-1}	$1.57\exp\left(\frac{-41.47}{T}\right)-1.29\times10^{4}\exp\left(\frac{-431.4}{T}\right)$
$C_{X,\,max}$	Biomass concentration when cell growth ceases	kg/m^3	$-0.3279\,T^2+18.484\,T-191.06$
$C_{B,\,max}$	Product concentration when cell growth ceases	kg/m^3	$-0.4421\,T^2+26.41\,T-279.75$
Y_X	Limit cellular yield	Kg/kg	$2.704\exp(-0.1225\,T)$
Y_{BX}	Yield of product based on cell growth	Kg/kg	$0.2556\exp(0.1086\,T)$
K_A	Substrate saturation constant	Kg/m^3	4.1
K	Substrate inhibition coefficient	m^3/kg	$1.393\times10^{-4}\exp(0.1004\,T)$
m_B	Ethanol production associated with growth	Kg/kg h	0.1
m_X	Maintenance coefficient	Kg/kg h	0.2
m	constant	–	1.0
n	constant	–	1.5

density is high for ethanol fermentation, the growth conditions are hindered due to access to nutrients, space limitations, and cell interactions are reduced [35]. For the sake of simplicity, the following equations consider that all cells are viable.

The corresponding reaction rates are [34]:

$$r_X=\mu_{max}\frac{C_A}{K_A+C_A}\exp(-KC_A)\left(1-\frac{C_X}{C_{X,\,max}}\right)^m\left(1-\frac{C_B}{C_{B,\,max}}\right)^n C_X \tag{8.5}$$

$$r_B=Y_{BX}r_X+m_BC_X \tag{8.6}$$

$$r_A=\frac{r_X}{Y_X}+m_XC_X \tag{8.7}$$

The reaction rate for component C is the same as that for component B but the ratio of molecular weights between B and C has to be accounted since reaction B is given on a mass basis. The values of the parameters used for the simulation are presented in Table 8.2 together with yield and productivity of product B. Yield of component B has been calculated as the ratio of the total amount of B produced to the total amount of A initially present and fed to the bioreactor. On the other hand, the productivity of B is calculated dividing the total amount of B produced by the final bioreactor volume and the operating time.

For the pervaporation membrane, the following ideal equation is used:

$$J_i=\overline{P_i}C_{o,i} \tag{8.8}$$

In this equation the flux of a component through a membrane (J_i) depends on the concentration of component i ($C_{o,i}$) inside the fermenter and the membrane area (A). In pervaporation, the driving force for the transport of the components through the membrane is the difference in vapor pressure of each component between feed (fermenter broth) and permeate. Thus, a more rigorous model requires consideration of the activity coefficients in the liquid phase, the corresponding vapor pressure of each pure component, the concentrations in the permeate and the permeate pressure [36, 37]. The performance of the pervaporation membrane and the transport properties of the membrane have been theoretically and experimentally analyzed by several authors [38–40]. It must be sated that for the analysis presented here, these simplifications do not change the main conclusions.

8.4.1.2 Conventional fed-batch fermenter analysis

Figure 8.4 shows a simulation of a conventional fed-batch bioreactor using the parameters presented in Table 8.3. Concentration profiles of substrate (A), component (B), and biomass are presented. Substrate (A) is consumed during the fermentation process to produce product B and biomass (X). During a period of approximately 35 h (operating time), the substrate is practically consumed. The usual stages of exponential growth of biomass and substrate-limited growth, where growth balances the dilution, can be observed. Also, the rate of production of B is reduced as the

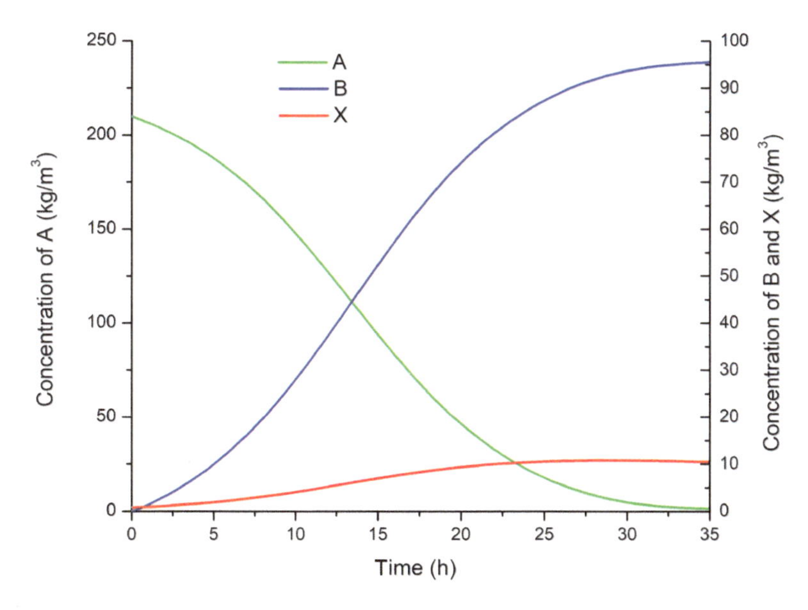

Figure 8.4: Concentration profiles as a function of time for a fed-batch reactor according to values presented in Table 8.3.

Table 8.3: Parameters used in a Fed-Batch reactor for ethanol production including calculated values for ethanol yield and productivity.

Parameter	Value
Initial concentrations in the bioreactor. C_A, C_B, C_C, X	[210, 0, 0, 0.9] (kg/m^3)
Initial reactor volume (V)	10 (m^3)
Feed concentrations.$C_{f,A} C_{f,B} C_{f,X}$	[210, 0, 0.9] (kg/m^3)
Operating temperature (T)	31 °C
Feed volumetric flow (F_f)	0.3 m^3/h
Outlet volumetric flow (F_o)	0.0 m^3/h
Operating time	35 h
Product (B) yield	0.456 kg B/kg A
Productivity of B	2.66 kg B/m^3 h

total amount of available substrate is depleted, but also due to product inhibition. Thus, it is expected that the reaction rate would be faster if the product concentration is lower within the fermenter.

8.4.1.3 Hybrid system – Effect of when the units are integrated

Since the reaction rate is limited by the product concentration, a pervaporation membrane is proposed for in situ product removal. The membrane area and permeability of B is assumed constant ($A = 1.7$ m^2 and $\bar{P}_B =$ 1 kg of B/m^2 h bar). Membrane selectivity in this case is infinite and only B permeates through the membrane. Figure 8.4 shows that the maximum production rate of B, associated with $P_{B,\ max}$, is reached at a concentration of B of around 50 kg/m^3 where the slope of the concentration of B as function of time is maximum. Thus, the question to solve is if the pervaporation membrane should operate from the beginning of this operation or when the concentration of B reaches 50 kg/m^3. So, from simulations the calculated values of operating time as well as yield and productivity of B, starting the pervaporation unit at several concentrations of B, are presented in Figure 8.5. For each simulation, the concentration profiles are similar to the ones presented in Figure 8.4 (not shown). The total operating time here is defined as the time when the substrate concentration falls beneath 1 kg/m^3. However, it is expected that the total operating time will change depending on the moment that the pervaporation membrane starts to remove product (B).

For all simulations shown in Figure 8.5, a hybrid fed-batch and pervaporation bioreactor has a productivity higher than 3.8 kg B/m^{3}h, which is also higher than for a conventional bioreactor (2.66 kg B/m^{3}h). Besides, the best moment to start the product removal is from the beginning of the fed-batch operation, since both high

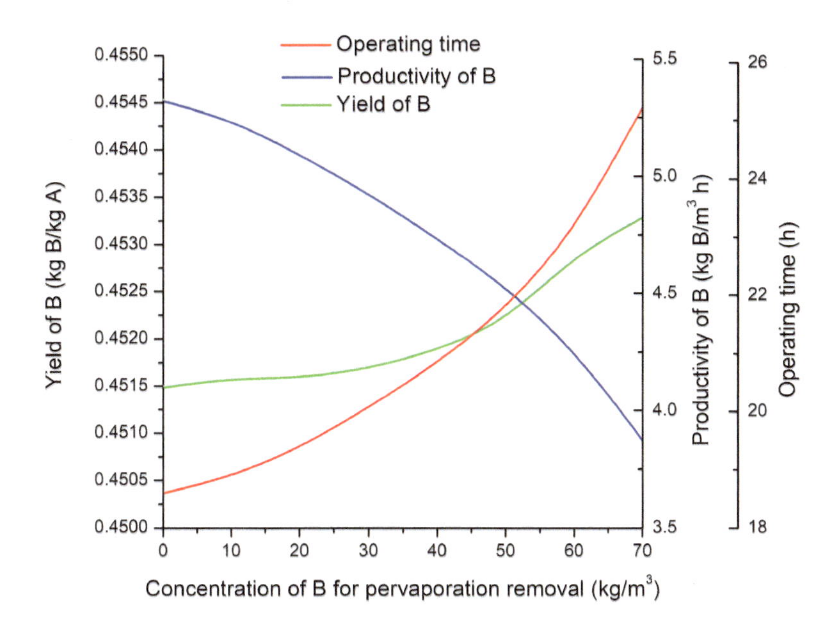

Figure 8.5: Influence of concentration of B reached for starting the B removal by pervaporation on (a) Operating time and productivity (b) Yield of B. $\overline{P_B} = 1$ kg of B/m^2 h bar. $A = 1.7$ m^2.

productivity (5.3 kg B/m^{3}h) and low operating time (18.5 h) can be achieved. Removal by pervaporation depends on the driving force of the component that is to be removed. Thus, at the beginning of the fed-batch operation, the concentration of B is low and so is its removal rate by pervaporation.

A continuous removal of B enhances the rate of production of B and the biomass generation and with a higher cell density (X reaches 12 kg/m^3), so the reaction is faster than in a conventional bioreactor. The higher cell density generates a slightly lower B yield value (Hybrid = 0.452 vs Conventional = 0.456 kg of B/kg of A) because a slightly higher fraction of substrate is used to produce biomass as compared to the conventional fed-batch bioreactor (Figure 8.4). Nevertheless, a higher biomass fraction increases the amount of B that is produced in a given period and so the reactor productivity rises. By using a hybrid fermentation – pervaporation system it is possible then to increase by almost twofold the productivity of B using the same reactor volume (in fact a bit lower because the product removal slightly reduces the final total volume).

8.4.1.4 Hybrid system – Effect of membrane selectivity

A point that is important to explore is the performance of the hybrid system in terms of membrane selectivity. This is relevant since scientific literature stresses the need for more selective membranes, but what is really significant is to define the selectivity.

In the previous section, it was found that productivity, in a hybrid fermenter – pervaporation system, is higher (5.3 kg B/m^3 h) compared to a conventional fed-batch bioreactor (2.66 kg B/m^3 h) using a membrane with an infinite selectivity toward B. Figure 8.6 shows simulation results on the effect of membrane selectivity on productivity, yield, and operating time. The membrane used for these simulations is selective for component B but also removes water. If membrane selectivity is reduced, the productivity can be increased significantly (with a maximum of 9 kg B/m^{3}h). There are two reasons for this behavior: (a) Biomass concentration and (b) substrate concentration. As the membrane is able to remove water, besides product B, the concentration of biomass and substrate are increased and thus the performance of the hybrid system is improved due to a higher reaction rate. With a high cell density and a high substrate concentration, the production rate of B rises and so its concentration. Therefore, the driving force for removal of B by pervaporation is also increased. It is important to remark that the concentration of B increases by removing water but this concentration is not high enough to induce inhibition.

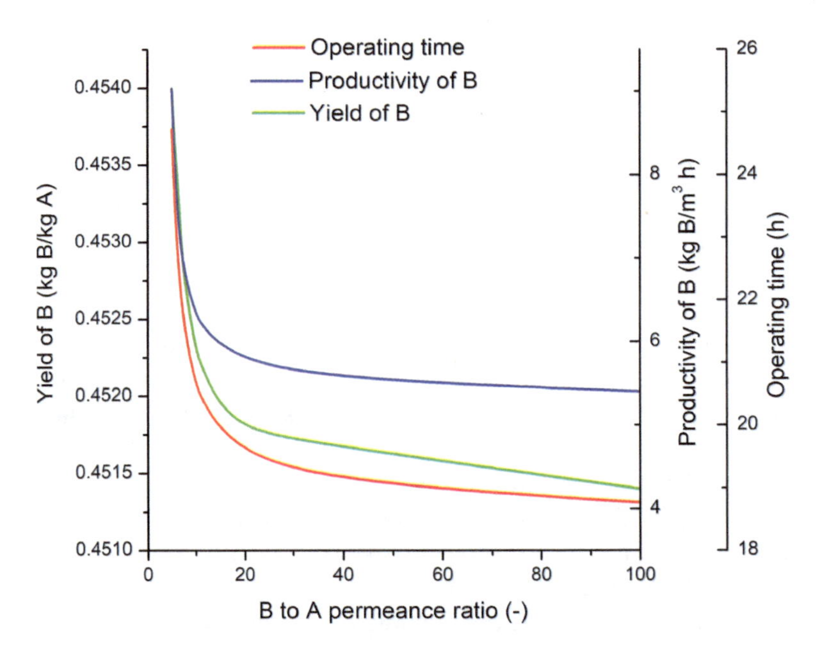

Figure 8.6: Influence of pervaporation membrane selectivity on (a) Operating time and productivity (b) Yield of B. Permeance of B = 1 kg of B/m2 h bar. Membrane area 1.7 m^2.

The yield of B slightly increases (Figure 8.6) as the membrane selectivity is reduced because the biomass concentration increases due to water removal but it does not due to substrate intake. Thus, lesser substrate is used for biomass production and the higher the yield of B. However, this improved Yield of B is not so high and is similar to the value shown in the previous section. It is important to mention that an extra benefit is expected from the hybrid system. When a submerged membrane configuration is used, it is possible to increase the cell density but with less shear rate stress on the biomass compared to an external recycling system. This enhances cell viability and thus productivity.

Figure 8.6 also shows that there is a limiting selectivity where the bioreactor can operate. If the membrane selectivity is lower than this value, the reactor performance is unstable and B is completely removed from the bioreactor before the substrate is completely consumed.

However, there is a downside of using low selective membranes. For relatively low selective membranes the operating time is increased and the concentration of the recovered component B is reduced. Operating costs are directly related to these two factors. High operating times increase the operating cost and reduce the total number of batches per day. On the other hand, low product concentrations increase the downstream processing. Consequently, there is an optimal point for membrane selectivity where productivity of component B, operating times, and final concentration of product B are economical balanced.

The following examples will show other alternatives and elements to take in mind for designing fermentation systems and bioreactors.

8.4.2 Hybrid fermentation and Donnan Dialysis

The objective of this case study is to identify the potential of a bioprocess intensification at an early stage of process design, where *in situ* product removal (ISPR) with ion exchange membranes is applied. A stepwise strategy to evaluate the design of the integrated system is based on a bottleneck identification. First, the processes are investigated separately and then the integrated system is evaluated. Initially, a continuous fermentation is assessed from productivity and substrate utilization point of view. In parallel, the membrane separation performance is determined at the operative window of the fermentation. Finally, the integrated system is investigated as a function of membrane system design. The case study uses a combined approach, where mathematical dynamic/static models and experimental evidence are exploited to investigate the system. Alongside this the solution, the design and operating challenges are discussed and potential modeling scenarios proposed for further investigation.

8.4.2.1 System description

As a case study, lactic acid fermentation is selected. Lactic acid is a commodity widely used in manufacture, food and pharmaceutical industries [41]. More recently, there is increasing interest in using this as a precursor for the sustainable production of polymers such as polylactic acid (PLA). Currently, PLA production is more expensive than its petrochemical based competitors. This has driven substantial research to improve the economic potential of lactic acid bioproduction at different levels such as: diversifying raw materials, microorganism engineering, fermentation intensification, and downstream process intensification.

Lactic acid can be produced from carbon sources using lactic acid bacteria. These microorganisms are a big group of gram-positive aero-tolerant bacteria characterized by similar properties and the ability to produce lactic acid as the main product. They obtain energy through the metabolism of sugars, which has been considered to be decoupled from growth, and therefore it is analyzed separately [42]. It is known this fermentation is impaired by product inhibition, meaning productivity is limited at low pH and high lactate concentrations, where even neutralized lactate impairs the microorganism [43]. This inhibition is driven by the acidification of the cytoplasm, affecting the microorganism transmembrane pH gradient and reducing the energy available for cell growth [44]. Therefore, it is natural to propose that *in situ* product removal and pH control benefits the fermentation yield and productivity.

In general, carboxylic acids separation from a dilute stream is not straightforward thus there is a substantial amount of research addressing this matter. Due to the relevance of this topic, it has been nicely reviewed in literature from different perspectives [30, 44–49]. Among the options, membrane-based separation processes are attractive due to high selectivity, capabilities of being operated aseptically, and can be considered without any by-products.

Donnan dialysis has been selected for lactate *in situ* product removal since simultaneously this allows inhibitor removal, has an intrinsic antifouling mechanism, facilitates pH control, and it can be further intensified [50–52]. A sketch of an integrated membrane bioreactor is shown in Figure 8.7.

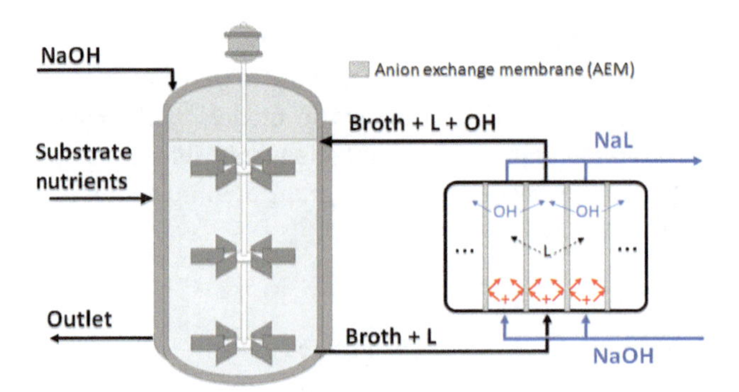

Figure 8.7: Sketch of an integrated bioreactor and Donnan Dialysis module for *in situ* organic acid removal, with pH control.

The Donnan Dialysis module uses only anion exchange membranes (AEM). This system is composed of multiple cells of feed and dialysate chambers (in Figure 8.7 only one cell is depicted). The fermentation broth is fed to every other channel. In the adjacent chambers, a concentrated sodium hydroxide solution is supplied. Due to the large hydroxide concentration gradient between the feed and dialysate chambers (evident in the pH difference: from approximately 6 to 11), the hydroxyl ions are transported through the membranes toward the feed channels. Hydroxyl ion flux induces an electric potential gradient across the membranes that drives lactate toward the dialysate channels. Both ion fluxes are coupled through Faraday´s law and the electroneutrality condition [53, 54]. Positively charged ions are practically retained in the chambers by the AEM due to their charge (i.e. electrostatic repulsion, referred to as Donnan exclusion). The main disadvantage of this membrane-based separation is the rather low ion fluxes accomplished. However, ion flux can be enhanced using a novel technology called Reverse Electro-Enhanced Dialysis (REED) [30, 55, 56].

8.4.2.2 System modeling

a) Fermentation model

Lactic acid bacteria kinetics have been extensively investigated. Despite the increasing understanding of the metabolism, the complexity of hundreds of reactions subject to regulatory mechanism means that unstructured or semi-structured models must be employed for process system engineering. A dynamic model is presented for substrate consumption, biomass growth and lactate production. In this particular application, the conventional fermentation model is extended to account for the dissociation of the main monoprotic acids, fermentation pH-buffer capacity, controlled pH, lactate ion removal, and hydroxyl ion addition.

The growth kinetics model for *Lactobacillus* sp. is based on the Luedeking-Piret model for growth associated and nongrowth associated lactic acid production, plus substrate, and product inhibition [57]. Conveniently, this model is capable of describing both batch and continuous experimental data. The kinetic model includes substrate limitation through a Monod expression, while a noncompetitive inhibition model represents substrate inhibition. Lactate inhibition is included using a linear model within a threshold defined by the minimum and maximum inhibition values. The biomass growth (q_x), product formation (q_L), substrate uptake (q_s), and species dissociation rates for lactic acid and undissociated protein are defined as: r_{HL}, r_{HP}. Notice that the original nomenclature was slightly modified in order to allow model extension.

$$q_x = \mu_{max}\left(\frac{S}{K_{sx}+S}\right)\left(\frac{K_{ix}}{K_{ix}+S}\right)\left(1-\frac{L-L_{ix}}{L_{mx}-L_{ix}}\right) \tag{8.9}$$

$$q_L = \alpha \frac{dX}{dt} + q_{L,max} \left(\frac{S}{K_{sL} + S} \right) \left(\frac{K_{iL}}{K_{iL} + S} \right) \left(1 - \frac{L - L_{iL}}{L_{mL} - L_{iL}} \right) \tag{8.10}$$

$$q_s = q_{s,max} \left(\frac{S}{K_{ss} + S} \right) \left(\frac{K_{is}}{K_{is} + S} \right) \left(1 - \frac{L - L_{is}}{L_{ms} - L_{is}} \right) \tag{8.11}$$

$$r_{HL} = k_2[L] - k_1[HL][OH] \tag{8.12}$$

$$r_{HP} = k_4[P] - k_3[HP][OH] \tag{8.13}$$

In order to account for the fermentation pH buffer capacity, hypothetical species referred to as “protein” (HP and P) are proposed to represent a wide range of molecules. Due to the lack of understanding of the source of the pH buffer capacity, it is assumed a protein production rate that keeps the total protein concentration constant, and thus the buffer capacity.

Dynamic mass balances for a stirred bioreactor are proposed for: lactate (L), lactic acid (HL), hydroxyl ion (OH), sodium (Na), dissociated protein (P), undissociated protein (HP), substrate (S), and biomass (X). The generic mass balances apply for batch, fed-batch, and continuous operation according to the volumetric flow rates. Besides, they include the base addition for pH control and the ionic species removal/addition.

$$\frac{dC_k}{dt} = \frac{q_{feed}}{V} (C_{in,k} - C_k) + \frac{q_{base}}{V} (C_{base,k} - C_k) + R_k + J_k \tag{8.14}$$

$$\frac{dV}{dt} = q_{feed} + q_{base} - q_{prod} \tag{8.15}$$

$$q_{base} = K_p \left((pH_{set} - pH) + \frac{1}{\tau_I} \int_0^t (pH_{set} - pH) dt \right) \tag{8.16}$$

Where C_k is the reactor concentration for specie k, V is the reactor operative volume, q_{feed} is the inlet flow rate, $C_{in,k}$ is the inlet concentration of k, q_{base} is the base inlet flow rate (for pH control), q_{prod} is the outlet flow rate,$C_{base,k}$ is the species concentration in the base inlet stream, R_k is the net reaction rate of specie k, J_k is the specie removal/addition rate, K_p is the pH controller gain, τ_I is the integral time of the pH controller and pH_{set} is the desired pH set point.

The model parameters are depicted in Table 8.4 and operational reactor values for continuous operation at constant volume are shown in Table 8.5.

b) Donnan Dialysis model

In the scientific literature, different types of models have been used to describe ion transport through convection, diffusion, and migration in dialytic systems. These approaches include static black-box models to dynamic first principles models [30, 50, 54, 56, 58]. Increasing the model complexity, more parameters are required

Table 8.4: Model parameters and extra relations.

Parameter	Value	Units	Parameter	Value	Units	Parameter	Value	Units
μ_{max}	1.10	h^{-1}	$q_{S,max}$	3.42	$gg^{-1}h^{-1}$	α	0.39	gg^{-1}
K_{sx}	1.32	gl^{-1}	K_{ss}	2.05	gl^{-1}	$q_{L,max}$	3.02	$gg^{-1}h^{-1}$
K_{ix}	304	gl^{-1}	K_{is}	140	gl^{-1}	K_{sL}	2.05	gl^{-1}
L_{ix}	1.39	gl^{-1}	L_{is}	47.1	gl^{-1}	K_{iL}	140	gl^{-1}
L_{mx}	49.9	gl^{-1}	L_{ms}	95.5	gl^{-1}	L_{iL}	47.1	gl^{-1}
						L_{mL}	95.5	gl^{-1}
pK_{HL}	3.89	–	pK_{HP}	5	–	$K_d = K_a/K_w$*		

* K_d corresponds to the dissociation constant, K_a is the acid dissociation constant and K_w ionic product of water.

Table 8.5: Operative values for the pH-controlled bioreactor.

Input variable	Value	Units
V	5	l
pH_{set}	5.75	–
K_p*	360	lh^{-1}
τ_I*	100	–
$C_{in,S}$	2-100	gl^{-1}
$C_{base,OH}$	2	M
q_{feed}	0.5040–5.0400	lh^{-1}
q_{prod}	$q_{feed} + q_{base}$	lh^{-1}

* Parameters not optimized since only static analysis is performed.

but they are also able to provide a deeper system understanding. Since the fermentation dynamics are slower than the membrane separation, the dynamics of the membrane separation system is neglected and a simple semi-structured static model is proposed to evaluate the membrane performance.

During static Donnan Dialysis, a dominating resistance at the boundary layer, or the membrane itself, can limit the lactate flux through the membrane. At low hydroxyl ion concentrations, the system behaves under boundary layer control. There, the concentration gradients within the membrane are negligible and the lactate flux is proportional to the hydroxyl ion concentration in the dialysate channel. The proportionality coefficient is a function of the ions' diffusivity and the boundary layer thickness. By increasing the hydroxyl ion concentration, the transport resistance within the membrane becomes increasingly important until the point it dominates the ion transport. At these conditions, steep concentration profiles are generated within the membrane and the ion transport becomes independent on the hydroxyl ion concentration in the dialysate channel. The ion fluxes are a function of their diffusivities and the membrane thickness [50]. A dual transport mechanism was proposed to model the experimental observations, combining a Langmuir function for reaction-diffusion mechanism and a proportional function for a solution-diffusion mechanism. The lactate flux can be represented by Eq. 8.17.

$$J_L = J_{max}\left(\frac{kC_{OH}}{1+kC_{OH}}\right) + P_{sd}C_{OH} \quad (8.17)$$

Where J_L is the lactate flux through the anion exchange membrane, $J_{\max}$ is the maximum lactate flux achievable under sole reaction-diffusion mechanism, $1/K$ is equivalent to the Langmuir saturation constant, and P_{sd} is the proportionality constant for the solution-diffusion mechanism. The Donnan Dialysis model parameters for lactate transport through the Neosepta-AMH membrane are depicted in Table 8.6 (recalculated from [53]). It should be mentioned that this flux model was estimated for a constant lactate concentration at a feed channel in the neighborhood of expected values from a fermentation broth (1 M).

Table 8.6: Donnan Dialysis model parameters (estimated from experimental data reported by [50]).

Input variable	Value	Units
V	5	l
pH_{set}	5.75	–
K_p*	360	$l\,h^{-1}$
τ_I*	100	–
$C_{in,S}$	2–100	$g\,l^{-1}$
$C_{base,OH}$	2	M
q_{feed}	0.5040–5.0400	$l\,h^{-1}$
q_{prod}	$q_{feed} + q_{base}$	$l\,h^{-1}$

8.4.2.3 Results and discussions

Although the proposed methodology is really performed starting with the fermentation analysis toward the integrated system investigation, herein, the Donnan Dialysis system analysis is depicted first and the fermentation plus the integrated system is analyzed afterward using the same figures. Notice that the Donnan Dialysis model units are in molar base, this was done in order to compare the values with the literature. In the implementation, the models' consistency must be carefully evaluated.

a) Donnan Dialysis

In order to determine the required membrane area for the fermentation, the lactate flux should be estimated. In Figure 8.8, experimental data and the model prediction for lactate flux through the membrane are plotted as a function of the hydroxyl ion concentration in the dialysate channel.

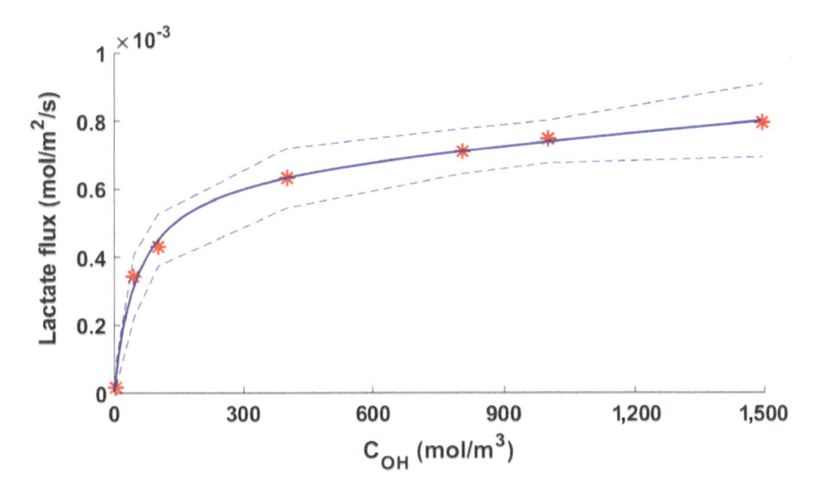

Figure 8.8: Lactate flux prediction versus experimental data [50]. Simultaneous functional confidence intervals for the predictor are shown (95% confidence interval). The adjusted correlation coefficient for the estimation is $r^2_{adj} = 0.991$.

From the flux predictions, at low hydroxyl ion concentration in the dialysate channel, it is advantageous to increase it (boundary layer-controlled transport). However, lactate flux tends to saturate at hydroxyl ion concentrations higher than 0.9 M (membrane-controlled transport). From the design point of view, it is expected to have the economic trade-off in the threshold between controlling regions. For the integrated system design, a membrane flux of $J_L = 7.412 \times 10^{-4}$ mol/m^2/s is selected (corresponding to a $C_{OH} = 1M$). From an operative point of view, it is interesting that there is almost one order of magnitude difference on the lactate membrane flux as a function of the hydroxyl ion concentration in the dialysate channel. It indicates that this variable could be used for control purposes [59].

A limitation of the presented model is the inability of predicting the hydroxyl ion flux toward the feed channel, which is necessary to evaluate the pH evolution in the fermenter. Using a first principles model, it has been shown the high efficiency of Donnan exclusion. Then, the hydroxyl ion flux is assumed to equal the lactate flux [53].

Finally, there is a remaining issue in the membrane system design. How to calculate the influence of the lactate inlet concentration on the membrane flux?

Experimentally, it has been shown that acetate flux increases linearly by increasing the acid inlet concentration up to 1M [50]. In order to quantify this increase for lactate, another black box model should be coupled or a first principle model derived, which is beyond the scope of this exercise. Therefore, the influence of the inlet lactate concentration is neglected. This might not be a problem in an early stage of process design since at higher lactate concentrations, larger ion fluxes are expected and thus lower membrane area requirements.

b) Continuous fermentation and integrated systems analysis

It should be stated that the proposed model uses differential equations for all the species. Since dissociation reactions are very fast compared to the fermentation reaction rates, it is expected that the system of equations is very stiff. As a consequence, simulation times can be very large and eventually the simulation may not converge. Therefore, an appropriate ODE solver for stiff equations must be used. Alternatively, the dissociation reactions and the electro-neutrality condition can be used as algebraic equations. Then, the system of differential algebraic equations (DAE) must be solved as it has been done for instance for the anaerobic digestion model (ADM1) implementation within the benchmark sludge model (BSM2) platform [60].

For the investigation of the continuous fermentation with pH control, a static analysis is performed to evaluate the lactate and biomass productivities as a function of the inlet substrate concentration and dilution rate, making the analysis scale independent. The results correspond to the surfaces depicted in Figure 8.9. As expected due to the proposed inhibition terms, the fermentation is characterized by having an optimal combination of dilution rate and inlet substrate concentration to obtain the largest either lactate or biomass productivities.

The selection of the optimal operating points based on productivity might not be very convenient from a plantwide design perspective. The intermediate product concentrations generate larger operating costs for the downstream processing (results not shown). This issue can be handled in the detailed engineering design using a multi-objective optimization or through a full economic plant optimization. Nevertheless, optimal productivities at this stage are useful to provide system understanding toward the first designs of an integrated process.

It can be seen from Figure 8.9a and 8.9b around the optimal dilution rates, there are not productivity improvements by increasing the inlet substrate concentration. This is mainly due to the product inhibition and thus, the extra substrate fed is wasted in the outlet stream. Substrate inhibition is not representative in this operating region due to the high K_{is}, K_{ix}, and K_{iL} values. The product inhibition and substrate availability are the motivation to investigate the potential productivity intensification by integrating the Donnan Dialysis module.

The intensified system performance, as depicted in Figure 8.7, is evaluated by increasing the membrane area from $A_m = 0$ to $A_m = 0.24\,m^2$. For the sake of simplicity,

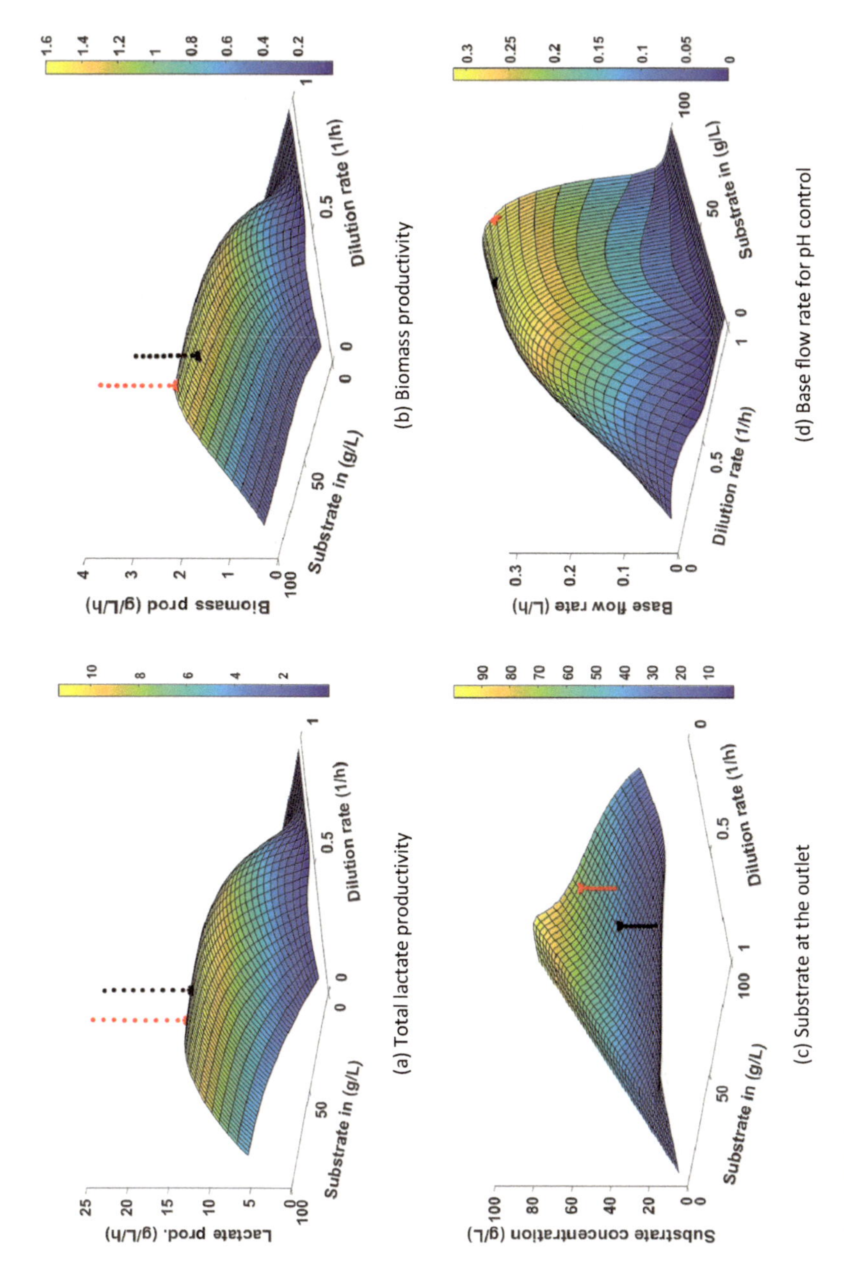

Figure 8.9: Fermentation key performance indexes and potential optimal enhancement by coupling the Donnan Dialysis module. Black points correspond to the optimal operating point for lactate productivity and Red points correspond to the optimal operating point for biomass productivity.

the membrane area interval corresponds to a region where the module facilitates pH control in the fermenter and does not fully substitute it, otherwise another control structure should be implemented [59]. In Figure 8.9, the performance of the intensified process is shown by dots when the membrane area is linearly increased. The highest enhancements obtained at the largest module are summarized in Figure 8.10.

The *in situ* lactate removal from the fermentation broth has a substantial effect on the lactate and biomass productivities in both optimal operating points from 90% to 108%. When the membrane is coupled, the lactate removal enhances biomass growth and substrate uptake (46% and 21%). Therefore, more lactate is produced until a new equilibrium point is reached where lactate production rates matches the membrane removal. That point is very close to the initial optimum lactate concentration in the fermentation broth as seen in Figure 8.9. Additionally, there is no significant change in the base consumption (for the pH control) even though there is an increase in the lactic acid production rate. This is because the extra acid produced is

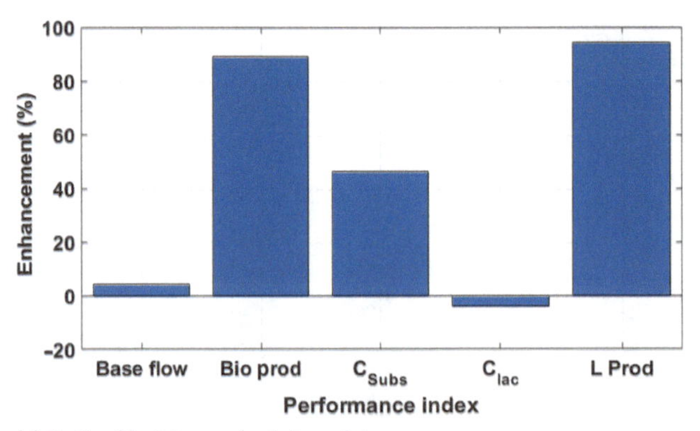

(a) Optimal lactate productivity point

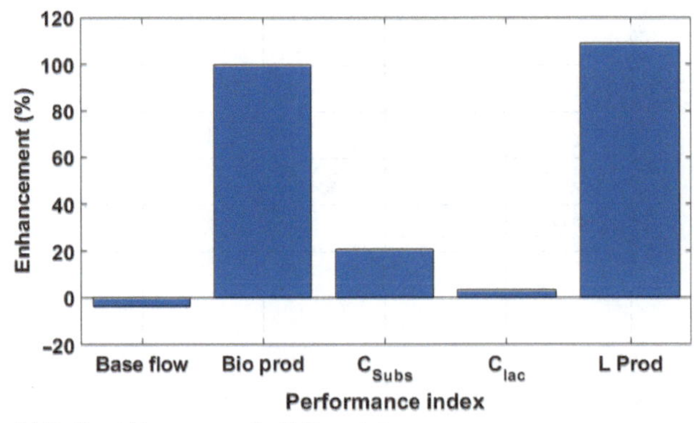

(b) Optimal biomass productivity point

Figure 8.10: Key performance indexes enhancement by integrating the Donnan Dialysis module to the fermentation.

neutralized by the hydroxyl ion flux, avoiding extra neutralization in the intensified process.

In summary, this case study allows highlights that:
- Membrane technology plays an important role in developing hybrid technologies for bioprocesses.
- Process understanding from experimental evidence plus models is vital for bottleneck identification and quantification, which is the genesis of process intensification.
- Structural changes by bioprocess intensification brings substantial performance improvements, but it does not necessarily reach the orders-of-magnitude, as stated in early literature on process intensification. In our experience, one order of magnitude is the limit.
- It is important to identify dynamic system constraints at an early stage of process design in order to foresee and assess controllability issues.
- Process system engineering tools are relevant for integrated processes design and operation.

c) Remaining Challenges

Accounting for the remaining substrate available in the reactor, there is still room for further process intensification by using cell confinement which will allow bioreactor operation at higher dilution rates than the specific growth rate [43, 61]. Besides, larger lactate removal can be investigated. However, interesting challenges remain prior to further design intensification.

From a more general perspective, the integrated system operation is a challenge including the pH control structure design. pH regulation is not straightforward due to the highly nonlinear behavior and fast reaction rates. In particular, the used base concentration could be a problem since low inlet concentrations result in a dilution effect and high concentrations could be a significant disturbance. Although only static results are discussed, it is important to validate the dynamic simulations to avoid large pH disruptions. In this case, there is a small influence of base addition on the dilution rate (6% of the total flow rate) and the pH controller avoids large and prolonged pH disturbances.

It is relevant to account for dynamic constraints in this application especially for the pH behavior. The system startup is a concern since the membrane unit should cope with the time variant lactic acid production. This could induce pH disturbances that offset the fermentation. From an operational perspective, it has been shown that Donnan Dialysis has a big operating window, however this could be insufficient and a more advanced dialysis operation must be used to favor the size of the operational window, such as REED technology [55, 56, 62]. Nevertheless, the intensified REED technology brings new process control challenges due the periodic operation [59]. Finally, a clear limitation is the partially decoupled process and control design strategy since the system controllability is tightly related to the process design.

8.5 Future perspectives

It is clear that raw biomaterial availability and how to transform it into energy and higher value products requires bioprocessing toward a circular economy. It has been estimated that 50% of the products in current use can be obtained through bioprocessing [1]. However, the inherent process limitations have hindered bioprocesses development.

Hybrid operations are a powerful intensification approach to overcome some of the challenges of biotransformations and downstream processes. The two examples presented in this chapter illustrate very well these concepts. Deliberately, both examples have been focused on the integration of the fermentation step with product removal (so called *in situ* product removal). Nevertheless, in the future other options for the application of hybrid intensification technologies will need to be explored. For example, many enzymes are used outside the cell in order to avoid metabolic constraints. Doing so enables increased flexibility through the use of alternative concentrations and conditions for a given enzyme. Indeed, pathways of enzymes are increasingly used to assist in organic synthesis [63]. However, in some cases it will be advantageous to combine enzyme steps together (especially for cases where cofactor regeneration is required to shift thermodynamics), and in other cases there may be potential yield improvements by combining reaction steps without intermediate product isolation. The same approach discussed in this chapter can also be used in these cases to make hybrid reaction-reaction intensified processes. One of the obvious advantages of combining enzyme steps to create pathways in this way is that most enzymes work under relatively similar conditions, meaning few compromises. Meanwhile organic chemists are also interested in combing the selectivity of enzymes together with the speed on conventional metal catalysts. Such chemo-enzymic processes are also an area where bioprocess intensification will play an increasingly important role.

From recent contributions, there is a clear trend improving the predictive power of hybrid bioprocess mathematical descriptions. This is possible due to the increasing understanding of the biochemical, chemical, and physical mechanisms of the processes using better monitoring techniques, modeling approaches, process simulators, and computational power. As a consequence, model-based frameworks for intensified process design are also evolving and speeding up the hybrid process development, migrating from relatively slow trial and error, and simplified, approaches toward more systematic methodologies. Hybrid bioprocess development brings new problems and thus challenges how engineers conceive of, and design, process alternatives. The relevance of modeling was stressed in the overview of the methodologies and the simple case studies, where process understanding is the basis for designing the hybrid technology while identifying process constraints and potential implementation issues. Forthcoming model-based methodologies will be robust and will address more complexity, from specific technology issues, to plant-wide operability. Of course, all this will be possible through investment in complementary modeling efforts while building demonstration facilities and including hybrid technologies within process simulators.

Nomenclature

F: Total volumetric flow
ρ: Bulk density
J_i: Mass flux of component i
A: Membrane area
t: Operating time
nc: Total number of components in the fermentation broth
C_i: Mass concentration of component i per volume unit
r_i: Production reaction rate of component i
V: Reactor volume
$m_{p,i}$: Total weight of component i removed by the membrane as permeate
$\overline{P_i}$: Mass permeance of component i through the membrane.

Subscripts

f: Inlet stream
o: Outlet stream
p: Permeate stream
X: Biomass

Bibliography

[1] A. Gorak and A. Stankiewicz, "Research agenda for process intensification – towards a sustainable world of 2050," 2011.

[2] O. A. Prado-Rubio, R. Morales-Rodríguez, P. Andrade-Santacoloma, and H. HernÁndez-Escoto, *Process intensification in biotechnology applications*. 2016.

[3] A. Bar-Even, E. Noor, Y. Savir, W. Liebermeister, D. Davidi, T. RS, and R. Milo, "The moderately efficient enzyme: Evolutionary and physicochemical trends in shaping enzyme parameters," *Biochemistry*, vol. 50, pp. 4402–4410, 2011.

[4] H. Ling, W. Teo, B. Chen, S. Su Jan Leong, and M. Wook Chang, "Microbial tolerance engineering toward biochemical production: from lignocellulose to products," *Curr. Opin. Biotechnol.*, vol. 29, pp. 99–106, 2014.

[5] F. Carstensen, A. Apel, and M. Wessling, "In situ product recovery: Submerged membranes vs. external loop membranes," *J. Memb. Sci.*, vol. 394–395, pp. 1–36, 2012.
[6] J. P. M. Sanders, J. H. Clark, G. J. Harmsen, H. J. Heeres, J. J. Heijnen, S. R. A. Kersten, W. P. M. van Swaaij, and J. A. Moulijn, "Process intensification in the future production of base chemicals from biomass," *Chem. Eng. Process. Process Intensif.*, vol. 51, pp. 117–136, 2012.
[7] J. J. Klemeš and P. S. Varbanov, "Process intensification and integration: An assessment," *Clean Technol. Environ. Policy*, vol. 15, no. 3, pp. 417–422, 2013.
[8] W. T. Hecke, G. Kaur, and H. DeWever, "Advances in in-situ product recovery (ISPR) in whole cell biotechnology during the last decade," *Biotechnol. Adv.*, vol. 32, pp. 1245–1255, 2014.
[9] J. T. Dafoe and A. J. Daugulis, "In situ product removal in fermentation systems: Improved process performance and rational extractant selection," *Biotechnol. Lett.*, vol. 36, no. 3, pp. 443–460, 2014.
[10] V. Rangarajan and K. G. Clarke, "Process development and intensification for enhanced production of Bacillus lipopeptides," *Biotechnol. Genet. Eng. Rev.*, vol. 31, no. 1–2, pp. 46–68, 2015.
[11] J. Woodley, "Bioprocess intensification for the effective production of chemical products," *Comput. Chem. Eng.*, vol. 105, pp. 297–307, 2017.
[12] Y. Satyawali, K. Vanbroekhoven, and W. Dejonghe, "Process intensification: The future for enzymatic processes?," *Biochem. Eng. J.*, vol. 121, pp. 196–223, 2017.
[13] K. Tadele, S. Verma, M. A. Gonzalez, and R. S. Varma, "A sustainable approach to empower the bio-based future: Upgrading of biomass via process intensification," *Green Chem.*, vol. 19, no. 7, pp. 1624–1627, 2017.
[14] H. Vaghari, M. Eskandari, V. Sobhani, A. Berenjian, Y. Song, and H. Jafarizadeh-Malmiri, "Process intensification for production and recovery of biological products," *Am. J. Biochem. Biotechnol.*, vol. 11, no. 1, pp. 37–43, 2015.
[15] A. Górak and A. Stankiewicz, Eds., *Intensification of Biobased Processes*. The Royal Society of Chemistry, 2018.
[16] T. Van Gerven and A. Stankiewicz, "Structure, energy, synergy, time: the fundamentals of process intensification," *Ind Eng Chem Res*, vol. 48, no. 5, pp. 2465–2474, 2009.
[17] J. Arizmendi-Sánchez and P. Sharratt, "Phenomena-based modularisation of chemical process models to approach intensive options," *Chem. Eng. J.*, vol. 135, pp. 83–94, 2008.
[18] P. Philip Lutze, R. Gani, and J. M. Woodley, "Process intensification: A perspective on process synthesis," *Chem. Eng. Process.*, vol. 49, pp. 547–558, 2010.
[19] A. Stankiewicz and J. A. Moulijn, "Process Intensification: Transforming Chemical Engineering," *Chem. Eng. Prog.*, vol. 96, no. 1, pp. 22–34, 2000.
[20] W. L. Luyben, *Chemical Reactor Design and Control*. John Wiley & Sons, Inc., 2007.
[21] M. Baldea, "From process integration to process intensification," *Comput. Chem. Eng.*, vol. 81, pp. 104–114, 2015.
[22] S. E. Demirel, J. Li, and M. M. F. Hasan, "Systematic process intensification using building blocks," *Comput. Chem. Eng.*, vol. 105, pp. 2–38, 2017.
[23] S. Choi, C. W. Song, J. H. Shin, and Y. L. Sang, "Biorefineries for the production of top building block chemicals and their derivatives," *Metab. Eng.*, vol. 28, pp. 223–239, 2015.
[24] F. Cavani, S. Albonetti, F. Basile, and A. Gandini, *Chemicals and Fuels from Bio-Based Building Blocks*, vol. 1. 2016.
[25] M. N. H. Z. Alam, M. Pinelo, K. Samanta, G. Jonsson, A. Meyer, and K. V. Gernaey, "A continuous membrane microbioreactor system for development of integrated pectin modification and separation processes," *Chem. Eng. J.*, vol. 167, no. 2–3, pp. 418–426, 2011.

[26] A. Pohar, P. Žnidaršič-Plazl, and I. Plazl, "Integrated system of a microbioreactor and a miniaturized continuous separator for enzyme catalyzed reactions," *Chem. Eng. J.*, vol. 189–190, pp. 376–382, 2012.
[27] P. Lutze, D. K. Babi, J. M. Woodley, and R. Gani, "Phenomena based methodology forprocess synthesis incorporating process intensification," *Ind Eng Chem Res*, vol. 52, no. 22, pp. 7127–7144, 2013.
[28] N. Anantasarn, U. Suriyapraphadilok, and D. K. Babi, "A computer-aided approach for achieving sustainable process design by process intensification," *Comput. Chem. Eng.*, vol. 105, pp. 56–73, 2017.
[29] P. Lutze, A. Roman Martinez, J. Woodley, and R. Gani, "A systematic synthesis and design methodology to achieve process intensification in (bio) chemical processes," *Comput. Chem. Eng.*, vol. 36, pp. 189–207, 2012.
[30] O. A. Prado-Rubio, "Integration of Bioreactor and Membrane Separation Processes: A model-based approach," Technical University of Denmark, 2010.
[31] O. A. Prado-Rubio, S. B. Jørgensen, and G. Jonsson, *Control system development for integrated bioreactor and membrane separation process*, vol. 28, no. C. 2010.
[32] H. J. Noorman and J. J. Heijnen, "Biochemical engineering's grand adventure," *Chem. Eng. Sci.*, vol. 170, pp. 677–693, 2017.
[33] R. van René and B. Annevelink, "Status Report Biorefinery 2007," Wageningen, 2007.
[34] D. I. Atala, A. C. Costa, R. Maciel, and F. Maugeri, "Kinetics of ethanol fermentation with high biomass concentration considering the effect of temperature.," *Appl. Biochem. Biotechnol.*, vol. 91–93, pp. 353–65, Jan. 2001.
[35] A. 6 Jarzebski, J. J. Malinowski, and G. Goma, "Modeling of Ethanol Fermentation at High Yeast Concentrations," vol. 34, pp. 1225–1230, 1989.
[36] R. W. Baker, *Membrane Technology and Applications*, Second. West Sussex: John Wiley & Sons, Ltd., 2004.
[37] R. W. Baker, J. G. Wijmans, and Y. Huang, "Permeability, permeance and selectivity: A preferred way of reporting pervaporation performance data," *J. Memb. Sci.*, vol. 348, no. 1–2, pp. 346–352, Feb. 2010.
[38] S. Yadav, G. Rawat, P. Tripathi, and R. K. Saxena, "Dual substrate strategy to enhance butanol production using high cell inoculum and its efficient recovery by pervaporation.," *Bioresour. Technol.*, vol. 152, pp. 377–383, Jan. 2014.
[39] O. J. Jaramillo-Pineda, M. Á. Gómez-García, and J. Fontalvo, "Removal of ethanolic fermentation inhibitors using polydimethylsiloxane (PDMS) membranes by pervaporation," *Rev. Ión*, vol. 25, no. 1, pp. 51–59, 2012.
[40] D. M. Aguilar-Valencia, M. A. Gómez-García, and J. Fontalvo, "Effect of pH, CO2, and High Glucose Concentrations on Polydimethylsiloxane Pervaporation Membranes for Ethanol Removal," *Ind. Eng. Chem. Res.*, vol. 51, no. 27, pp. 9328–9334, Jul. 2012.
[41] C. Åkerberg and Z. G., "An economic evaluation of the fermentative production of lactic acid from wheat flour," *Bioresour. Technol.*, vol. 75, no. 2, pp. 119–126, 2000.
[42] G. Stephanopoulos, A. Aristidou, and J. Nielsen, *Metabolic Engineering. Principles and Methodologies*. Academic Press, 1998.
[43] J. Nielsen, J. Villadsen, and G. Lidén, *Bioreaction Engineering Principles*, 2nd ed. Kluwer Academic/Plenum Publishers, 2003.
[44] M. Othman, A. B. Ariff, L. Rios-Solis, and M. Halim, "Extractive Fermentation of Lactic Acid in Lactic Acid Bacteria Cultivation: A Review," *Front Microbiol.*, vol. 8, p. 2285, 2017.
[45] T. Ghaffar, M. Irshad, Z. Anwar, T. Aqil, Z. Zulifqar, A. Tariq, M. Kamran, N. Ehsan, and S. Mehmood, "Recent trends in lactic acid biotechnology: A brief review on production to purification," *J. Radiat. Res. Appl. Sci.*, vol. 7, no. 2, pp. 222–229, 2014.

[46] L. Qian-Zhu, J. Xing-Lin, F. Xin-Jun, W. Ji-Ming, S. Chao, Z. Hai-Bo, X. Mo, and L. Hui-Zhou, "Recovery Processes of Organic Acids from Fermentation Broths in the Biomass-Based Industry," *J. Microbiol. Biotechnol.*, vol. 26, no. 1, pp. 1–8, 2016.
[47] N. Murali, K. Srinivas, and B. K. Ahring, "Biochemical Production and Separation of Carboxylic Acids for Biorefinery Applications," *Fermentation*, vol. 3, p. 22, 2017.
[48] A. Komesu, M. R. Wolf Maciel, and R. Maciel Filho, "Separation and purification technologies for lactic acid – A brief review," *BioRes*, vol. 12, no. 3, pp. 6885–6901, 2017.
[49] H. D. Lee, M. Y. Lee, Y. S. Hwang, Y. H. Cho, H. W. Kim, and H. B. Park, "Separation and Purification of Lactic Acid from Fermentation Broth Using Membrane-Integrated Separation Processes," *Ind. Eng. Chem. Res*, vol. 56, no. 27, pp. 8301–8310, 2017.
[50] A. Zheleznov, D. Windmöller, S. Körner, and K. Böddeker, "Dialytic Transport of Carboxylic Acids through an Anion Exchange Membrane," *J. Memb. Sci.*, vol. 139, pp. 137–143, 1998.
[51] O. A. Prado Rubio, S. B. Jørgensen, and G. E. Jonsson, *Lactic Acid Recovery in Electro-Enhanced Dialysis: Modelling and Validation*, vol. 26. 2009.
[52] O. A. Prado-Rubio, S. B. Jørgensen, and G. Jonsson, "Reverse Electro-Enhanced Dialysis for lactate recovery from a fermentation broth," *J. Memb. Sci.*, vol. 374, no. 1–2, 2011.
[53] O. A. Prado-Rubio, M. Møllerhøj, S. B. Jørgensen, and G. Jonsson, "Modeling Donnan dialysis separation for carboxylic anion recovery," *Comput. Chem. Eng.*, vol. 34, no. 10, 2010.
[54] H. Strathmann, *Ion-Exchange Membrane Separation Processes*, Membrane S. Elsevier, 2004.
[55] A. Garde, "Production of Lactic Acid from Renewable Resources using Electrodialysis for Product Recovery," Technical University of Denmark, 2002.
[56] J. Rype, "Modelling of Electrically Driven Processes," Technical Uniersity of Denmark, 2003.
[57] M. Boonmee, N. Leksawasdi, W. Bridge, and P. Rogers, "Batch and Continuous Culture of Lactococcus lactis NZ133: Experimental Data and Model Development," *Biochem. Eng. J.*, vol. 14, pp. 127–135, 2003.
[58] V. Fila and K. Bouzek, "A Mathematical Model of Multiple Ion Transport Across an Ion-Selective Membrane under Current Load Conditions," *J. Appl. Electrochem.*, vol. 33, pp. 675–684, 2003.
[59] O. A. Prado-Rubio, S. B. Jørgensen, and G. Jonsson, "pH control structure design for a periodically operated membrane separation process," *Comput. Chem. Eng.*, vol. 43, 2012.
[60] C. Rosen, D. Vrecko, K. V. Gernaey, and U. Jeppsson, "Implementing ADM1 for benchmark simulations in Matlab/Simulink," 2006.
[61] O. A. Prado-Rubio, D. Rodriguez-Gomez, and R. Morales-Rodriguez, "Model-Based Approach to Enhance Configurations for 2G Butanol Production through ABE Process," *Recent Innov. Chem. Eng.*, 2018.
[62] O. A. Prado-Rubio, S. B. Jørgensen, and G. Jonsson, "Model based investigation of the potential lactate recovery using electro-enhanced dialysis - Static analysis," *Sep. Purif. Technol.*, vol. 78, no. 2, 2011.
[63] S. France, L. Hepworth, N. Turner, and S. Flitsch, "Constructing biocatalytic cascades: In vitro and in vivo approaches to de novo multi-enzyme pathways," *ACS Catal.*, vol. 7, pp. 710–724, 2017.

Juan Álvaro León, Oscar Andrés Prado-Rubio and Javier Fontalvo

9 Design of hybrid distillation and vapor permeation or pervaporation systems

Abstract: It is accepted that conventional distillation is one of the most used separation processes in industry. This versatile and relatively simple technology can produce high-purity products. Despite considered a mature technology, one of the main disadvantages of distillation still resides in its high energy demand which conflicts with current and future energy utilization constraints. New approaches to increase energy efficiency have been explored for several decades where process intensification concepts are inspiring and have produced substantial improvements. For example, recent technologies have been proposed such as: dividing wall columns, Higee distillation, cyclic distillation, heat-integrated distillation, reactive distillation, and reactive dividing-wall columns. However, some opportunities remain for mixtures with azeotropes, which are especially energy intensive to separate by distillation. Therefore, despite the technological advances, new concepts are required combining distillation with more energetically efficient separation technologies.

Among alternatives, hybrid distillation systems with pervaporation or vapor permeation are especially interesting. Thus, they have been extensively studied over the last couple of decades. The purpose of this chapter is to show the main configuration of these hybrid systems, some examples and several design tools for hybrid distillation-membrane processes. Besides, it presents the main aspects of pervaporation and vapor permeation technology. Different discussions are presented not only for systems design but also for membrane producers in order to aim for more suitable materials and performance.

Keywords: Pervaporation, azeotropes, energy consumption, process design

9.1 Introduction

Conventional distillation is a mature technology widely applied in industry to separate multicomponent mixtures. It is well known that separation capabilities of distillation processes are limited by the mixture liquid-vapor equilibria and energy demand. Then, the separation must be performed by enhanced processing schemes such as extractive or azeotropic distillation to overcome thermodynamic limitations in particular mixtures (i.e., azeotropes and/or distillation boundaries) [1, 2]. From

---

Juan Álvaro León, Oscar Andrés Prado-Rubio, Javier Fontalvo, National University of Colombia, Department of Chemical Engineering, Laboratory of Process Intensification and Hybrid Systems

https://doi.org/10.1515/9783110596120-009

process intensification perspectives, different approaches have been proposed to obtain substantial increase in the separation performance such as dividing-wall columns [3], internal heat-integrated distillation columns (HIDiCs) [4], Higee distillation [5], cyclic distillation [6], reactive distillation [7], and reactive dividing-wall columns [8]. Despite the efforts, energy demand in distillation is still considered high.

Recently, distillation-pervaporation hybrid systems have been proposed as an alternative for limited separations, to reduce energy consumption or improve operating conditions [9–15]. Membrane-assisted distillation is promising as the separation by pervaporation is not limited by the phase equilibrium [16], allowing to split nonideal mixtures beyond their azeotropic compositions while simultaneously reducing the energy consumption [15]. In a conventional distillation-pervaporation hybrid system, the pervaporation module is externally connected to the distillation column (externally connected distillation-pervaporation, ECDP) [17]. The mixture separation is carried out by the two units, commonly integrated by a recycle stream of the membrane retentate or permeate to the distillation column [18]. Energy and separation cost savings between 20% and 60% have been reported as compared to systems where just distillation or azeotropic distillation is used [15]. Distillation is a unit operation with lower capital costs than membrane technology. Thus, a hybrid distillation and membrane system exploits the advantages of both operations while reduces their disadvantages.

9.2 Basic configurations of hybrid distillation and membrane systems

Hybrid distillation-membrane configurations used for process intensification are presented in Figure 9.1. The membrane unit used in those configurations can be either pervaporation or vapor permeation [2, 19]. In pervaporation the feed is a liquid, while in vapor permeation it is a vapor [15]. In both cases, the permeate stream, constituted by the compounds that pass through the membrane, is a vapor phase at low pressure. Thus, in Figure 9.1 the permeate stream (P) that is returned to the column is usually: condensed at low pressure, pressurized as a liquid to the operating pressure of the column, heated to the required temperature, and finally fed back to the column. Notice that all these steps are not shown in the diagram. Figure 9.1a shows a hybrid distillation and membrane system where the membrane is used as a pretreatment stage. In this configuration, there is not an integration as such between the two processes [20]. For binary mixtures, configuration presented in Figure 9.1a is not very attractive since the feed flow to the membrane and the required component removal is higher than for instance in configuration 9.1b or 9.1c, thus increasing the membrane area [21]. However, using a sequential configuration the number of stages, energy consumption

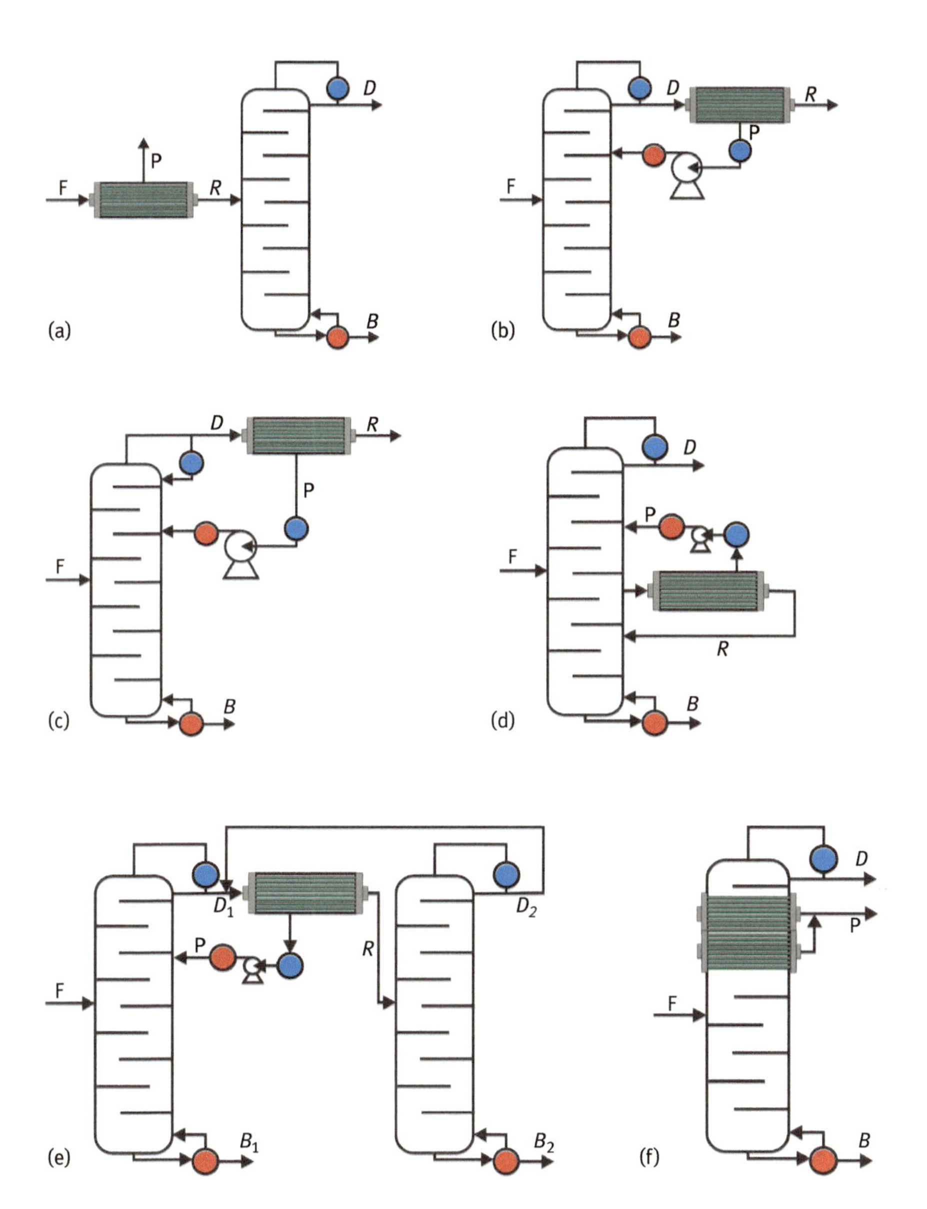

Figure 9.1: Configurations of hybrid distillation – membrane (pervaporation or vapor permeation) processes. Feed (F), distillate (D), permeate (P), and bottom (B) streams are shown.

and reflux ratio can be reduced in the distillation column. Therefore, an economic analysis is required to determine the configuration convenience.

For multicomponent mixtures, the sequential configuration is interesting if it is not possible to reach a desired composition in the product streams from the distillation column with a given feed (F). Then, the membrane can be used to adjust the inlet composition to a more suitable one even overcoming a distillation boundary prior the distillation unit. In addition, for a mixture of three components, the membrane can be used to remove one of the components while the distillation column concentrates the other two components [20]. As for binary mixtures, the membrane previous to the distillation column could reduce the number of stages and reflux ratio of the distillation column.

In Figures 9.1b and 9.2c, the membrane is used as a refining step to adjust the distillate (an analogous configuration can be used for the bottom stream). For distillation of components with similar boiling points, these configurations reduce the number of distillation stages or the reflux ratio of the distillation column. Also for systems with azeotropes, the membrane can overcome the azeotrope composition because it is not limited by the liquid-vapor equilibria. The advantage of using vapor permeation (Figure 9.1c) is that the membrane unit does not require additional energy because the vapor is already available from the top of the distillation column [22]. From an energy point of view, Figure 9.1c could require less energy in the reboiler than Figure 9.1b. However, by using pervaporation (Figure 9.1b), it is possible to pressurize the feed to the membrane unit to enhance the flux, and as a consequence, the required membrane area for this hybrid system is reduced [15]. Notice that the pervaporation unit still requires an energy supply. Thus, an economic analysis is important to assess these two hybrid processes.

For binary and multicomponent mixtures, the number of stages and the reflux ratio in the distillation unit can be decreased by using a membrane for separation of a column side stream (Figure 9.1d) [23, 24]. The membrane performs a part of the separation and reduces the work performed by the distillation column in terms of number of stages.

A more integrated process is shown in Figure 9.1e where the membrane is used to overcome a distillation boundary after a first column, and a second distillation carries on the refining step. In this case, the first distillation column will produce a stream close to azeotropic conditions. Subsequently, a membrane unit is used to overcome the azeotrope so the final separation takes place in a second column from a different region of the compositional diagram [15, 18]. For a binary mixture with a low boiling point azeotropic mixture, the purified components are obtained at the bottom of the distillation columns.

An externally connected hybrid system normally requires a lower energy consumption to separate nonideal organic mixtures in comparison to pressure swing distillation or azeotropic distillation, obtaining products with the same composition requirements [25]. Additionally, operation costs are reduced by an ECDP system since the thermal energy requirement is lower than in an azeotropic distillation without the use of an entrainer (or solvent). As counterpart, an external membrane unit imposes the operative limitations of the membrane unit to the hybrid distillation-membrane system. Mass and heat transfer resistance build up, known in membrane technology as concentration and temperature polarization, occur inside the membrane unit and decrease the performance of the unit [16, 26–28]. On top of that, an inter-heating system is required between the pervaporation modules to avoid a significant temperature drop and maintain the membrane flux high [16]. Due to these considerations, it has been proposed to have a highly integrated system where the membrane is placed within the distillation column (Figure 9.1f) [29, 30]. Integrating the membrane into the column creates a system that can overcome the distillation boundaries while both mass and energy transfer adjacent to the membrane and the transmembrane driving force are improved. Consequently, the required membrane area is reduced compared to ECDP and there is no need of interstage heating.

In this chapter several examples, based on Figure 9.1, for designing hybrid distillation-pervaporation systems are presented for binary and multicomponent mixtures. The corresponding design methodologies are explained in detail. In order to exploit the insights presented there in, it is advised to have some basic knowledge on design strategies of distillation columns for binary and multicomponent mixtures.

9.3 Basic approaches for designing hybrid distillation and membrane systems

Tools for designing hybrid distillation and membrane systems are rather similar to those used in distillation: conceptual design, rigorous design, and optimization. Thus, shortcut methods have been developed for binary and multicomponent mixtures based on graphical techniques. For binary mixtures, the design of the hybrid system is based on the McCabe-Thiele diagram [23]. Due to the interaction of the membrane unit with the distillation column, new operating lines are obtained for the distillation column. With the new operating lines, the number of stages or the reflux ratio can be calculated analogously to single distillation columns with multiple side streams [31]. Once the distillation column in the hybrid system has been designed, the membrane area can be calculated as a function of the mass transfer driving force, as it is shown below (Eqs. 9.1 and 9.9). An example of this approach is presented in Section 9.8. For multicomponent mixtures, the conceptual design is based on what is called membrane residual curves [32–34]. Then, on ternary diagrams, a conceptual design can be proposed based on the distillation lines (or distillation residue curves) and membrane residue curves. The membrane residue curves are explained in Section 9.5 while three examples of the design approach are presented in Sections 9.6, 9.9, and 9.10.

Rigorous design of hybrid distillation and membrane systems have been performed by using commercial software for the distillation column in combination with user models for the membrane [15, 29, 35–37]. From this simulation, it is possible to obtain energy consumption, separation costs, membrane area, membrane energy consumption and operating conditions for both units. On the other hand, a superstructure approach has been used to design and optimize the hybrid system by placing the membrane unit in various configurations [38]. This approach is based on a combination of a superstructure for distillation processes with a superstructure for gas permeation networks. In this way, it is possible to find an optimized structure and parameter for the hybrid system. Section 9.7 presents an example of a rigorous approach where ASPEN is used in combination with MATLAB for designing a hybrid process where pervaporation is used as a final dewatering stage of the process [35].

9.4 Pervaporation and vapor permeation

Pervaporation and vapor permeation are promising membrane technologies that can be integrated with separation processes and reactive systems for process. In pervaporation, a liquid is placed in contact with a selective membrane while vacuum is applied in the permeate side to remove and evaporate some components (Figure 9.2). The removed components are transferred through the membrane, vaporized, and received in the permeate side. The driving force for this process is the difference in vapor partial pressure of the components that are transported. This difference in vapor partial pressure is produced by the chemical potential difference between the feed and permeate channels. Because the components that are transported through the membrane are also vaporized, the energy required for this evaporation process comes from the liquid internal energy, and consequently, the liquid cools down. Thus, in pervaporation, several modules in series are used where the retentate, fed to a subsequent membrane module, has to be reheated. If the liquid is not reheated, the partial pressure is reduced and so the driving force for the

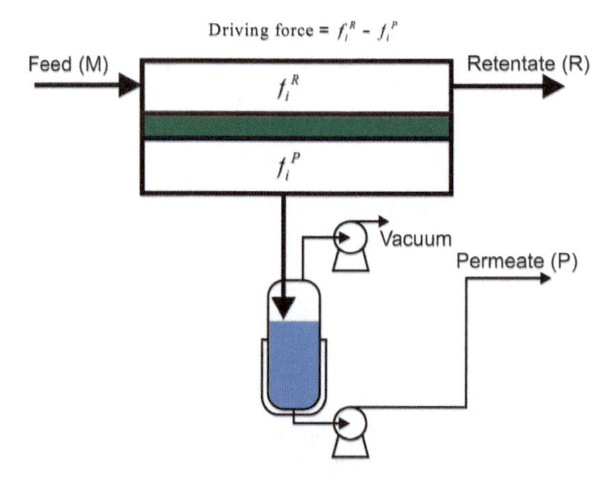

Figure 9.2: Sketch of a pervaporation or vapor permeation membrane.

transport. Always, in pervaporation, a high driving force is desired to have high fluxes in low membrane areas.
In vapor permeation, the membrane unit is fed with vapor phase, and thus, there is no phase change across the membrane. As a consequence, the retentate temperature does not drop compared to pervaporation. Analogous to pervaporation, the driving force of a given component is the difference in partial pressure between permeate and retentate sides of the membrane.

Polymeric and ceramic membranes are used in pervaporation systems. Usually, dense polymeric membranes are used and the transport is based on a "solution-diffusion" mechanism [39]. In this mechanism, the components in the liquid phase are transported first to the membrane surface, they are solubilized in the polymeric layer, and transported by diffusion through the membrane to the permeate side. The transported components are then desorbed and vaporized in the permeate side. In porous membranes, the mechanism is based on diffusion inside the pores [40]. For high selective membranes, mass transport is based on activated transport or surface diffusion [41]. In both dense and porous membranes, the component transport improves as temperature rises.

Polymeric and ceramic membranes are prepared as a composite material with several layers. First, a support that consists of several layers of polymer or ceramic materials [16, 42–44]. These layers are porous to provide structural support to the selective layer and reduce additional mass transport resistance [45]. Second, several polymeric or ceramic layers are placed on top of the support, which provide the membrane separating characteristics. Usually, several layers of the membrane are required to reduce defects or holes on the final separating layer. These holes reduce the performance of the membrane dramatically in terms of selectivity and are accompanied with high permeances. Ceramic separating layers are used with ceramic supports [44] but polymeric separating layers can be used with either polymeric [16] or ceramic [43, 46] supports.

9.5 Membrane residue curve maps

The flux of a given component i through a pervaporation or a vapor permeation membrane can be calculated with the following equation:

$$J_i = \overline{P_i}\left(f_i^R - f_i^P\right) \tag{9.1}$$

where, f_i is the fugacity or partial pressure of component i in retentate (R) or permeate (P) and $\overline{P_i}$ is its permeance through the membrane. A component permeance depends on intrinsic properties of the membrane material and its interaction with the components that are transported. For pervaporation the partial pressure in the liquid is:

$$f_i^R = \gamma_i x_i p_i^o \tag{9.2}$$

While for vapor permeation, the corresponding equation for the vapor, which is similar to the fugacity in the permeate side (Eq. 9.4), is:

$$f_i^R = x_i p^R \tag{9.3}$$

$$f_i^P = y_i p^P \tag{9.4}$$

For pervaporation and vapor permeation, the membrane selectivity of a component i related to component j is calculated as a ratio of permeances by:

$$S_{ij} = \frac{\overline{P_i}}{\overline{P_j}} \tag{9.5}$$

Analogous to distillation, the separation factor is used to characterize the membrane performance as shown in Eq. 9.6. According to Eq. 9.6, separation factor depends on retentate and permeate concentrations while selectivity, as defined in Eq. 9.5, depends on intrinsic membrane properties ($\overline{P_i}$).

$$\alpha_{ij} = \frac{y_i/(1-y_i)}{x_i/(1-x_i)} \tag{9.6}$$

A sketch of a membrane module suitable for mass balances is presented in Figure 9.3. The permeate flux (J) on a local position on the membrane is mixed with the bulk permeate stream ($P|_A$) that is flowing at that specific place on the permeate side. The molar balances, on a differential membrane area (ΔA), are:

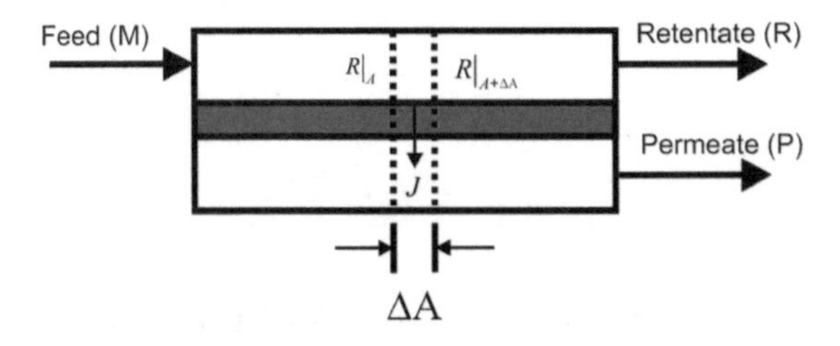

Figure 9.3: Mass transfer on a differential element of membrane area for pervaporation or vapor permeation.

$$R|_A = J\Delta A + R|_{A+\Delta A} \tag{9.7}$$

$$(Rx_i)|_A = Jy_i^* \Delta A + (Rx_i)|_{A+\Delta A} \tag{9.8}$$

By combining Eqs. 9.7 and 9.8 and for an infinitesimal differential membrane area, it is obtained:

$$\frac{d(Rx_i)}{dA} = -J_i \tag{9.9}$$

$$\frac{dx_i}{dA'} = (x_i - y_i^*) \tag{9.10}$$

Where A' is a dimensionless membrane area described by:

$$A' = \ln\left[\frac{M}{R|_A}\right] \tag{9.11}$$

It is necessary to have a relation between the retentate and permeate compositions to use Eq. 9.10. Thus, the flux per component through the membrane is given by:

$$J_i = J\, y_i \tag{9.12}$$

In this last equation, the corresponding flux per component (J_i) can be calculated from Eq. 9.1. Consequently, Eq. 9.10 is also solved by using Eqs. 9.1 and 9.12 for all the components in the mixture, except for a reference component. It is important to mention that for Eq. 9.10 the sum of molar fractions in retentate (x_i) or permeate (y_i) is the unity.

Based on the permeance ratios, the selectivity order of this membrane is component 1 > component 2 > component 3. An expected behavior is presented in Figure 9.4a for a selective vapor permeation membrane to component 1. The membrane residue curve (MRC) presents the composition profile in the retentate (Figure 9.3) and these profiles moves away from component 1, for which the membrane is more selective. Thus, the MRC moves away from component 1 and finishes in the component 3 vertex. Consequently, the last component to be removed from the retentate side is the one which permeability is the lowest (component 3). The local permeate compositions (y_i^*), or compositions on top of the membrane surface at the permeate side, are also shown

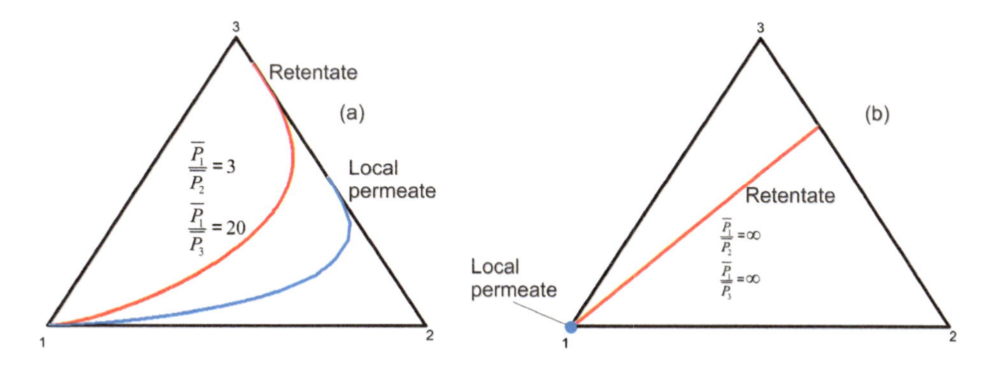

Figure 9.4: Membrane residue curves for a selective membrane (a) and a very high selective membrane (b).

in Figure 9.4a. These local compositions move in the same direction that the retentate compositions and, thus for a given composition in the retentate, the permeate has a larger content of component 1 and 2.

Figure 9.4b shows a membrane that has an infinite selectivity to component 1, so none of the other components are transported through the membrane. In this case, the MRC is a straight line that moves away from component 1 and the permeate local and accumulative compositions correspond to pure component 1.

The following sections present some examples where hybrid distillation-membrane processes, based on the block diagrams presented in Figure 9.1, are designed.

9.6 Case studies

9.6.1 Sequential membrane and distillation units

A mixture of isopropyl alcohol (IPA), water, and acetone has a couple of binary azeotropes and a distillation boundary shown in Figure 9.5 along the distillation residue curves. If a feed stream composition is located below the distillation boundary,

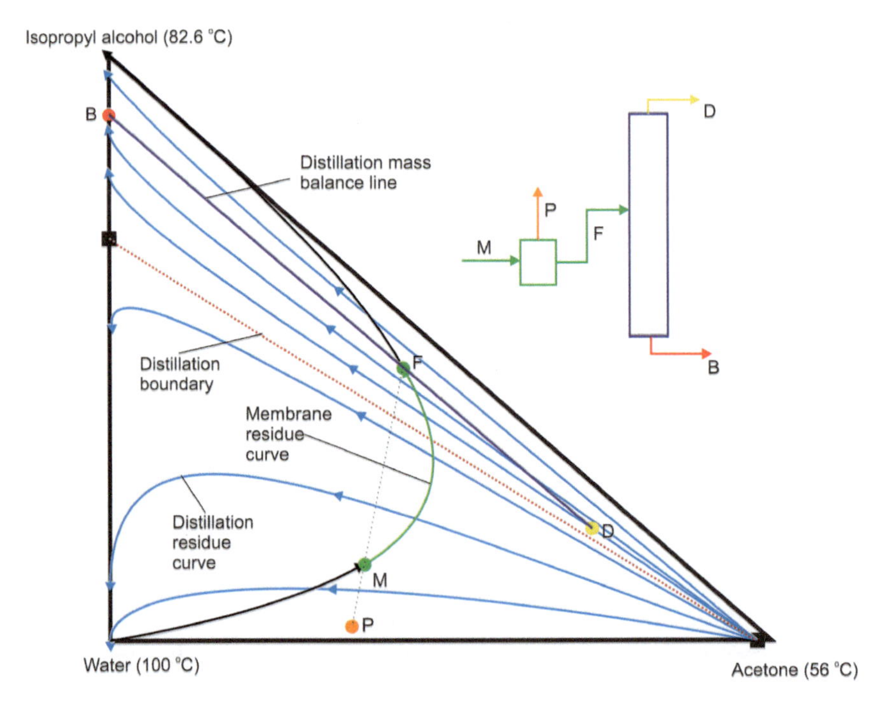

Figure 9.5: Design lines for a hybrid distillation–pervaporation process. The pervaporation unit is used to pretreat the feed to the distillation column.

point M in Figure 9.5, it is not possible, with a single column, to obtained pure IPA. However, if there is a membrane that has an infinity or a high selectivity to water, it is possible to pretreat stream M and to obtain a retentate to feed the distillation column (F) with a composition located above the distillation boundary. Within this region, it is possible by distillation to purify IPA. The corresponding MRC is shown in the figure with a membrane with a permeance order water > acetone > IPA. The conceptual design of this kind of systems can be performed graphically by using ternary diagrams with residue curves for distillation (DRC) and the membrane (MRC).

From the feeding point M, the corresponding MRC is used to cross the distillation boundary to reach a convenient feeding point for the column at the upper region, for instance, point F. Using point F, it can be drawn the corresponding global mass balance or the DRCs to produce a distillate (D) and bottom (B) streams. The permeate composition (P) depends on membrane selectivity. Also, the final permeate composition will be located in a straight line connecting F and M that are the outlet retentate and membrane feed streams compositions, respectively. Notice that, depending on membrane selectivity, the permeate stream will require additional purification or can be fed back to a suitable place in the chemical plant. The membrane area for this process can be calculated by integrating Eq. 9.8a and a mass balance in the permeate side of this membrane. Membrane area is a direct function of the total feed flow rate to the membrane for the same compositions presented in Figure 9.5.

9.6.2 Hybrid distillation-pervaporation system using pervaporation as a final distillate separation step

A configuration similar to the one shown in Figure 9.1b is presented in Figure 9.6 for purification of ethanol from a distillation column. In this configuration, a pervaporation unit will be used as a final separation step at the top of the column. The pervaporation unit can be simulated in MATLAB to find the total membrane area, the outlet permeate composition, and the outlet flows of retentate and permeate streams for a given composition of a single compound in the retentate stream. Equation 9.10 has been used for the retentate in combination with Eqs. 9.1 and 9.2. Also, to have the molar fractions in the permeate, a molar balance in the permeate has to be used which is similar to the one in the retentate (Eq. 9.10). This molar fractions (y_i) are required in order to calculate the flux through the membrane according to Eqs. 9.1 and 9.4

The behavior of the membrane area as a function of the membrane selectivity will be analyzed before designing the hybrid distillation-pervaporation system. The membrane selectivity is an important design parameter, specially since it is usually assumed that a very selective membrane is convenient for a given process. Table 9.1 presents the parameters used for simulating the performance of a water-selective

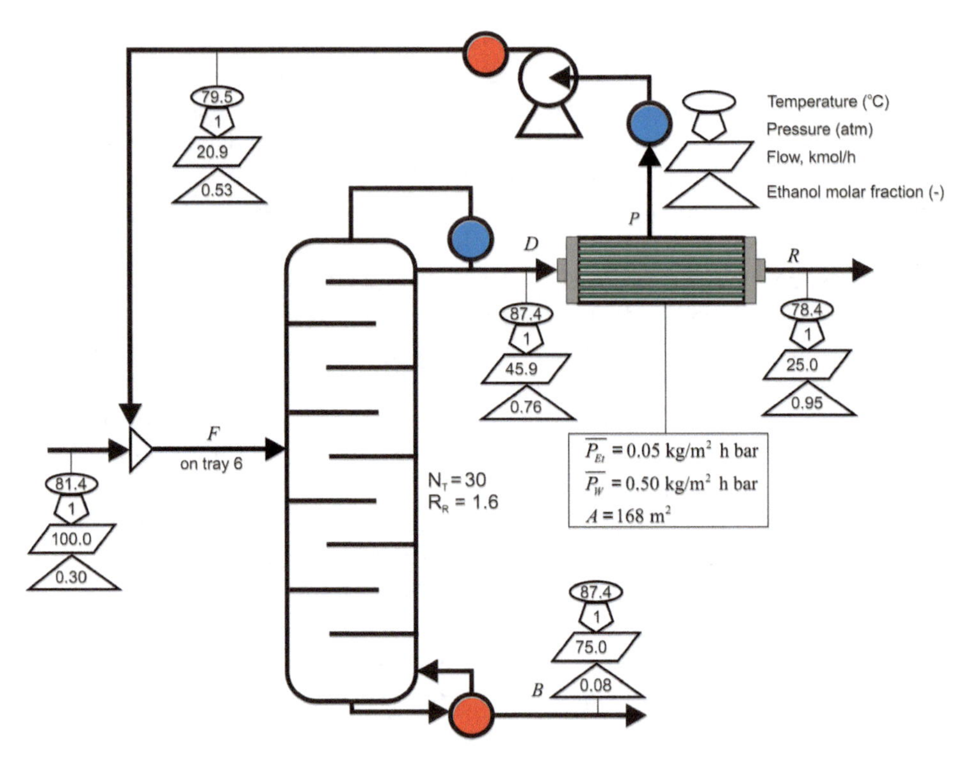

Figure 9.6: Hybrid distillation–pervaporation system for ethanol dehydration. The pervaporation unit is used as a final purification step.

Table 9.1: Parameters used for simulation (Figure 9.7) of a pervaporation module using a water-selective membrane.

Parameter or operating condition	Value
Water feed molar fraction	0.24
Feed temperature [K]	351.55
Feed flow [kmol/h]	37
Final water molar fraction in the retentate	0.1
Permeate pressure [bar]	0.1
Ethanol permeance [kmol / m^2 h bar]	It changes according to Figure 9.7
Water permeance [kmol / m^2 h bar]	0.5

pervaporation module for separation of an ethanol-water mixture. The corresponding results for the calculated membrane area as a function of the ethanol permeance are presented in Figure 9.7. A highly selective membrane has low ethanol permeance. Membrane area is reduced if a low selective membrane is used, which will also reduce capital cost for the process. Interestingly, a membrane with a very

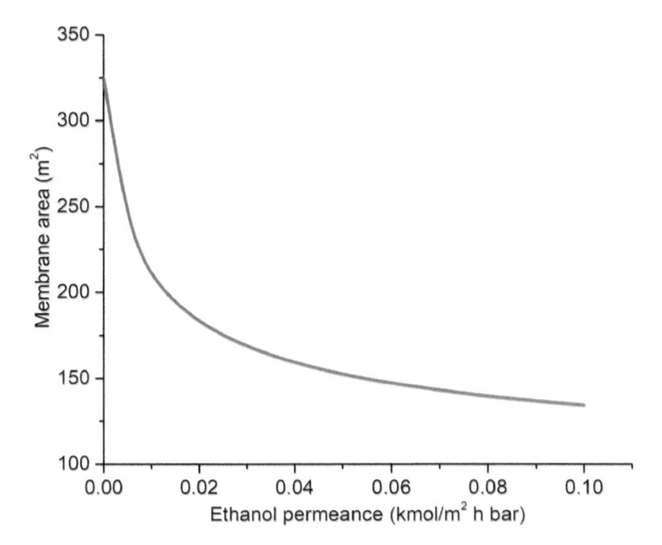

Figure 9.7: Effect of ethanol permeance on membrane area for dehydration of ethanol by pervaporation.

good selectivity toward water will require a considerable larger membrane area than a membrane with a selectivity of just 5 (Eq. 9.4). If membrane selectivity is very high, the water concentration in the permeate will also be very high, so the driving force decreases (according to Eqs. 9.1 and 9.4). Consequently, when the driving force decreases the corresponding membrane area required for removing a specific amount of water is increased. If permeate pressure is very low, the permeate concentration will not be important on the driving force and the influence of membrane selectivity on the membrane area is reduced. However, for industrial application the permeate pressure cannot be very low because the idea is to use a cooling service fluid of low cost, for instance, water.

The economic implications of this behavior are important to select a specific membrane for a hybrid process. These implications are not just based on membrane area, shown here, but also on separation costs and energy aspects related with the obtained permeate stream. These will be partially analyzed below.

As a final refining step, pervaporation can overcome the azeotropic composition of the ethanol-water mixture and produce anhydrous ethanol (Figure 9.6). The obtained permeate can be recycled to the distillation column or the feed. Recycling the permeate to the column is more convenient from the exergy point of view of [47], but the second option is used in this example for illustrative purposes. This process was simulated in ASPEN V7.2 integrated with a MATLAB (R2013a) subroutine for the pervaporation module and using MS Excel as a link [35]. Table 9.1 presents the information used for this simulation with an ethanol permeance of 0.05 kmol/m^2 h. Permeate stream is recycled at the column pressure (1.013 bar) and heated to its bubble point.

The calculated recycle flow is closed to 20% of the feed flow to the complete system (100 kmol/h). One option to reduce the membrane area of this process is by increasing the temperature and pressure of the stream fed to the pervaporation unit (results not shown). In this way, the corresponding driving force for the pervaporation process will increase and so the membrane area is reduced. An economic assessment has to be performed to evaluate if these changes are convenient for the process.

For this process, the influence of ethanol membrane selectivity on the membrane area and reboiler duty is shown in Figure 9.8. Very selective membranes require a high membrane area but reduce the energy consumption in the distillation tower. With high selective membranes, the ethanol concentration in the permeate increases and so the concentration of the feed stream to the distillation column reducing the reboiler duty. On the other hand, low selective membranes require a low membrane area but the corresponding energy consumption of the distillation tower is high.

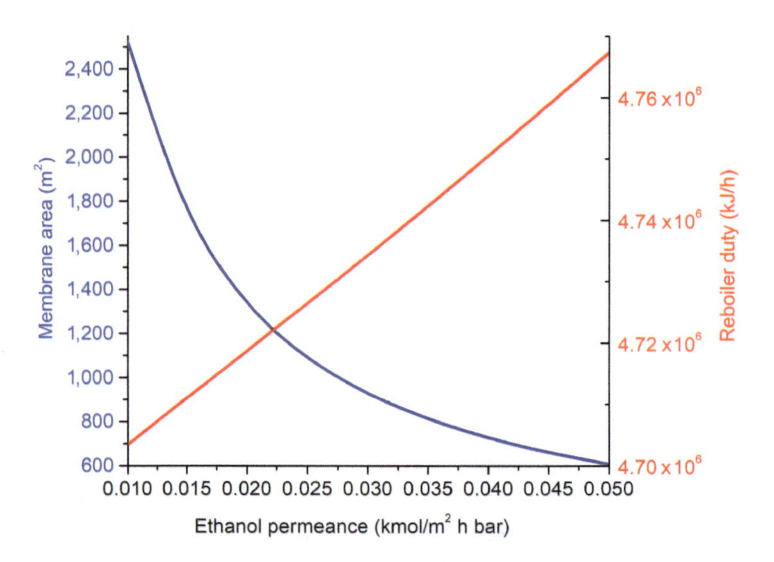

Figure 9.8: Effect of ethanol permeance on membrane area and reboiler duty for a hybrid distillation pervaporation process for ethanol production of Figure 9.6.

There is an economic trade-off regarding the membrane selectivity, where the costs of membrane area and energy are balanced. In general, very selective membranes are convenient to use if their cost is relatively low, which is not common in the market. Usually, a more selective membrane also has a higher price as compared with membranes with lower selectivity but the same lifetime.

Regarding systems that are already installed, although this is not always the case, it is expected that the membrane selectivity decreases with time. If this is the case, according to Figures 9.7 and 9.8, the required membrane area will be lower. Then, the installed membrane area in a chemical plant will be sufficient to continue with the process. However, the energy consumption will increase accordingly and there will be a point where the operating cost is not attractive, then the membranes must be replaced with new ones.

9.6.3 Hybrid distillation and membrane system for reducing the number of stages

The separation by distillation of some mixtures of interest for industry usually requires a number of stages and high external reflux ratios. Any reduction in these aspects has a significant impact on capital and operating costs. In that regard, pervaporation or vapor permeation can be used to reduce both of these parameters.

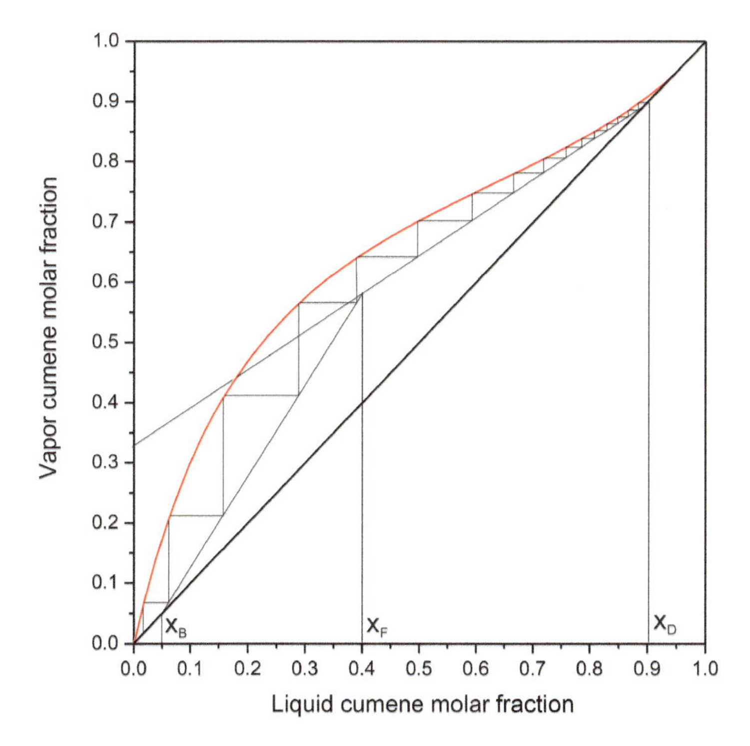

Figure 9.9: Number of ideal stages for purification of cumene from phenol with the McCabe-Thiele method.

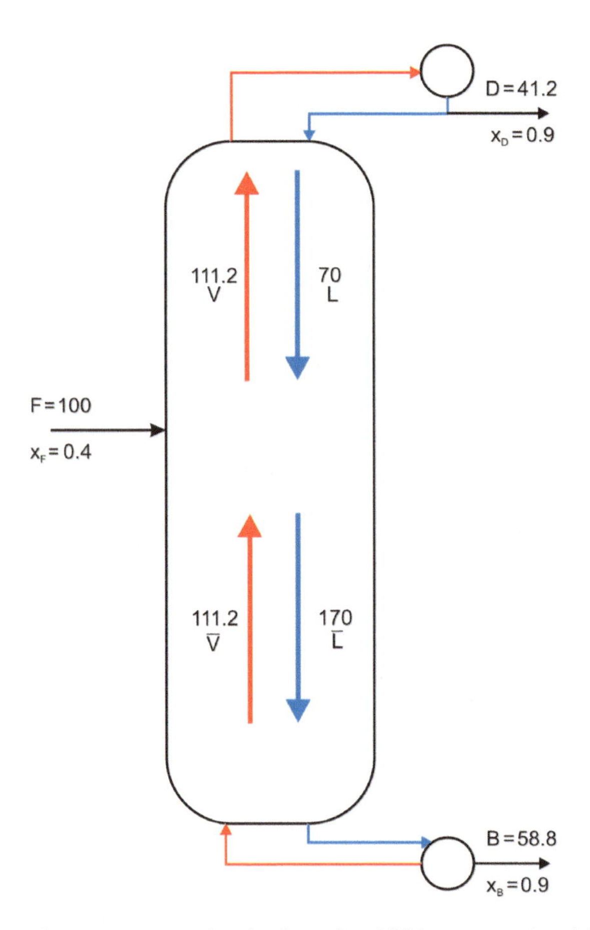

Figure 9.10: Internal molar flows (kmol/h) in a conventional distillation column for cumene purification from phenol of Figure 9.9. Cumene molar fraction, *x*.

For example, a distillation column design using a McCabe-Thiele approach is depicted in Figure 9.9. The figure shows that for a moderate separation to obtained cumene, with a concentration of 90% molar fraction, a total number of 15.2 total ideal stages (including reboiler) are required. The corresponding external reflux ratio is 1.7. A molar flow balance inside of a conventional distillation column produces the flows presented in Figure 9.10.

Using the configuration shown in Figure 9.1d, a pervaporation unit can be used to split a side stream from the column into two streams: P and R, as depicted in Figure 9.11. In this way, a higher concentration of cumene and phenol can be achieved. The side stream could be taken from the section of the column where the total liquid flow is the highest, which in this case corresponds to the stripping section.

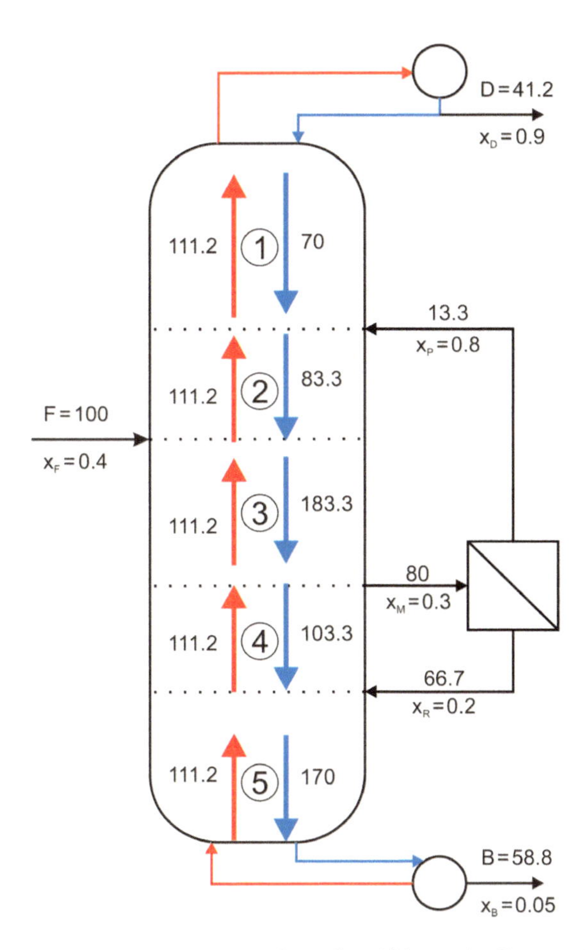

Figure 9.11: Internal molar flows (kmol/h) in a distillation column for cumene purification from phenol for a hybrid distillation–pervaporation system. Cumene molar fraction, *x*.

For instance, consider as a case study, a side stream of 80 kmol/h that is withdrawn from the column and fed to the pervaporation module to produce permeate and retentate streams with molar fractions of cumene of 0.8 and 0.2, respectively (Figure 9.12). Operating lines in this case can be calculated from the mass balances of the streams inside the column by assuming constant molar flow. The cumene molar fractions of the side stream and the retentate are known and, for this example, they are 0.3 and 0.2, respectively. The cumene molar fraction in the permeate depends on the membrane selectivity but in this case, a value of 0.8 is used. For the molar flow balance inside the column, all side streams will be assumed as saturated liquids.

For instance, for Section 9.2 the liquid in this section is:

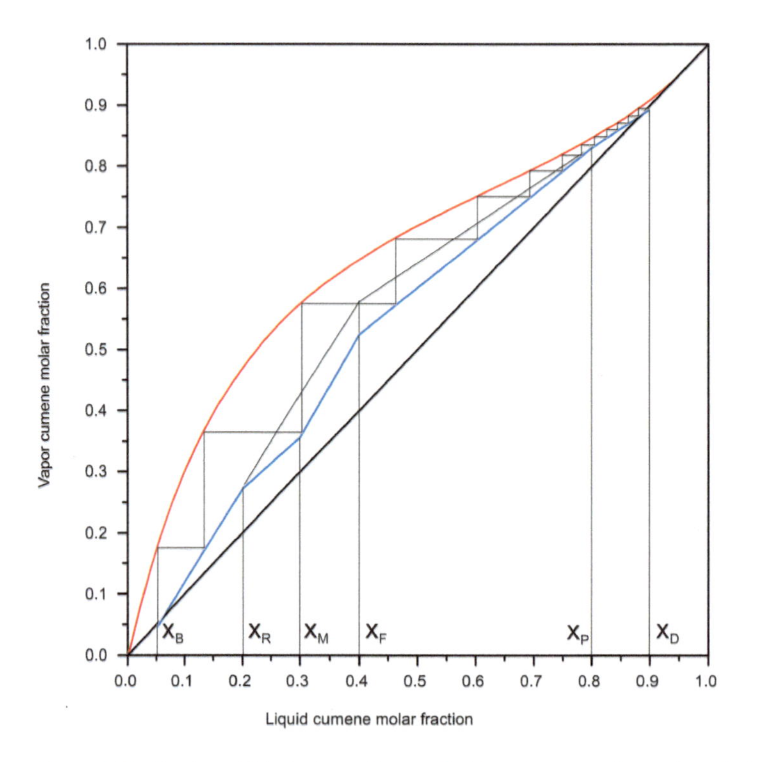

Figure 9.12: Number of ideal stages for purification of cumene from phenol with the McCabe-Thiele method for a hybrid distillation–pervaporation system presented in Figure 9.11. The blue line corresponds to the operating line of the hybrid distillation–pervaporation system.

$$L^2 = L^1 + P = 70 + 13.33\frac{kmol}{h} = 83.33\frac{kmol}{h} \tag{9.13}$$

Using similar balances to the one presented in Eq. 9.13, the calculated flows can be seen in Figure 9.12. With these values, it is possible to calculate the slope of the distillation operating lines (L/V) and also the number of ideal stages. Notice that the number of ideal stages in this hybrid process is reduced to 13 (including reboiler). Including the membrane unit, the operating lines are placed farther away from the liquid-vapor equilibrium compared to conventional distillation (Figure 9.9) and so the number of stages in the column is reduced. From a systematic point of view, it has been shown that is also possible to use analytical expressions for the operating lines, and then minimize the area between them and the diagonal of the *xy* diagram which at the end minimizes the number of theoretical stages in the column [23, 24].

9.6.4 Pervaporation or vapor permeation for multicomponent distillation

In some cases, a feasible separation cannot be achieved by a conventional distillation column, even if the mass balance is met. The separation feasibility depends on the product composition, azeotrope formation, and operating parameters as reflux or boil-up ratio. To achieve a feasible separation in distillation, the composition profiles of a section must intercept the composition profiles of the immediately subsequent section. Thus, the composition profiles of the rectifying and stripping sections must intercept each other. An ECDP column can be used to generate a feasible separation when it is not feasible with conventional distillation, since an ECDP column allows modifying the composition profiles inside the distillation column.

Figure 9.13 shows a general scheme of the streams involved in an ECDP system. The column is divided into several sections according to the feed location and the input/output streams of the pervaporation unit. Section 9.1 corresponds to the rectifying section, where the distillate is produced. The pervaporation unit is located in Section 9.2 and immediately below Section 9.1. There, a fraction of the liquid stream is fed to the pervaporation module. The retentate is then recycled to the distillation column above Section 9.2, while the permeate is obtained as a product stream. Section 9.3 corresponds to the stripping section of the ECDP system. In this configuration, a continuous composition profile must be generated by Sections 9.1, 9.2, and 9.3 to achieve a feasible separation.

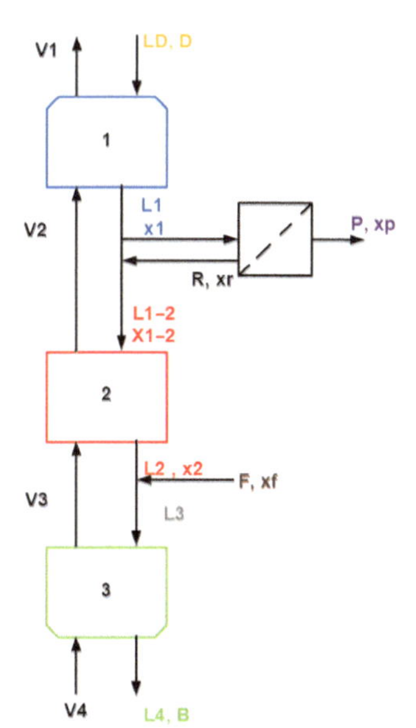

Figure 9.13: Block diagram of an externally connected distillation pervaporation (ECDP) system [20] where (1) is a rectifying section, (2) intermediate rectifying section subsequent to the pervaporation unit, and (3) stripping section. Condenser and reboiler are not shown.

The composition profile for the liquid and vapor phases in every section can be determined by a difference point equation (DPE) that applies to any section of the distillation column in the block diagram presented in Figure 9.13 [20, 48]. In this case, constant molar flow is assumed through the specific section. However, the molar flow between sections can change based on the point of mass addition or withdraw. The general DPE for a given section is [20]:

$$\frac{dx_i}{dn} = \left(\frac{1}{R_\Delta} + 1\right)(x_i - y_i) + \frac{1}{R_\Delta}(X_\Delta - x_i) \tag{9.14}$$

where:

$$X_\Delta = \frac{Vy_i^T - Lx_i^T}{\Delta}; \quad R_\Delta = \frac{L}{\Delta}; \quad \Delta = V - L \tag{9.15}$$

Equations 9.14 and 9.15 for the rectifying and stripping section (Sections 9.1 and 9.3) are analogous to balance equations derived by Barbosa and Doherty [49]. For every section of the distillation column (shown in Figure 9.13) it is possible to plot a liquid composition profile in a mass balance triangle (MBT) for ternary mixtures.

As an example, a conventional distillation column and an ECDP column are evaluated for the separation of a mixture of acetone-isopropanol-water. As aforementioned, the azeotropes formation produces two regions in the composition space. Due to the distillation boundary, conventional distillation is constrained to operate in one of the two regions. In Figure 9.14a, it is shown that an unfeasible distillation column with feed and product compositions located inside region 2 for a distillate to feed molar flow ratio of D/F = 0.2. Although the mass balance is met and the feed and

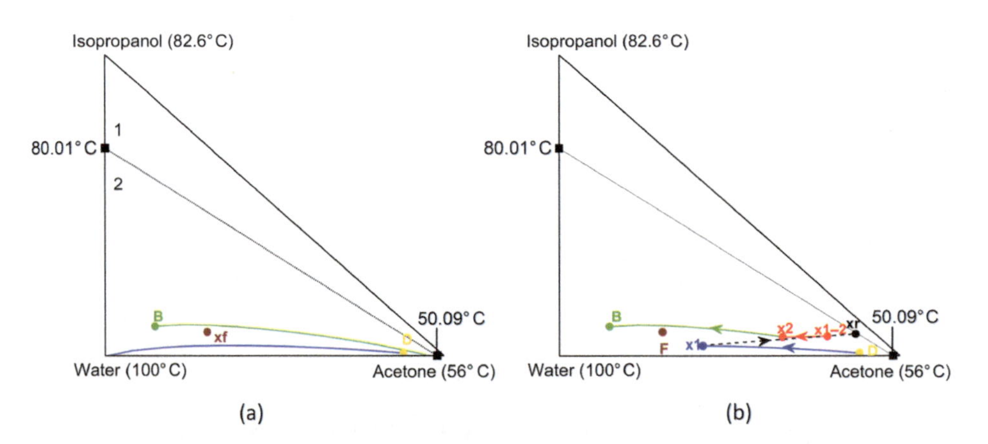

Figure 9.14: Column profiles for separation of a mixture of isopropanol, acetone and water. (a) Conventional distillation and (b) externally connected distillation pervaporation (ECDP) system according to Figure 9.13 using a water-selective pervaporation membrane (Note: nomenclature according to Figure 9.13).

product compositions are located in the same region, the separation cannot be achieved. A feasible separation could be achieved by increasing the isopropanol composition in D or reducing it in B, so both composition profiles intercept.
On the other hand, an ECDP column can achieve a feasible separation with the same D/F ratio, feed composition, and product specifications (for D and B) shown in Figure 9.14a. A selective water membrane (infinite selectivity) is used to dehydrate a side stream with a composition x_1 (Figure 9.13), and 90% of L_1 is fed to the pervaporation membrane. For the sake of simplicity, a 95% water removal from the pervaporation feed is specified, which results in a retentate with a low water concentration (x_r). The retentate with a composition x_r, is fed back to the distillation column on top of Section 9.2 and mixed with the remaining liquid that comes from Section 9.1 with a composition x_1. The final composition of the liquid inlet stream to Section 9.2 ($L_{1\text{-}2}$) is $x_{1\text{-}2}$, which is located on a straight line between x_1 and x_r (dashed line in Figure 9.14b). Table 9.2 shows the product compositions presented in Figure 9.13a and that also correspond to those presented in Figure 9.14b. The composition profiles are presented in Figure 9.14b for Section 9.1, 9.2, and 9.3 in blue, red, and green, respectively. The dashed line describes the influence of the pervaporation unit on the liquid concentration profile in the column.

Since Section 9.2 is also a rectifying section, the liquid phase L_2 is slightly enriched with water (Figure 9.14b). This latter arises possibilities of achieving an interception between the liquid composition profiles of rectifying and stripping sections, in the point x_2. In the stripping section, acetone is separated until it reaches the product composition B. The pervaporation unit provides a "bridge" to connect the liquid composition profiles of stripping and rectifying sections that are shown in Figure 9.14b by the red and dashed lines. This "bridge" depends on membrane selectivity and the fraction of liquid that is taken from the column as a side stream to the pervaporation unit. A similar design can be implemented using vapor permeation instead of pervaporation.

Table 9.2: Molar fractions of streams of Figures 9.13 and 9.14b for an externally connected distillation and pervaporation (ECDP) system.

Stream	Molar compositions		
	Acetone	Isopropanol	Water
F	0.3	0.08	0.62
D[1]	0.88	0.01	0.11
Xr	0.87	0.073	0.057
x1-2	0.786	0.065	0.149
x2	0.653	0.064	0.283
B	0.15	0.1	0.75

[1]Consider D is the liquid produced in the partial condenser and recycled to the column, which is in equilibrium with the distillate at the top of the column (vapor).

9.6.5 Hybrid distillation and pervaporation systems in a single column

In a hybrid distillation-pervaporation system in a single unit (DPSU), the membrane is located inside the distillation column. Figure 9.15a shows a set of membrane tubes with several layers that are perpendicular to each other to be placed within the column. The selective layer of membrane is placed on their external surface. Thus, liquid and vapor flow in countercurrent and externally to the membrane tubes while the permeate is removed from the internal lumen of the tubes by a vacuum jacket (Figure 9.15b).

Unlike ECDP column, in a DPSU column, the pervaporation and distillation mechanisms occur simultaneously in the same section. Therefore, an alternative

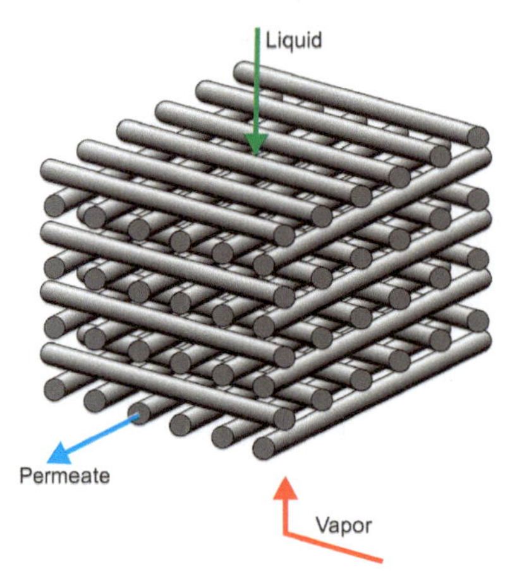

Figure 9.15a: Set of membrane tubes used inside a distillation pervaporation column in a single Unit (DPSU).

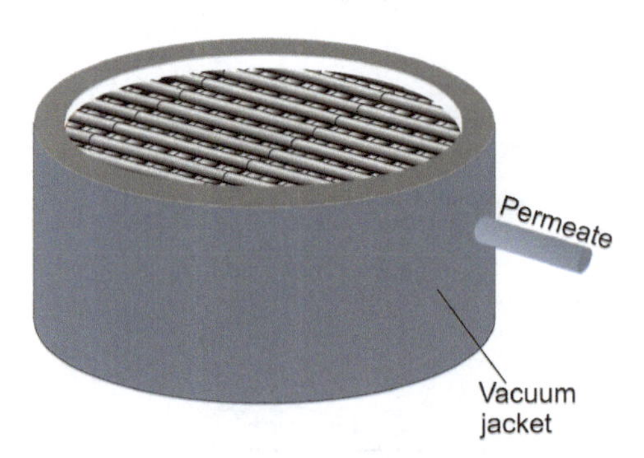

Figure 9.15b: Set of membrane tubes with its jacked for permeate removal used inside of a distillation pervaporation column in a single unit (DPSU).

mathematical model for the DPSU column is proposed to introduce the membrane flux to a distillation process applying transformed compositions.

An important tool that is used for mass balances is the transformed composition. Commonly, the transformed compositions are applied for the graphical interpretation of a mixture separation, where more than four components are involved. Additionally, they allow reducing the degrees of freedom with several simultaneous operations. The transformed compositions have been extensively used in reactive distillation to develop simplified models based on phase equilibrium, avoiding to introduce mass and energy transfer rates in the equations [50, 51].

The number of moles of any component in a mixture of C components in a pervaporation system is:

$$n_i = n_i^0 - N_i(t), \qquad i = 1, \ldots, C \tag{9.16}$$

where (t) is the time in a batch pervaporation or membrane area in a steady-state pervaporation. t is an independent variable in the system. Besides, n_i^0 is the initial number of moles of component i in the retentate, $n_i$is the remaining number of moles of i in the retentate for a value of $t \geq 0$ and $N_i(t)$is the number of moles i that have been permeated as a function of t. Also, $N_i(t)$ is defined by the total membrane flux and permeate composition:

$$N_i(t) = \sum_{t=0}^{t} x_{i,P}(t)J(t)A\Delta t \tag{9.17}$$

Where $x_{i,p}(t)$ is the permeate composition in function of t, $J(t)$ is the total membrane flux, A is the total membrane area, and $\Delta t$is the time difference. For a selective membrane, the permeate composition can be assumed constant. Then, Eq. 9.17 can be written as:

$$N_i(t) = x_{i,P} \sum_{t=0}^{t} J(t)A\Delta t \tag{9.18}$$

By a simplification in Eq. 9.18, $\beta = \beta(t) = \sum_{t=0}^{t} J(t)A\Delta t$.

Substituting β in Eq. 9.16:

$$n_i = n_i^0 - x_{i,P}\beta \tag{9.19}$$

The total number of moles in the retentate as function of t is:

$$n_T = n_T^0 - \beta \tag{9.20}$$

Combining Eqs. 9.19 and 9.20, the mole fraction in the retentate can be determined:

$$x_i = \frac{n_i}{n_T} = \frac{n_i^0 - x_{i,P}\beta}{n_T^0 - \beta} = \left[\frac{\frac{n_i^0}{n_T^0} - x_{i,P}\left(\frac{\beta}{n_T^0}\right)}{1 - \left(\frac{\beta}{n_T^0}\right)}\right] \tag{9.21}$$

In Eq. 9.21, the followed simplification can be introduced $\phi = (\beta/n_T^0)$, such that:

$$x_i = \frac{x_i^0 - x_{i,P}\phi}{1-\phi} \tag{9.22}$$

According to Eq. 9.22, *C+1* degrees of freedom are generated due to the introduction of the removed fraction ϕ and the mole fraction in permeate $x_{i,P}$. If the functions $x_{i,P} = f(x_i)$ and $\sum x_{i,P} = 1$ are specified, only a degree of freedom remains. A reference component k must be introduced to eliminate ϕ. The mole fraction in the retentate for a reference component k with $\alpha = 1-\phi$ is:

$$x_k = \left(\frac{x_k^0 - x_{k,P}\phi}{\alpha}\right) \tag{9.23}$$

Reorganizing Eq. 9.23:

$$\frac{\phi}{\alpha} = x_{k,P}^{-1}\left(\frac{x_k^0}{\alpha} - x_k\right) \tag{9.24}$$

Substituting Eqs. 9.24 in 9.22 and rearranging:

$$x_i - x_{i,P}x_{k,P}^{-1}x_k = \frac{x_i^0 - x_{i,P}x_{k,P}^{-1}x_k^0}{\alpha} \tag{9.25}$$

Reorganizing Eq. 9.24, ϕ can be defined:

$$\phi = x_{k,P}^{-1}\left(x_k^0 - \alpha x_k\right) \tag{9.26}$$

Substituting ϕ in Eq. 9.26 with $\alpha = 1-\phi$, and rearranging:

$$\alpha = \frac{1 - x_{k,P}^{-1}x_k^0}{1 - x_{k.P}^{-1}x_k} \tag{9.27}$$

Then, Eq. 9.27 is replaced in 9.25 and rearranged to obtain:

$$X_i = \frac{x_i - x_{i,P}x_{k,P}^{-1}x_k}{1 - x_{k,P}^{-1}x_k} = \frac{x_i^0 - x_{i,P}x_{k,P}^{-1}x_k^0}{1 - x_{k,P}^{-1}x_k^0} \tag{9.28}$$

where X_i is the transformed composition for independent *C – 2* components in the mixture, analogous to a chemical reaction [51]. The mole fractions in the right side of Eq. 9.28 are constant.

For the dependent component:

$$\sum_{i=1}^{C-1} X_i = 1 \tag{9.29}$$

Besides, for the reference component k the transformed composition is $X_k = 0$. For pervaporation membranes, the permeate composition is [34]:

$$x_{i,P} = \frac{S_{ij} f_i^R}{\sum_{\substack{i=1 \\ i \neq j}}^{C} S_{ij} f_i^R} \tag{9.30}$$

where S_{ij} is the membrane selectivity of i in comparison to j and f_i^R is the fugacity of i in the retentate.

Figure 9.16 shows a diagram for a DPSU column considering a membrane located inside the rectifying section. The following assumptions are considered for a mathematical model of a DPSU column with a hybrid rectifying-pervaporation section (R-MS):

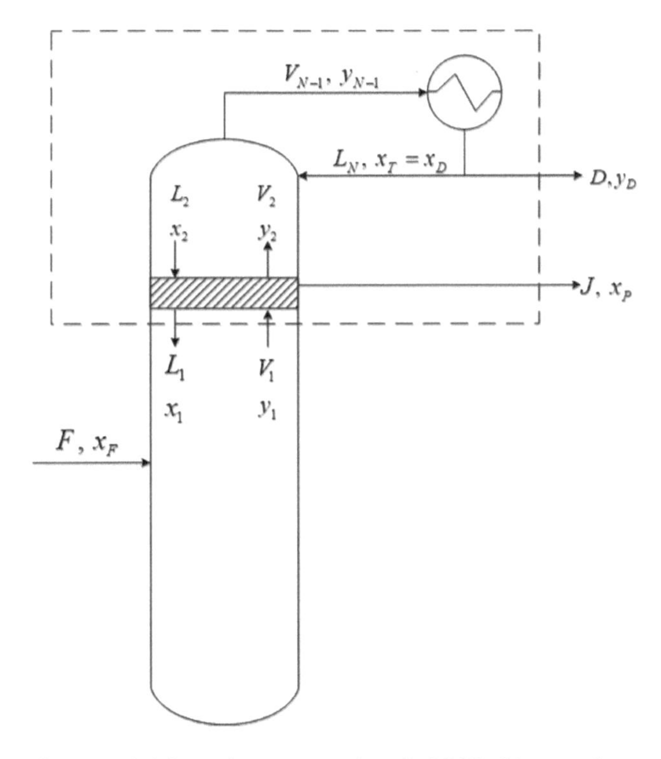

Figure 9.16: Schematic representation of a DPSU with a membrane section located in the rectifying section.

- The simplified model is an equilibrium model (EQ). Therefore, no mass transfer rates are considered.
- Equimolar mass transfer between the liquid and vapor phase.
- Only pervaporation is the separation mechanism carried out by the membrane. The membrane walls are completely wet by the internal liquid stream.
- Constant molar overflow (CMO) is not assumed for the liquid phase, due to the removal by the pervaporation.
- Constant molar overflow (CMO) is assumed for the vapor phase.
- Pervaporation and distillation mechanisms occur simultaneously inside the hybrid rectifying-pervaporation section.
- Partial condenser is assumed.

The mass balance around the control volume in Figure 9.16 for the component i is:

$$V_1 y_{i,1} = L_1 x_{i,1} + D y_{i,1} + J x_{i,P} \tag{9.31}$$

For the reference component k:

$$V_1 y_{k,1} = L_1 x_{k,1} + D y_{k,1} + J x_{k,P} \tag{9.32}$$

The Eq. 9.32 can be rearranged:

$$J = \frac{V_1 y_{k,1} - L_1 x_{k,1} - D y_{k,1}}{x_{k,P}} \tag{9.33}$$

The total membrane flux J is common for the component i and k; then it can be eliminated by combining Eqs. (9.31) with (9.33):

$$V_1\left(y_{i,1} - x_{i,P} x_{k,P}^{-1} y_{k,1}\right) = L_1\left(x_{i,1} - x_{i,P} x_{k,P}^{-1} x_{k,1}\right) + D\left(y_{i,D} - x_{i,P} x_{k,P}^{-1} y_{k,D}\right) \tag{9.34}$$

The terms inside the brackets can be replaced by Eq. (9.28). Also, for the vapor phase, transformed compositions are used:

$$V_1\left(1 - x_{k,P}^{-1} y_{k,1}\right) Y_{i,1} = L_1\left(1 - x_{k,P}^{-1} x_{k,1}\right) X_{i,1} + D\left(1 - x_{k,P}^{-1} y_{k,D}\right) Y_{i,D} \tag{9.35}$$

Equation 9.35 is the operation line in transformed variables of a hybrid rectifying-pervaporation section of a DPSU. The internal transformed reflux ratios can be defined as:

$$\bar{R}_n = \frac{L_1\left(1 - x_{k,P}^{-1} x_{k,1}\right)}{D\left(1 - x_{k,P}^{-1} x_{k,D}\right)} = \frac{\bar{L}_1}{\bar{D}} \tag{9.36}$$

The vapor stream across the R-MS is constant and the overall mass balance in transformed variables based on Eq. 9.35 is:

$$V_1\left(1-x_{k,P}^{-1}y_{k,1}\right)=L_1\left(1-x_{k,P}^{-1}x_{k,1}\right)+D\left(1-x_{k,P}^{-1}y_{k,D}\right)\rightarrow\bar{V}_1=\bar{D}+\bar{L}_1 \quad (9.37)$$

Combining Eqs. 9.35 to 9.37, the operation equation for a R-MS is:

$$X_{i,1}=\left(\frac{\bar{R}_n+1}{\bar{R}_n}\right)Y_{i,1}-\left(\frac{1}{\bar{R}_n}\right)Y_{i,D} \quad (9.38)$$

Due to $X_1-X_2=\Delta X$, thus:

$$X_{i,1}-X_{i,2}=\left(\frac{\bar{R}_n+1}{\bar{R}_n}\right)Y_{i,1}-X_{i,2}-\left(\frac{1}{\bar{R}_n}\right)Y_{i,D} \quad (9.39)$$

Approximating Eq. 9.39 by the first derivative $(dX/dh'_R)_{X_2}$and inverting the sign for an incremental integration from the top to the feed:

$$\frac{dX_i}{dh'_R}=X_i-\frac{\bar{R}_n+1}{\bar{R}_n}Y_i+\frac{1}{\bar{R}_n}Y_{i,D} \quad (9.40)$$

The initial conditions of Eq. 9.40 are:

$$h'_R=h'_{R,0};\ X_i(h'_{R,0})=X_{i,D}$$

An overall mass balance around the partial condenser is:

$$V_1\left(1-x_{k,P}^{-1}y_{k,N-1}\right)=L_N\left(1-x_{k,P}^{-1}x_{k,D}\right)+D\left(1-x_{k,P}^{-1}y_{k,N-1}\right) \quad (9.41)$$

If Eq. 9.41 is used to eliminate V_1 from Eq. 9.37:

$$D\left(\frac{1-x_{k,P}^{-1}y_{k,D}}{1-x_{k,P}^{-1}y_{k,N-1}}\right)+L_N\left(\frac{1-x_{k,P}^{-1}x_{k,D}}{1-x_{k,P}^{-1}y_{k,N-1}}\right)=D\left(\frac{1-x_{k,P}^{-1}y_{k,D}}{1-x_{k,P}^{-1}y_{k,1}}\right)+L_1\left(\frac{1-x_{k,P}^{-1}x_{k,1}}{1-x_{k,P}^{-1}y_{k1}}\right) \quad (9.42)$$

Rewriting Eq. 9.42:

$$\frac{L_1\left(1-x_{k,P}^{-1}x_{k,1}\right)}{D\left(1-x_{k,P}^{-1}y_{k,D}\right)}+1=\left(\frac{1-x_{k,P}^{-1}y_{k,1}}{1-x_{k,P}^{-1}y_{k,N-1}}\right)\left[1+\frac{L_N}{D}\left(\frac{1-x_{k,P}^{-1}x_{k,D}}{1-x_{k,P}^{-1}y_{k,D}}\right)\right] \quad (9.43)$$

The transformed external reflux is:

$$\bar{R}_{Ex}=\frac{L_N}{D}\frac{1-x_{k,P}^{-1}x_{k,D}}{1-x_{k,P}^{-1}y_{k,D}}=R_{Ex}\frac{1-x_{k,P}^{-1}x_{k,D}}{1-x_{k,P}^{-1}y_{k,D}} \quad (9.44)$$

Replacing Eqs. 9.36 and 9.44 in 9.43, the transformed internal reflux ratio is obtained:

$$\bar{R}_n + 1 = \frac{1 - x_{k,P}^{-1} y_{k,1}}{1 - x_{k,P}^{-1} y_{k,N-1}} \left(\bar{R}_{Ex} + 1\right) \tag{9.45}$$

For a conventional distillation, the liquid composition profile in the rectifying section is:

$$\frac{dx_i}{dh_R} = x_i - \frac{R_{Ex} + 1}{R_{Ex}} y_i + \frac{1}{R_{Ex}} y_{i,D} \tag{9.46}$$

With the initial condition:

$x_i(h_R = h_{R,0}) = x_{i,D}$ and $y_{i,D} = K_i x_{i,D}$

Also, for a conventional distillation the corresponding liquid composition profile in the stripping section is:

$$\frac{dx_i}{dh_S} = \frac{S}{S+1} y_i - x_i + \frac{1}{S+1} x_{i,B} \tag{9.47}$$

With the initial condition:

$$x_i(h_S = h_{S,0}) = x_{i,B}$$

Case study: acetone-isopropanol-water mixture

The same mixture to the one presented in Section 9.9 is used in this section (Figure 9.14a). According to the liquid-vapor equilibria shown in Figure 9.14a, a conventional distillation with feed composition located in region 1 would be able to produce a distillate rich in acetone. Due to the distillation boundary, it is not

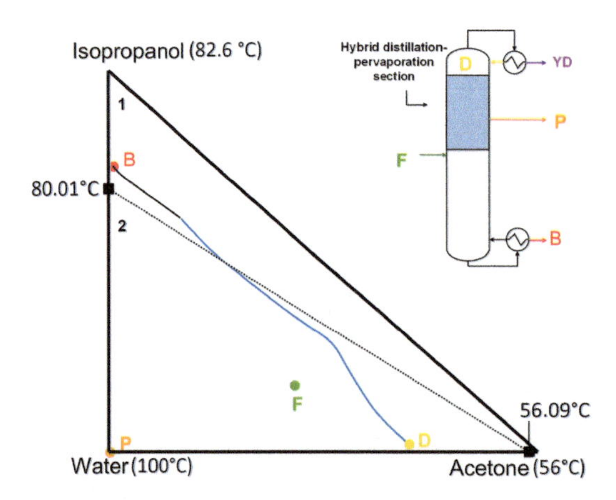

Figure 9.17: DPSU column with D in region 1 and B in region 2 at R = 5 and S = 7. Data presented in Table 9.3.

possible to achieve a feasible separation with D and B located in different regions. A high separation of the acetone-isopropanol-water mixture requires a sequence of distillation columns for instance by azeotropic distillation or ECDP.

Using a DPSU column, the liquid composition is controlled by the liquid-vapor equilibria but also by the pervaporation process. This example shows that it is possible to overcome the distillation boundary using a water-selective membrane in a DPSU column, as Figure 9.17 shows. In the rectifying section, the liquid phase is enriched in isopropanol, which is the component with intermediate volatility in region 2. Since water is removed by pervaporation (heavy component in region 2 according to Figure 9.14a), the liquid composition does not achieve the pure water vertex through the rectifying section. On the contrary, it tends to achieve the pure isopropanol vertex, even if this is located beyond the distillation boundary. Isopropanol is the heavy component in region 1. Although it is not shown in Figure 9.17, the distillation lines in the rectifying section follow analogous trajectories to ideal homogeneous mixtures where not azeotropes are present [52]. Therefore, it is possible to obtain a feasible separation with product compositions located in two regions using a water-selective membrane in a DPSU column. After the liquid composition achieves region 1, a conventional stripping section can be used to produce a bottom stream that is rich in isopropanol. The distillation boundary can be overcome depending on membrane selectivity and also the volatility order.

For the type of topology presented in Figure 9.14a, the distillation boundary can be overcome by removing through the pervaporation of the heaviest component in the region. In this way, the stable node of one region (e.g., isopropanol) can be turned into a stable node of the whole topology and water is no longer a stable node of region 2. On the contrary, if the component with intermediate volatility in region 2 (isopropanol) is removed the stationary point in that region is not shifted. Unlike to ECDP columns, the distillation boundary is overcome by a continuous composition profile.

As an example, Table 9.3 shows the flows and compositions for a DPSU column using a water-selective membrane. For this PDSU configuration, 99% of acetone and 97% of isopropanol are recovered in the distillation and bottom streams, respectively.

Table 9.3: Mass balance of the output and input streams for a DPSU column with a water-selective membrane.

Stream	Flow (kmol/h)	Molar compositions		
		Acetone	Isopropanol	Water
Feed	100	0.437	0.174	0.389
Distillate	50	0.87	0.016	0.114
Bottoms	22.56	0.01	0.75	0.24
Permeate	27.44	<0.001	< 0.001	>0.999

The fraction of the feed removed by the membrane and the membrane area depend explicitly on the feed composition. In this case study, the feed composition is located in region 2 relatively away from the distillation boundary (Figure 9.17), and thus, 27% of the feed flow requires to be removed by pervaporation for overcoming the distillation boundary. In general, to overcome the distillation boundary, the closer is the feed (located in region 2) to the distillation boundary, the lower the amount of water to be removed by pervaporation will be. Therefore, the membrane area is reduced by decreasing the water composition in the feed.

9.7 Nomenclature

A: Membrane area [m^2]
B: Bottom stream [$kmol\ h^{-1}$]
D: Distillate stream [$kmol\ h^{-1}$]
$\bar{D}$: Distillate stream [$kmol\ h^{-1}$]
F: Feed stream [$kmol\ h^{-1}$]
h_R': Dimensionless column height of R-MS
h_S: Dimensionless column height of SS
J: Membrane total flux [$kmol\ h^{-1}$]
K: Distribution coefficient
L: Liquid stream inside the column [$kmol\ h^{-1}$]
$\bar{L}$: Transformed liquid stream inside the column [$kmol\ h^{-1}$]
N: Number of moles removed by pervaporation
n: Number of moles in the retentate
R_{Ex}: External reflux L_N/D
R_n: Internal reflux L_n/D
$\bar{R}_{Ex}$: External transformed reflux
$\bar{R}_n$: Internal transformed reflux
R_Δ: Reflux ratio
S: Boil-up ratio
S_{ij}: Membrane selectivity
t: Time [h]
V: Vapor stream inside the column [$kmol\ h^{-1}$]
$\bar{V}$: Transformed vapor stream inside the column [$kmol\ h^{-1}$]
X: Transformed mole composition in the liquid
x: Mole composition in the liquid
x_P: Mole composition in permeate
x_i^T: Molar fraction of the liquid at the top of the section
X_Δ: Difference point
y_i^T: Molar fraction of the vapor at the top of the section
Y: Transformed mole composition in the vapor
y: Mole composition in the vapor

Symbols

α: Fraction of initial number of moles remaining in the retentate
ε: Reaction advance [kmol]
ϕ: Fraction of the initial number of moles permeated by pervaporation $\phi = 1 - \alpha$
γ: Activity coefficient in the liquid
ν: Stoichiometric coefficient
Δ: Net molar flow in a section

Subscript

i: Chemical specie i
N: Position on top of the R-MS height
n: Position on the R-MS height
$N-1$: Position before condenser
o: Initial
T: Total

9.8 Exercises

1. Discuss situations or applications where the hybrid membrane-distillation processes presented in Figure 9.1 are convenient.
2. Figure 9.1c presents a hybrid distillation and membrane process where the membrane unit is used as a final treatment step. Discuss advantages and disadvantages of using pervaporation or vapor permeation as a membrane technology in the hybrid configuration.
3. List all the capital and operating cost associated with a pervaporation system according to Figure 9.2. Prepare the same list for a vapor permeation system. What are the differences between these two membrane technologies regarding operating and capital costs?
4. Design a hybrid distillation-pervaporation system for separating an equimolar mixture of water and acetonitrile using the hybrid configuration presented in Figure 9.1e and a McCabe-Thiele approach. The pervaporation membrane is water selective with permeances reported in [15].
5. Ethyl acetate-ethanol-water mixture presents three distillation regions. Using a triangular diagram present a conceptual design for the purification of the mixture if the feed is located in one of the distillation regions. Develop conceptual designs for a feed located in each distillation region. Use the membranes that are selective to the component that you consider more convenient for every case.

6. Design a hybrid pervaporation-distillation system for separating a mixture of methanol, 1-butene and methyl t-butyl ether (MTBE) with molar fractions of 0.45, 0.09, and 0.46, respectively. The hybrid configuration is presented in Figure 9.13. Use a membrane with an infinite selectivity to methanol. Calculate a suitable distillation ratio and the liquid composition profiles for the three sections of the distillation column to have a feasible operation. The expected product streams are high purity MTBE and a butene-methanol mixture close to the azeotropic composition.
7. Considering the acetone-isopropanol-water mixture shown in Figure 9.17, use a DPSU column with a water-selective membrane to achieve a feasible separation. Use R and S > 10.
8. Based on the example shown in Section 9.10, it is possible to achieve a feasible separation with an isopropanol selective membrane in a DPSU column. Explain the reason.
9. In what region should be located D and B in the example of Section 9.10 to achieve a feasible separation with an isopropanol selective membrane? What should be the feed composition? Determine the composition profiles of the entire DPSU column with R = 5 and S = 10. What is the fraction of isopropanol fed that should be removed by pervaporation?
10. How the volatility order affects the separation in a DPSU column? Use Figure 9.17 for your analysis.

References

[1] Seader JD, Henley EJ. Separation Process Principles. New York, USA, John Wiley & Sons, 1998.
[2] Skiborowski M, Harwardt A, Marquardt W. Conceptual Design of Distillation-Based Hybrid Separation Processes. Annu Rev Chem Biomol Eng 2013, 4, 45–68.
[3] Madenoor Ramapriya G, Tawarmalani M, Agrawal R. A systematic method to synthesize all dividing wall columns for n -component separation-Part I. AIChE J 2018, 64(2), 649–59.
[4] Matsuda K, Iwakabe K, Nakaiwa M. Recent Advances in Internally Heat-Integrated Distillation Columns (HIDiC) for Sustainable Development (Journal Review). J Chem Eng Japan 2012, 45(4/7), 363.
[5] Cortes Garcia GE, van der Schaaf J, Kiss AA. A review on process intensification in HiGee distillation. J Chem Technol Biotechnol 2017, 92(6), 1136–1156.
[6] Bîldea CS, Pătruţ C, Jørgensen SB, Abildskov J, Kiss AA. Cyclic distillation technology - a mini-review. J Chem Technol Biotechnol 2016, 91(5), 1215–1223.
[7] Harmsen GJ. Reactive distillation : The front-runner of industrial process intensification A full review of commercial applications, research, scale-up, design and operation. Chem Eng Process 2007, 46, 774–780.
[8] Weinfeld A, Owens SA, Eldridge RB. Reactive dividing wall columns : A comprehensive review. Chem Eng Process Process Intensif 2018, 123(November 2017), 20–33.
[9] Daviou M, Hoch PM, Eliceche AM. Design of membrane modules used in hybrid distillation / pervaporation systems. Ind Eng Chem Res 2004, 43, 3403–3412.

[10] Roizard D, Favre E, Horbez D. Improved Energy Efficiency of a Hybrid Pervaporation/ Distillation Process for Acetic Acid Production: Identification of Target Membrane Performances by Simulation. Ind Eng Chem Res 2014, 53(7768), 7779.
[11] Fontalvo J. A hybrid distillation - pervaporation system in a single unit for breaking distillation boundaries in multicomponent mixtures. In: DECHEMA, editor. 10th International Conference on Distillation & Absorption. Friedrichshafen (Germany), DECHEMA, 2014, 1–9.
[12] Ahmad SA, Lone SR. Hybrid Process (Pervaporation-Distillation): A Review. Int J Sci Eng Res 2012, 3(5), 1–5.
[13] Van Hoof V, Van den Abeele L, Buekenhoudt A, Dotremont C, Leysen R. Economic comparison between azeotropic distillation and different hybrid systems combining distillation with pervaporation for the dehydration of isopropanol. Sep Purif Technol 2004, 37(1), 33–49.
[14] Fontalvo J. Separation of multicomponent mixtures using an integrated distillation-pervaporation system in a single column. In: International Scientific Conference on Pervaporation, Vapor Permeation and Membrane Distillation. Torun (Poland), 2013.
[15] Fontalvo J, Cuellar P, Timmer JMK, Vorstman MAG, Wijers JG, Keurentjes JTF. Comparing pervaporation and vapor permeation hybrid distillation processes. Ind Eng Chem Res 2005, 44(14), 5259–5266.
[16] Baker RW. Membrane Technology and Applications. Second. West Sussex, John Wiley & Sons, Ltd., 2004.
[17] Lipnizki F, Field RW, Ten PK. Pervaporation-based hybrid process: a review of process design, applications and economics. J Memb Sci 1999, 153, 183–210.
[18] Pressly TG, Ng KM. A break – Even analysis of distillation-membrane hybrids. AIChE J 1998, 44(1), 93–105.
[19] Caballero J a., Grossmann IE, Keyvani M, Lenz ES. Design of Hybrid Distillation–Vapor Membrane Separation Systems. Ind Eng Chem Res 2009, 48(20), 9151–9162.
[20] Peters M, Kauchali S, Hildebrandt D, Glasser D. Separation of Methanol/butene/MTBE using hybrid distillation-membrane processes. IChemE Symp Ser 2006, 152, 152–161.
[21] Steegs PFM. Hybrid distillation / membrane unit processes for solvent dehydration. 2001,
[22] Sato K, Aoki K, Sugimoto K, et al. Dehydrating performance of commercial LTA zeolite membranes and application to fuel grade bio-ethanol production by hybrid distillation/vapor permeation process. Microporous Mesoporous Mater 2008, 115(1–2), 184–188.
[23] Stephan W, Noble RD, Koval CA. Design methodology for a membrane/distillation column hybrid process. J Memb Sci 1995, 99(3), 259–272.
[24] Pettersen T, Argo A, Noble RD, Koval CA. Design of combined membrane and distillation processes. Sep Technol 1996, 6, 175–187.
[25] Naidu Y, Malik RK. A generalized methodology for optimal configurations of hybrid distillation–pervaporation processes. Chem Eng Res Des 2011, 89(8), 1348–1361.
[26] Fontalvo J, Vorstman M a. G, Wijers JG, Keurentjes JTF. Heat supply and reduction of polarization effects in pervaporation by two-phase feed. J Memb Sci 2006, 279(1–2), 156–164.
[27] Fontalvo J, Fourcade E, Cuellar P, Wijers J, Keurentjes J. Study of the hydrodynamics in a pervaporation module and implications for the design of multi-tubular systems. J Memb Sci 2006, 281(1–2), 219–227.
[28] Favre E. Temperature polarization in Pervaporation. Desalination 2003, 154, 129–138.
[29] Fontalvo J, Keurentjes JTF. A hybrid distillation–pervaporation system in a single unit for breaking distillation boundaries in multicomponent mixtures. Chem Eng Res Des 2015, 99, 158–164.

[30] Fontalvo J, Keurentjes JTF, Wijers JG, Vorstman MAG. Pervaporation process and apparatus for carrying out same. 2007, WO/2007/03(EP1762295).
[31] Seader JD, Henley EJ. Separation process principles. New York, USA, John Wiley & Sons, 2006.
[32] Huang Y-S, Sundmacher K, Qi A, Schlünder E-U. Residue curve maps of reactive membrane separation. Chem Eng Sci 2004, 59, 2863–2879.
[33] Peters M, Kauchali S, Hildebrandt D, Glasser D. Derivation and properties of membrane residue curve maps. Ind Eng Chem Res 2006, 45, 9080–9087.
[34] Peters M, Kauchali S, Hildebrandt D, Glasser D. Application of membrane residue curve maps to batch and continuous processes. Ind Eng Chem Res 2008, 47(7), 2361–2376.
[35] Fontalvo J. Using user models in Matlab® within the Aspen Plus® interface with an Excel® link. Ing e Investig 2014, 34(2), 39–43.
[36] Hommerich U, Rautenbach R. Design and optimization of combined pervaporation / distillation processes for the production of MTBE. J Memb Sci 1998, 146, 53–64.
[37] Eliceche AM, Carolina Daviou M, Hoch PM, Ortiz Uribe I. Optimisation of azeotropic distillation columns combined with pervaporation membranes. Comput Chem Eng 2002, 26(4–5), 563–573.
[38] Kookos IK. Optimal design of membrane distillation column hybrid processes. Ind Eng Chem Res 2003, 42(8), 1731–1738.
[39] Wijmans J. The solution-diffusion model: a review. J Memb Sci 1995, 107(1–2), 1–21.
[40] Benes NE. Mass transport in thin supported silica membranes [Internet]. 2000,
[41] Choi JG, Do DD, Do HD. Surface diffusion of adsorbed molecules in porous media: Monolayer, multilayer, and capillary condensation regimes. Ind Eng Chem Res 2001, 40, 4005–4031.
[42] Duque Salazar AC, Gómez García MÁ, Fontalvo J, Jedrzejczyk M, Rynkowski JM, Dobrosz-Gómez I. Ethanol dehydration by pervaporation using microporous silica membranes. Desalin Water Treat 2013, 51(10–12), 2368–2376.
[43] Aguilar-Valencia DM, Gómez-García MÁ, Fontalvo J. Effect of pH, CO2, and High Glucose Concentrations on Polydimethylsiloxane Pervaporation Membranes for Ethanol Removal. Ind Eng Chem Res 2012, 51(27), 9328–9334.
[44] Peters TA, Fontalvo J, Vorstman MAG, et al. Hollow fibre microporous silica membranes for gas separation and pervaporation. Synthesis, performance and stability. J Memb Sci 2005, 248, 73–80.
[45] Bode E, Hoempler C. Transport resistances during pervaporation through a composite membrane: experiments and model calculations. J Memb Sci 1996, 113(1), 43–56.
[46] Peters TA, Benes NE, Keurentjes JTF. Hybrid ceramic-supported thin PVA pervaporation membranes: Long-term performance and thermal stability in the dehydration of alcohols. J Memb Sci 2008, 311(1–2), 7–11.
[47] Sankaranarayanan K, van der Kooi HJ, de Swaan Arons J. Efficiency and Sustainability in the Energy and Chemical Industries: Scientific Principles and Case Studies. 2nd Ed. New York, CRC Press, 2010.
[48] Tapp M, Holland ST, Hildebrandt D, Glasser D. Column Profile Maps. 1. Derivation and Interpretation. Ind Eng Chem Res 2004, 43, 364–374.
[49] Barbosa D, Doherty MF. The simple distillation of homogeneous reactive mixtures. Chem Eng Sci 1988, 43(3), 541–550.
[50] Ung S, Doherty M. Synthesis of reactive distillation systems with multiple equilibrium chemical reactions. Ind Eng Chem Res 1995, 2555–2565.
[51] Ung S, Doherty MF. Vapor-Liquid phase Equilibrium in System with Multiple Chemical Reactions. Chem Eng Sci 1995, 50(1), 23–48.

Paola Ibarra-Gonzalez and Ben-Guang Rong

10 Lignocellulosic biofuels process synthesis and intensification: Superstructure-based methodology

Abstract: Advanced biofuels from lignocellulosic biomass have been presented as a promising alternative to transportation fuels due to their many advantages over fossil fuels and first-generation biofuels. Some of these advantages are lower GHG emissions and minimal negative impacts on food production. However, nowadays, fossil fuels are still being considered the dominant source of transportation energy. This is because the production of advanced biofuels from lignocellulosic biomass is still in early stages of research and development and their success depends on the technology and total production costs. Synthesis and integration of new production facilities can reduce the production costs of biofuels and increase their viability. In this chapter, a systematic methodology framework based on rigorous simulations and a Mixed Integer Non-Linear Programming (MINLP) model for the conversion of lignocellulosic biomass to liquid (BtL) transportation fuels is presented. First, five process routes including thermochemical conversion, upgrading, and separation technologies are proposed. Then, the process simulator Aspen Plus V8.8 is used to perform rigorous simulation of the five process routes. The simulations and experimental data taken from the literature are used to predict conversion and separation factors, and capital and energy costs of unit operations. From the simulation results, the possibility of combining unit operations between the thermochemical routes, as well as mass and energy integration, are explored. Thereafter, the five process routes are interconnected and transformed into a processing superstructure. The superstructure is defined as a MINLP problem coded in GAMS 24.5.6, which sets the objective to minimize the total annual cost (TAC) of BtL fuels under different cases and integration scenarios. Under different product profile constraints, two network flowsheets are identified as optimal technology routes for the conversion of lignocellulosic biomass to biofuels, which are then rigorously simulated for benchmark purposes. From the two optimal case scenarios, different upgrading and separation configuration alternatives as well as process intensification possibilities are proposed. The results demonstrate that this methodology can explore and generate optimal total biofuels production processes.

Keywords: lignocellulosic biomass, biofuel, optimal synthesis, superstructure, MINLP optimization, integration, intensification

Paola Ibarra-Gonzalez, Ben-Guang Rong, Department of Chemical Engineering, Biotechnology and Environmental Technology, University of Southern Denmark, 5000 Odense, Denmark

https://doi.org/10.1515/9783110596120-010

10.1 Introduction

10.1.1 Worldwide transportation fuels consumption

Over the current decade, global transportation energy consumption has been dominated by two fuels: motor gasoline (including ethanol blends) and diesel (including biodiesel blends). Together, in 2012, these two fuels accounted for 75% of total delivered transportation energy use. From 2012 to 2040, the world transportation sector liquid fuels consumption is expected to grow by 36 quadrillion Btu, with diesel (including biodiesel) showing the largest gain (13 quadrillion Btu), jet fuel consumption increasing by 10 quadrillion Btu, and motor gasoline (including ethanol blends) increasing by 9 quadrillion Btu. However, fossil fuels share of total transportation energy is expected to decline from 96% in 2012 to 88% by 2040. Likewise, even though motor gasoline will remain the largest transportation fuel, its share of total transportation energy consumption is expected to decline from 39% in 2012 to 33% by 2040. On the other hand, the total transportation market share of diesel fuel (including biodiesel) will decline from 36% to 33% and the jet fuel share will increase from 12% to 14% by 2040 [1].

Moreover, it is crucial to analyze the impact of gasoline and diesel consumption in the transportation sector depending on the travel modes. For this matter, the transportation sector can be divided in two travel modes: passenger and freight modes. Motor gasoline is used primarily for the movement of people namely passenger modes, such as light-duty vehicles (cars, minivans, sport utility vehicles (SUVs), and commercial vehicles (LCVs)) and trucks, buses, 2- and 3-wheel vehicles, airplanes, and passenger trains [1, 2]. While diesel fuel is used primarily for the movement of raw, intermediate, and finished goods to consumers, i.e., freight modes. Freight modes include trucks (heavy-, medium-, and light-duty), marine vessels (international and domestic), rail, and pipelines [1].

In 2012, passenger-related fuel consumption accounted for 61% of total world transportation energy consumption. Among this mode of transport, light-duty vehicles accounted for 44% of total world transportation energy use, followed by aircraft at 11%. Buses, 2- and 3-wheel vehicles, and rail accounted for 6% of total world transportation energy use. Freight modes accounted for the other 39% of total world transportation energy consumption. Freight trucks made up by far the largest share (23%) of total transportation energy use, followed by marine vessels (12%) and rail and pipelines (4%). For passenger modes, total energy consumption rises by an average of 1.4%/year, while total freight-related energy consumption grows by an annual average of 1.5% [1].

Currently, worldwide, to provide the fuels needed by the transportation sector, petroleum and other liquid fuels such as natural gas to liquids, coal-to-liquids, and biomass-to-liquids are the dominant source of transportation energy. Petroleum products account for the largest share of transportation energy use by far, while non-petroleum fuels account for a small portion of the world energy mix. For instance, in 2017, petroleum products accounted for about 92% of the total US transportation sector energy use, as presented in Figure 10.1 [3]; while biofuels, such as ethanol and biodiesel, contributed only in about 5%. On the other hand, natural gas contributed nearly 3% and electricity provided less than 1% of total transportation sector energy use. Of the 5% accounted for biofuels in the transportation sector energy consumption, the ethanol's share was 4% and biodiesel's share was about 1% [4].

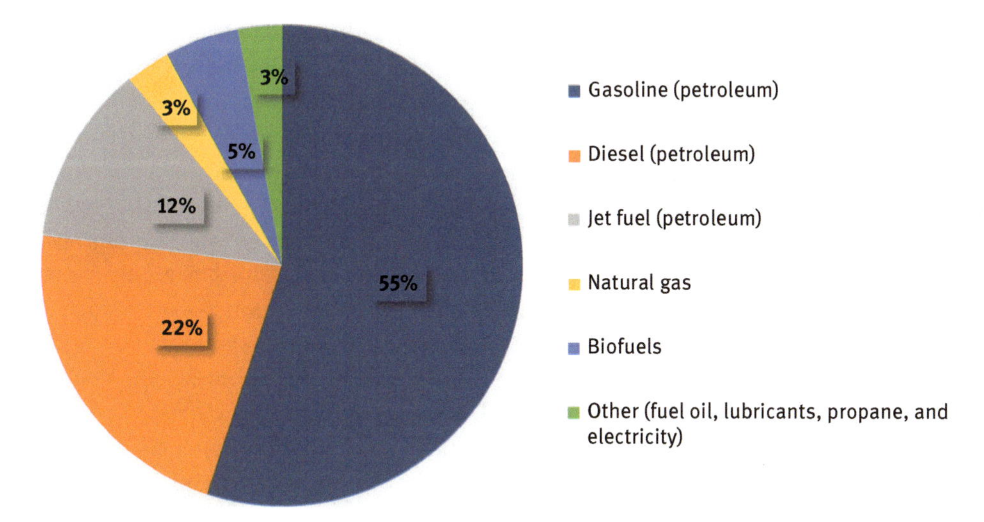

Figure 10.1: US transportation energy sources.

Furthermore, reducing the carbon intensity of the energy used to power vehicles is a long-term priority for the transportation sector. Governments around the world are adopting various policies to reduce energy consumption and greenhouse gas (GHG) emission from the transportation sector [5]. For example, countries like the USA and Mexico have both implemented fuel economy and GHG standards, which manufacturers must satisfy. These standards have focused on reducing GHG or carbon dioxide (CO_2) emissions, and on improving fuel economy (or reducing fuel consumption) [1]. However, in contrast to the proven ability of vehicle efficiency regulations to result in significant emission reductions, there is substantial uncertainty about the effectiveness of policies to accelerate the development and deployment of low-carbon biofuels [2].

10.1.2 Biofuels projection

While biofuel production has the potential to limit oil dependence, there is strong evidence that the first generation liquid biofuels that have achieved the greatest market penetration to date will not deliver substantial net carbon savings compared to fossil fuels [2]. Biofuels produced in the wrong way may require too much water, fertilizer, or land, and thus are not able to offset the GHG emissions. In addition, these biofuels may compete with food production and prevent conservation of the environment [6]. First generation biofuels production has led to concerns regarding the impact on the price and availability of the feedstocks as sources of food or animal feed [7]. Therefore, to ensure the feasibility of biofuels, several factors should be taken into consideration among which are energy security, food security, environmental and climate security, and sustainable development [6]. This means that the impact of the entire industrial biofuels production processes including feedstock production and handling, conversion technologies, and final applications should be considered.

Advanced biofuels produced from lignocellulosic biomass present advantages over conventional biofuels and fossil fuels. These can often be grown on lands that are unusable for food crops and are considered to have minimal negative impacts on food production [8]. Most of the technologies employed in the production of advanced biofuels already present a good potential to overcome technical bottlenecks by 2030, but they will not likely be cost-competitive with fossil fuels unless investments on research, development, and demonstration (RD&D) are performed [9]. Currently, the development of processes related to the production of advanced biofuels utilizing lignocellulosic biomass is still in the early stages of research and development [10]. RD&D is considered crucial to ensure technological advances and should be diversified among different technological processes. In addition, among RD&D, the development of new feedstocks, improvement of conversion processes and their efficiency, improvements in terms of feedstock pretreatment, advancement in the understanding of fuel properties, improvements in the design of machines, controlling mixed biomass feedstocks and co-processing of biomass and coal, and improvements on handling fuels and materials with different physical and chemical properties, as well as development of large-scale applications, construction of pilot projects, and process intensification should be explored to test their effects on the final production costs [9].

Therefore, the availability of biofuels in the future and the carbon benefits first generation fuels will deliver are unclear, and although significant progress has been made in the past years with respect to advanced biofuels with best environmental performance, they still present the least certainty about time to commercialization and rate of deployment [2, 9].

Thus, the development of advanced biofuel systems is critical to increase the potential of this resource in a sustainable and economic way. Synthesis and design of novel processes based on existing unit operations with a more efficient design is needed to reduce the high costs of these systems. For instance, the National Renewable Energy Laboratory

(NREL) has proposed and evaluated a thermochemical process for converting BtL transportation fuels via fast pyrolysis followed by hydroprocessing of pyrolysis oil using Aspen Plus as a simulation tool [11, 12]. Another study made by Betchel to support the NREL reports regarding the thermochemical conversion of BtL was developed with Aspen Plus process simulation model for a biomass-based gasification and Fischer Tropsch (FT) plant [13]. Moreover, Martin and Grossman proposed a superstructure for the production of diesel from switchgrass via gasification followed by FT. The optimization of the system was formulated as a MINLP problem for the optimal production of FT-diesel while minimizing the energy and hydrogen consumption [14]. Baliban et al. [15, 16] proposed an optimization-based process synthesis capable of analyzing distinct process designs of a hardwood BtL refinery considering the gasification of biomass and the FT hydrocarbons synthesis. Torres-Ortega et al. [17, 18] defined a synthesis network superstructure based on rigorous simulations and a MINLP-using-simplified model to optimize the integrated production of diesel and hydrotreated vegetable oil diesel.

Unfortunately, none of the previous works considered different thermochemical-based processes simultaneously. Moreover, the optimization-based works, used estimates from several literature sources to calculate the unit operation costs and energy costs. To the best of our knowledge, there are not studies that combine rigorous simulations and mathematical modeling approaches to explore several thermochemical processes to find the most economically and efficient design for the conversion of BtL transportation fuels. Additionally, besides Torres-Ortega et al. [17, 18] synthesis network approach, no paper uses rigorous simulations as performance and cost sources.

In light of the ongoing effort to improve products and processes, in this chapter, a systematic methodology including process synthesis, process modeling and simulation, process intensification and integration, and process evaluation and optimization is proposed for the conversion of lignocellulosic BtL transportation fuels. The systematic methodology framework combines rigorous simulations and a MINLP model for the synthesis of process routes for the conversion of lignocellulosic BtL transportation fuels. To achieve this, first, synthesis and design of five promising production processes for BtL fuels including thermochemical conversion, upgrading, and separation technologies are described. Then, the process simulator Aspen Plus V8.8 is used to perform rigorous simulation of the five process routes. Thereafter, the five process routes are transformed into a processing superstructure considering the data collected from the rigorous simulations. The superstructure is proposed and defined as a MINLP problem coded in GAMS 24.5.6, which sets the objective to minimize the TAC of BtL fuels under different cases and integration scenarios. Finally, for a set of given constraints, two optimal network flowsheets for specified product profiles are obtained and potential energy and/or mass integration are determined. Then, the optimal routes are rigorously simulated for benchmark purposes. As an addition, different upgrading and separation configuration alternatives for the optimal flowsheets as well as process intensification possibilities are proposed.

10.2 Lignocellulosic BtL transportation fuels production processes

10.2.1 Lignocellulosic biomass as source for BtL fuels

Moving the world market dependence away from fossil-based energy sources to renewable alternatives, such as lignocellulosic biomass, can be regarded as an important contribution toward the establishment of favorable conditions for the climate and sustainable economy [10]. Lignocellulosic biomass represents an extraordinarily large amount of renewable bioresource available in surplus on earth and is a suitable raw material for vast number of applications, such as resource for bioenergy production [19, 20].

There are indications that lignocellulosic biomass could well be the ultimate solution to raising the global demands for bioenergy industry. In addition, various assessments indicate that lignocellulosic biomass offers great potential as a resource for the provision of the future green transport fuels but also for the direct use in carbon sequestration in many parts of the world [21, 22].

Lignocellulosic biomass is the term used for biomass from woody or fibrous plant material being a combination of lignin, cellulose, and hemicellulose polymers interlinked in a heterogeneous matrix. The combined mass of cellulose and hemicellulose in the plant material varies with species but typically is around 50%–75% of the total dry mass with the remainder consisting of lignin. Lignocellulosic biomass is abundant, renewable, and comes from nonedible residues of food crop production or nonedible whole plant biomass. In general, lignocellulosic feedstocks are divided into three categories: agricultural residues, forest residues, and energy crops. These feedstocks include cereal, straw, wheat chaff, rice husks, corncobs, corn stover, sugarcane bagasse, nutshells, forest harvest, residue, wood process residues, and energy crops on marginal and degraded lands.

Second generation liquid biofuels from lignocellulosic biomass can be bred specifically for energy purposes, enabling higher production per unit land area, and a greater amount of aboveground plant material can be converted and used to produce biofuels. As a result, this will further increase land use efficiency compared to first-generation biofuels. Among other advantages, these fuels can come from a range of agricultural and wood-related residues without any direct claims on land. Land use efficiency is two to four times higher than first-generation biofuels. Likewise, a wider spectrum of land could be available for these feedstocks (grasslands, degraded land, marginal areas), which could lead to higher income for farmers [23]. Moreover, new and emerging biofuel technologies using lignocellulosic biomass have the potential to produce very low carbon biofuels.

10.2.2 Thermochemical conversion of lignocellulosic biomass

Thermochemical processes are of significant importance due to their ability to transform biomass into fluids, increase heating value, and enable easier handling, distribution, and storage. In recent years, research on using biomass for liquid fuels has been robust, ranging from studies of pyrolysis, hydrothermal liquefaction and gasification of lignocellulosic materials, and biomass-to-liquid technologies to upgrading processes [24].

The process selection is determined by the types of biomass used, the energy demand, and the applicability in either laboratory setups or industrial applications. For instance, a wide variety of lignocellulosic materials including agricultural wastes, forestry residues, grasses, and woody materials can be converted to biofuels through thermochemical conversion processes [25]. The thermochemical conversion of biomass to synthetic fuels is called biomass-to-liquid (BtL) where biomass is converted into synthesis gas and then to bio-oil. Currently, the major biofuels production processes are biomass conversion through pyrolysis and gasification followed by FT synthesis. These processes, besides biofuels production, can convert biomass into chemicals and power and therefore increase their feasibility. Furthermore, the thermochemical-based processes present many advantages over conventional biofuels and fossil fuels processes, among which are [26–28]:

- Faster conversion compared to biochemical conversions
- Physical pretreatment such as drying and grinding is needed, thus no requirement of chemical pretreatment
- Lower CO_2 emissions
- Zero emission of particulate matter
- Low NOx emissions
- Adjustable product quality (Properties such as octane and cetane number)
- Produce cleaner products and more efficient than the biomass they were derived from
- Based on well-established and mature commercial thermochemical conversion technologies.
- Able to convert types of lignocellulosic biomass (forestry and agricultural residues) that are difficult to handle using other conversion processes.

In general, the main objective of the thermochemical conversion technologies is to produce fuel components that are similar to those of current fossil-derived fuels and hence can be used in existing fuel distribution systems and with standard engines [29]. However, in order to be used as an alternative fuel, upgrading bio-oil is desirable.

10.2.3 Upgrading technologies to transportation fuels

The bio-oil produced from the thermochemical conversion has low energy density, high moisture content, high content of oxygen, and its physical form is not free flowing that creates a problem as a feedstock for reciprocating engines [26, 30]. Therefore, if bio-oil is to be used as an alternative fuel to conventional fossil fuels, performing an upgrading step is required to convert the bio-oil into a physically and chemically stable product with zero oxygen content and nearly chemically equivalent to petroleum [19]. Upgrading the bio-oil represents a challenge in finding a feasible path that involves proven technologies.

Upgrading technologies such as hydrotreating, hydrocracking, catalytic cracking, catalytic reforming, aromatic alkylation, olefin oligomerization, and so on can modify properties such as viscosity, density, heating value, oxygen, nitrogen and sulfur content, and chemical composition. Through these technologies, the products from the thermochemical conversion can be modified for meeting fuel standards and therefore be used pure or as blend in vehicles [27].

10.3 Systematic methodology framework: Synthesis of lignocellulosic BtL process routes

Process synthesis determines the optimal processing units and their interconnections, as well as the optimal design for the conversion of specific feedstocks into desired products. The synthesis of a system must be performed through a systematic process. In a systematic process, the feedstock and desired products are first defined. After that, the technologies for attaining the desired products are selected. Then, process design is carried out by identifying a set of interconnections between the feedstock, selected technologies, and products. Finally, from the different possible interconnections, several process routes that ensure the conversion of the feedstock to the desired products can be proposed.

The synthesis of a BtL production process can be performed by separating the synthesis problem in three detailed-level stages. In the first stage, individual technological routes are defined based on literature research and rigorous simulations. For the first stage, model compounds, stoichiometric reactions, yields, and operation conditions are used to evaluate the performance of the technological routes and to collect information regarding mass and energy balances, waste streams, possible mass and energy integrations, equipment sizing, and costs. Once sufficient information is collected, the second level stage can be performed. In the second stage, a superstructure optimization problem is defined including the most promising technological routes. The superstructure formulation follows the principle described at the beginning of the section, in which a set of interconnections between the technological routes are identified. Finally, the third stage uses the proposed superstructure as starting point and then uses rigorous models and sets of constraints to evaluate and find the network flowsheet that meets an objective function.

In addition, in each stage different synthesis tools can be considered. For example, in the second stage, different integration scenarios based on the waste streams information collected in the first stage can be included. In the third stage, possibilities of integration between units, i.e., intensification can be examined and then evaluated through rigorous models to confirm the feasibility and improvement of any structural changes in the processes.

For instance, the three stages systematic methodology, depicted in Figure 10.2, can be applied for the synthesis and evaluation of promising BtL process routes for the conversion of softwood biomass into transportation fuels.

In the case of study, different thermochemical-based process routes capable of converting spruce and pine forest residues into gasoline, diesel, and by-products (noncondensable gases, fuel gases, and aqueous products) are evaluated. First, the selection of the thermochemical conversion, upgrading, and separation technologies that enable the conversion of the softwood into the desired fuel products is performed. Then, the synthesis and design of the individual process routes is performed by sequencing or interconnecting the softwood biomass and desired fuel products with the thermochemical, upgrading, and separation technologies required for the conversion. From the different possible interconnections, process routes diagrams are proposed. The process simulator Aspen Plus V8.8 is used to perform rigorous simulation of the process routes. The information collected from the rigorous simulations (mass balances, reaction and separation performance, capital and utilities costs of processing blocks) is used to group the unit operations into reaction and separation blocks, leading to simplified processing blocks diagrams, which will facilitate the analysis of the possible combination networks between the thermochemical-based process routes. This is done by detecting which blocks perform the same tasks and present similar costs. The result is a superstructure for the conversion of softwood into gasoline, diesel, gases, and aqueous products. The final superstructure is formulated and defined as an MINLP problem coded in GAMS 24.5.6, which sets the objective to minimize the TAC's of BtL fuels under different constraints and integration scenarios. For the given constraints and objective function, optimal MINLP solutions were obtained and rigorously simulated for benchmark purposes. Finally, upgrading alternatives and possible intensifications are proposed for the optimal technological routes. The detailed systematic methodology is described in the next sections.

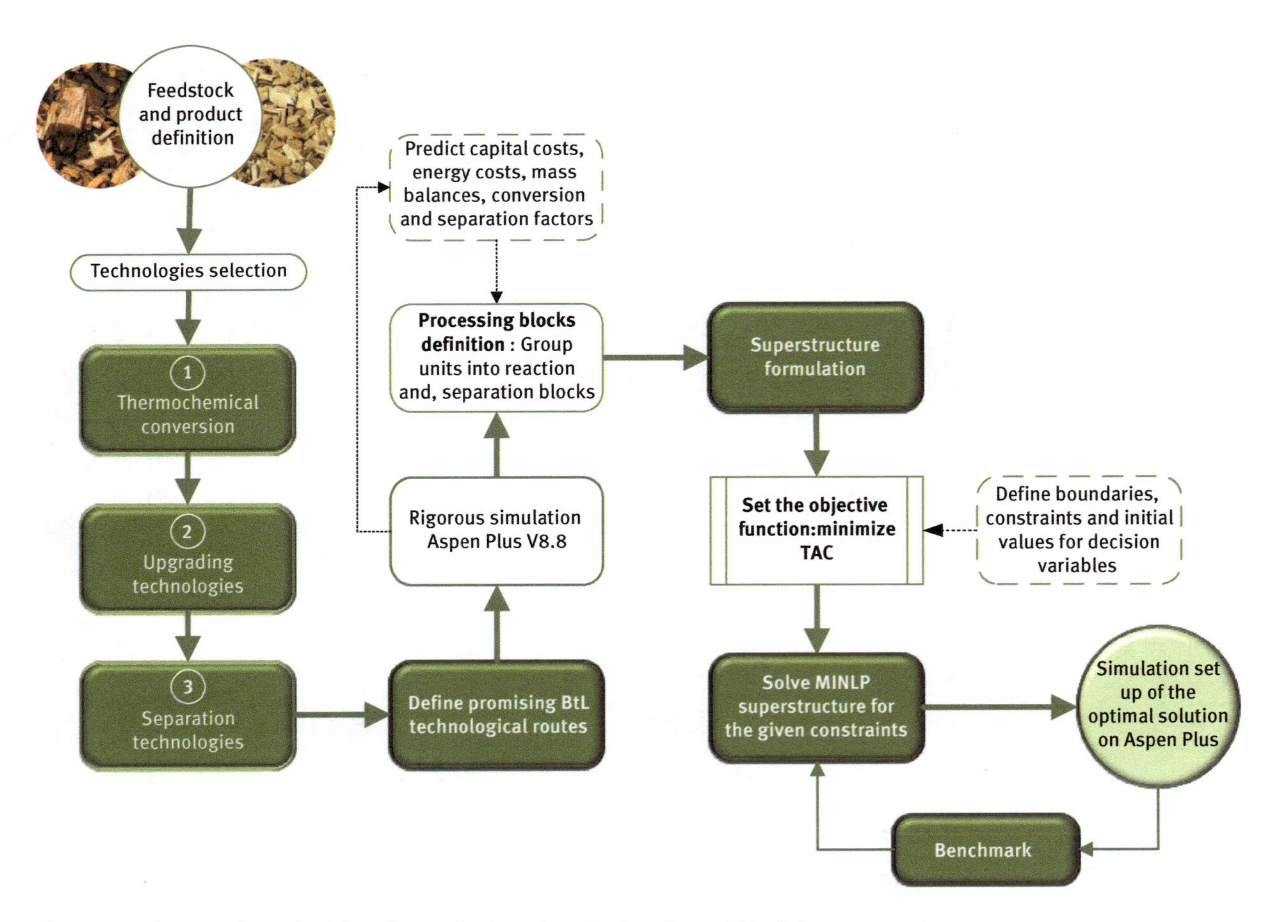

Figure 10.2: Systematic methodology for synthesis and evaluation of promising BtL process routes.

10.4 Synthesis, design, and evaluation of lignocellulosic BtL process routes

10.4.1 Synthesis of lignocellulosic BtL production process routes: Conceptual design

As explained before, the selection of feedstocks, products, and technologies is a major step in the synthesis and design of process routes. The feedstocks and the sequence of the technologies that converts them into the desired products is called process route. The process routes proposed in the case of study for the conversion of lignocellulosic biomass into biofuels are the result of the sequencing of three technological sections (Figure 10.3):

- *Thermochemical conversion*, gasification followed by low temperature and/or high temperature Fischer–Tropsch (LTFT and HTFT, respectively) or pyrolysis;
- *Upgrading technology*, FT syncrude fractional upgrading (aromatic alkylation, oligomerization, naphtha hydroisomerization, catalytic reforming, hydrotreating, hydrocracking, and hydrogenation), HTFT syncrude hydroprocessing, and pyrolysis oil hydroprocessing or catalytic cracking;
- *Separation technologies* such as distillation columns, solid separators, flash units, scrubbers, and fractionation columns.

The five process routes formulated by the combination of the different technological sections are;

- Gasification- HTFT-fractional upgrading-fractionation,
- Gasification-LTFT-fractional upgrading-fractionation,
- Gasification-HTFT-hydroprocessing,
- Pyrolysis-hydroprocessing-separation,
- Pyrolysis-catalytic cracking-fractionation.

The technological sections are initially combined for the synthesis of the five process routes and will be then interconnected for the superstructure formulation. The five process routes are described in the following sections.

10.4.1.1 Gasification- HTFT hydrocarbon production-FT hydrocarbon upgrading (PR-GHTUF)

In the gasification reactor, the previously dried and grinded biomass reacts with air, oxygen, or steam at temperatures of 800–1,000 °C, atmospheric pressure and residence time of 3–4 seconds [31]. The product raw gas is a gaseous mixture of CO, CO_2, H_2, CH_4, N_2, C_2H_4, C_2H_6, and ammonia, which is sent to a cyclone to remove solid particles. The raw gas is then cooled down in a scrubber by contact with a

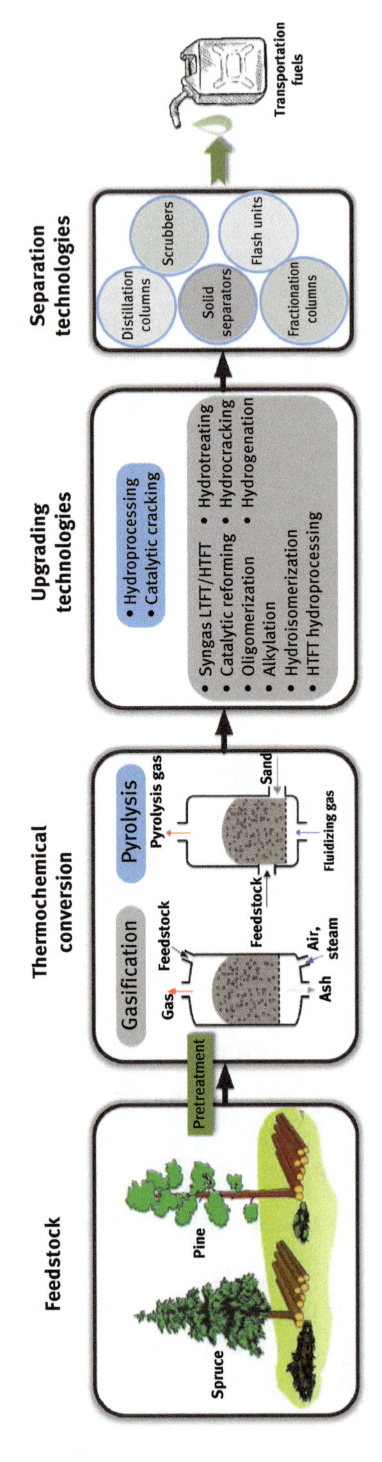

Figure 10.3: Technological sections for conversion of lignocellulosic BtL.

countercurrent stream flow of water, which also reduces the concentration of ammonia. After cleaning, the gas is sent to a water gas shift reactor (WGSR) to adjust the H_2/CO ratio [32] and the gas finally obtained is called syngas. The syngas can be used for heat and power production or can be upgraded to transportation fuels by adding processing steps, such as Fischer-Tropsch (FT) synthesis and product upgrading operations. With respect to FT, there are two working temperature ranges, known as the low temperature FT (LTFT) and high temperature FT (HTFT) regimes in which the syngas is by passed over supported metal catalysts (Fe, Co, Ru, Rh, and Ni) to produce hydrocarbons [33].

In the HTFT reactor the syngas is by passed over iron based catalyst such as Fe_2O_3/Cr_2O_3 to produce short chain hydrocarbons (mainly naphtha range products and gases) at reaction temperatures of 320–375 °C [33]. The FT synthesis produces a syncrude consisting of a mixture of linear hydrocarbons with similarities to crude oil. However, the syncrude is present as a multiphase mixture with up to four different phases and thus several separation units are required to recover the fractions. After fractions recovery, upgrading is required to convert them into a single phase "crude oil." The upgrading techniques considered in the present process route are aromatic alkylation of tail gas, liquefied petroleum gas (LPG) oligomerization, catalytic reforming of naphtha range products, and distillate and wax fraction hydroprocessing. On the other hand, the byproduct gases and aqueous fractions can be recovered and integrated to the process or considered as emissions.

10.4.1.2 Gasification- LTFT hydrocarbon production-FT hydrocarbon upgrading (PR-GLTUF)

In this process route, similarly to the Gasification-HTFT hydrocarbon production-FT hydrocarbon upgrading process route, the syngas is produced through gasification and is upgraded to hydrocarbons via FT reaction. The main difference relies on the type of FT reaction, which in this case is a LTFT reaction carried out at 200–250 °C over cobalt-based catalyst such as $Co\text{-}ZrO_2$. The temperature difference promotes the production of longer chain molecules in the distillate and wax range. Therefore, the product distribution is different and other upgrading units are required besides the upgrading technologies mentioned for the HTFT syncrude upgrading. For instance, the naphtha range product can be separated into C_5 and C_6–C_8 fractions. The C_5 fraction can be upgraded via hydroisomerization and the C_6–C_8 fraction via aromatization [33]. In addition, the amount of tail gas produced through this route is lower and therefore upgrading this fraction is not recommended.

10.4.1.3 Gasification- HTFT hydrocarbon production-FT hydrocarbon hydroprocessing (PR-GHTH)

The process route described in Section 10.4.1.1 shares with this process route the thermochemical conversion technology and the type of FT reactor. However, a different upgrading technique is considered. The FT product contains mainly three fractions: naphtha range product and distillate and wax products in lower percentages. Three individual hydroprocessing units to upgrade each of the fractions are needed: for hydrotreating the Fischer-Tropsch distillate and naphtha product and for hydrocracking the wax [13]. In the distillate hydrotreating unit, the FT distillate is catalytically hydrotreated to yield a high-quality diesel-blending component [34]. The naphtha product is catalytically hydrotreated under a hydrogen environment in the presence of a catalyst to yield saturated naphtha, which is further processed in a naphtha reformer to produce a high-octane gasoline-blending component. Finally, the wax is catalytically cracked in a hydrocracking unit under a high-pressure hydrogen environment to yield more desirable naphtha and distillate products. The naphtha obtained from hydrocracking is sent to an isomerization reactor to increase the stability of the products and to yield a high-octane gasoline blending component [13].

10.4.1.4 Pyrolysis-Hydroprocessing (PR-PHS)

In the pyrolysis reactor, the pretreated softwood is thermally decomposed in the absence of oxygen. The lignocellulosic biomass is heated to 500 °C by contact with hot sand and a fluidizing agent, the reaction is carried out in less than 2 seconds at atmospheric pressure [11]. The pyrolysis products: solid char, noncondensable gases (NCGs), and pyrolysis vapors are sent to a cyclone where solids removal is carried out. The gases are rapidly quenched in two quench columns and the pyrolysis oil is separated from the NCGs. The NCGs can be combusted and recycled back to the pyrolysis reactor to be used as fluidizing agent. The final pyrolysis oil, also called bio-oil, still contains some fine char that is inevitably carried over from the cyclone, and can only be removed by liquid filtration [35]. However, the final bio-oil is a complex mixture of water and hundreds of organic compounds such as acids, aldehydes, ketones, phenolics, alcohols, ethers, esters, anhydrosugars, furans, nitrogen compounds as well as large molecular oligomers. These characteristics make bio-oil a low-grade liquid fuel, highly oxygenated, acid and corrosive to common materials, thermally and chemically instable, as well as nonmiscible with petroleum fuels [36]. Therefore, the bio-oil needs to be upgraded before being considered as replacement for fossil fuels. Hydroprocessing rejects the oxygen as water by catalytic reaction with hydrogen. Multistage processing, where mild hydrotreating is followed by more severe hydrotreating, has been found to overcome the reactivity of the bio-oil and prevent catalyst coking [37, 38]. Therefore, pyrolysis bio-oil can be pretreated in a stabilization bed under relatively mild process conditions (140 to 180°C and 82 atm), followed by processing under more severe hydrotreating conditions in the first and second stage hydrotreating reactors (180 to 250 °C at 136 atm, and 350 to 425 °C at 136 atm, respectively) [11, 39]. The products from the last hydrotreating stage are gas, hydrotreated bio-oil, and an aqueous fraction. The compounds in the gas product are light hydrocarbons (CH_4, C_2H_6), carbon monoxide, and carbon dioxide. The aqueous fraction is mainly water with a low concentration of carbon dioxide. The upgraded oil obtained contains less than 2 wt% oxygen and is formed by hydrocarbons in the naphtha and diesel range, as well as a heavy fraction. The heavy fraction is sent to a hydrocracking reactor to be catalytically cracked to additional fuel. The product from hydrocracking is a mixture of liquids spanning the gasoline and diesel range and some byproduct gas [11].

10.4.1.5 Pyrolysis-Catalytic Cracking (PR-PCC)

The conversion of lignocellulosic biomass to pyrolysis oil and its recovery is carried out as described in the pyrolysis-hydroprocessing route. However, the upgrading of the highly oxygenated product is performed by zeolite cracking. In zeolite cracking, the oxygen is rejected as CO_2, yielding mainly aromatic hydrocarbons but with extensive coke deposition on the catalyst. Cracking and dehydration are the main reactions seen. The advantage of the use of zeolites relies on the production of aromatics at atmospheric pressures without H_2 requirements. The catalytic cracking is conducted at temperatures of 450 °C, atmospheric pressure, and reaction times of 15 minutes in the presence of N_2 gas to stabilize the product. Excessive carbon production is presented with yields of 18 % (wt) solids (coke, char, and tar) and thus catalyst coking is present. The other products are the upgraded oil, gases, and aqueous fraction, which are recovered by fractionation [40–42].

10.4.2 BtL process routes: Design and evaluation of process flowsheets

The process route selection is one of the main design decisions that needs to be taken during the preliminary stages of design and development of a biofuel production plant. However, in order to obtain the optimal design, other factors such as plant economics and performance should be considered and evaluated. Therefore, after collecting from the literature and experimental works all the necessary information related to the BtL process routes (operation conditions, type of reactors, stoichiometric reactions, yields, separation needs, and so on) it is required to continue with the design of the process flowsheet for each process route under consideration. The process flowsheets are representations of the process stages considering the main plant items connected by process streams using module blocks to represent the unit operations along with a description of them. To exemplify this step, the process flowsheet for the Pyrolysis-Hydroprocessing-Separation route is depicted in Figure 10.4.

10.4.2.1 Computational Studies: Aspen Analysis

The process flowsheets together with the data collected allow the use of computer tools to perform steady-state energy and mass balances, sizing, and economic calculations of the proposed BtL process routes. Process simulation is important to represent the process routes by mathematical models, which are then solved to obtain information about their performance.

10.4.2.2 Flowsheets setup

In the case of study, the rigorous simulation of the pyrolysis, gasification, upgrading and separation technologies was performed in the process simulator Aspen Plus V8.8. To create the process model, thermodynamic package selection, flowsheet setup, specification of chemical components, and operating conditions are required. For the setup of the process routes in Aspen Plus, each of the technological sections (Thermochemical conversion, upgrading and separation) was setup as a simulation flowsheet. A total of 15 simulations were performed. On the other hand, for setting up the energy balances, mass balances, and economic calculations, the following assumptions were given:

Mass balances

- The lignocellulosic feedstock (spruce and pine residues) and ash were specified as nonconventional components since they are not included in the Aspen database. For this, their ultimate analysis including C, H, O, N, S, Cl, and ash elements, and proximate analysis, presented in Table 10.1 [43, 44], were entered.
- The stream class MCINCPSD was selected, which is recommended when both conventional and nonconventional solids with particle size distribution are present.
- Char and sand were defined as conventional solid components. Char was assumed 100% carbon and sand was represented by silicon dioxide.
- The system is at steady-state, which means that the mass flow rates and compositions do not change with time and the accumulation term is equal to zero.
- The system does not leak: totaling up the mass entering and leaving the system.
- Yield, stoichiometric reactions, conversion, and separation factors assuming a 98% recovery of key components were fixed according to experimental data found in the literature.

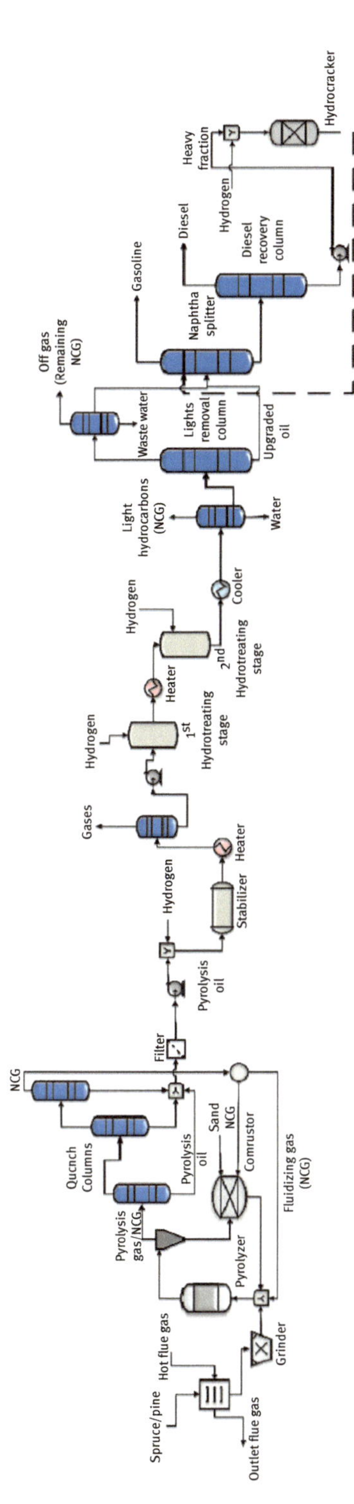

Figure 10.4: Pyrolysis-Hydroprocessing-Separation (PR-PHS) process flowsheet.

Table 10.1: Feedstock proximate and ultimate analyses.

	Proximate analysis (wt%, db)	Ultimate analysis (wt%, db)					Calorific values MJ/kg		
	Ash content	C	H	N	S	O	LHV	HHV	HHV_{Milne*}
Pine	1.33	51.30	6.10	0.40	0.02	40.85	19.34	20.67	20.59
Spruce	2.50	50.10	6.30	0.32	0.02	40.70	18.65	19.89	20.45

db- Dry base.
*Calculated from the elemental composition using the Milne formula.

- In the model to represent the product mixtures, model compounds were selected to represent each of the functional groups identified in the chemical characterizations found in the literature.
- Components not participating in the reactions were defined as inert components.

Energy balances

- Energy consumption of the unit operations was calculated using the thermodynamic package Peng-Robinson for the pyrolysis-based routes and Soave-Redlich-Kwong equation of state with Kabadi-Danner mixing rules for the gasification-based routes. The Peng-Robinson equation of state with the Boston-Mathias modifications and the electrolyte NRTL model with Redlich–Kwong equation of state were employed for the separation units.
- The biomass lower heating value (LHV) was also specified with the HCOALGEN and DCOALIGT property models chosen to estimate the biomass enthalpy of formation and specific heat capacity based on the ultimate and proximate analyses.
- No thermal losses were considered for the unit operations.

Economic evaluations

- The capital and energy costs (electricity, heating, and cooling) of the process routes were calculated for a fixed capacity of 500 kg/h of spruce and pine residues.
- Aspen Process Economic Analyzer V8.8 was used to calculate unit operation costs and utilities costs for fixed flowrates.
- Capital costs of the reactors were calculated by Aspen considering a base residence time (τ_{base}) of 300 s, which were then adjusted according to the residence time (τ_{adj}) reported in the literature for each specific reaction. The adjustment of the capital cost is depicted in Eq. 10.1.

$$\text{cost}_{\text{adj}} = \text{cost}_{\text{base}} \left(\frac{\tau_{\text{adj}}}{\tau_{\text{base}}} \right)^{0.5} (1) \tag{10.1}$$

10.4.2.3 BtL process routes: process flowsheets simulation and evaluation

From the simulations, product properties (density, viscosity, specific gravity, molecular weight, boiling point, flash point, cetane and octane numbers, and heat of combustion), capital costs, energy costs, mass balances, waste streams, conversion, and separation performance factors for all the unit operations were collected. The results from the rigorous simulations have been previously reported by Ibarra-Gonzalez et al. [45], where the five process routes were compared to determine the process for liquid fuel production that presents the minimum total annual cost and the highest conversion of softwood to liquid fuels. It was found that the Gasification-LTFT-fractional upgrading-fractionation process route (PR-GLTUF) was the most cost-effective among all the process routes when simultaneous diesel and gasoline production is preferred. In addition, the results showed that change and/or improvement of one technological section will significantly affect the performance of the other process sections and thus the total process routes. The major findings are summarized in Figures 10.5 and 10.6.

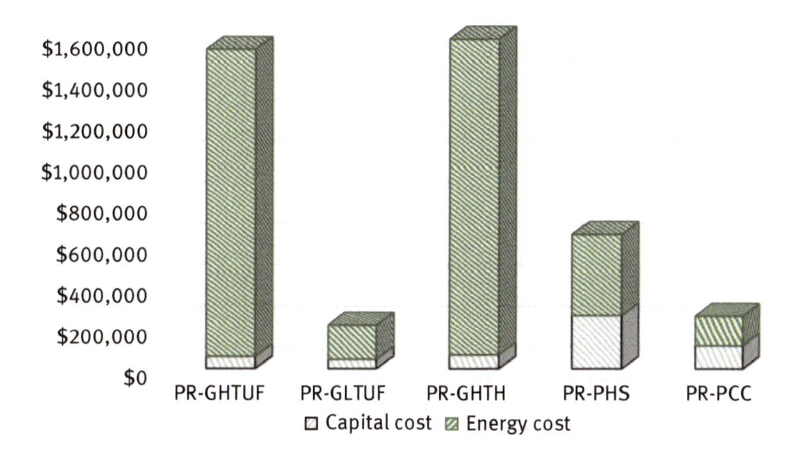

Figure 10.5: Total annual cost of the process routes (USD/year).

Moreover, with the data collected from the simulations is possible to proceed with the formulation of a processing superstructure to explore the distinct thermochemical-based process designs and propose new optimal total biofuels production processes that minimize the TAC of BtL fuels systems and increase the biofuels production. Likewise, with the liquid, gas, and solid emissions quantified from each process route, mass and energy integration scenarios can be considered.

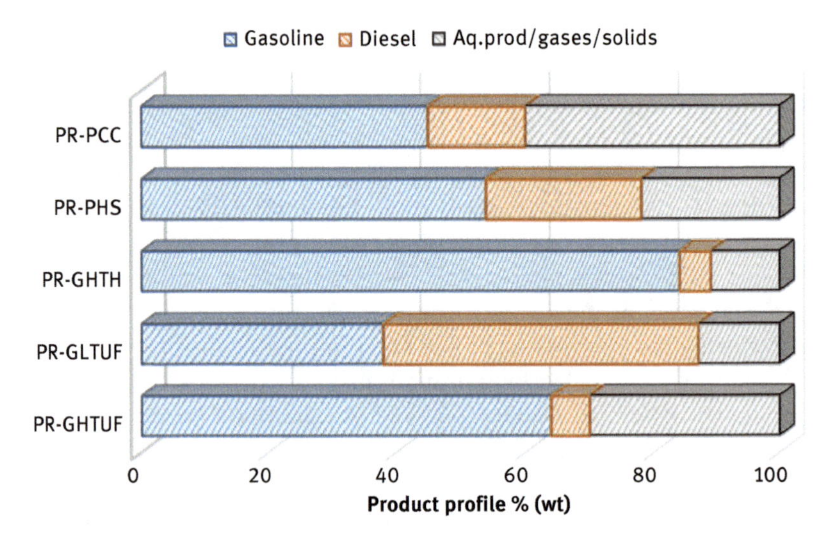

Figure 10.6: Product profiles of the BtL processes routes.

10.5 Integrated superstructure synthesis

10.5.1 Superstructure formulation: Processing blocks definition

A biomass network superstructure refers to a group of dependent and interconnected processes that utilize one or more biomass resources that lead to the production of single or multiple bioproducts [46]. The BtL processing superstructure relates the biomass resources to their products, available processes, and possible future processes.

In the case of study, the process routes that will be considered in the superstructure formulation are composed of many unit operations and, therefore, the analysis and modeling of the possible interconnections between them is possible but complex. To simplify the procedure, the unit operations from each process route can be grouped into processing blocks: *reaction* and *separation*. A processing block is considered as a set of units where the mass composition of the feed stream changes. The blocks definition is done by grouping mixers, pumps, and heat exchangers with the reactors and separation units associated to them, and by assigning them a specific tag. The reaction blocks were tagged as *RXN(i,j)* and the separation blocks as *SEP(i,j)*. The index *i* refers to the processing block number and *j* to the technology option (1-feedstock selection, 2-reaction block or 3-separation block). To exemplify the procedure, the formulation of reaction and separation blocks is illustrated in Figure 10.7.

The unit operations' information collected from the simulations (capital costs, energy costs, conversion, and separation factors) should be considered in the block

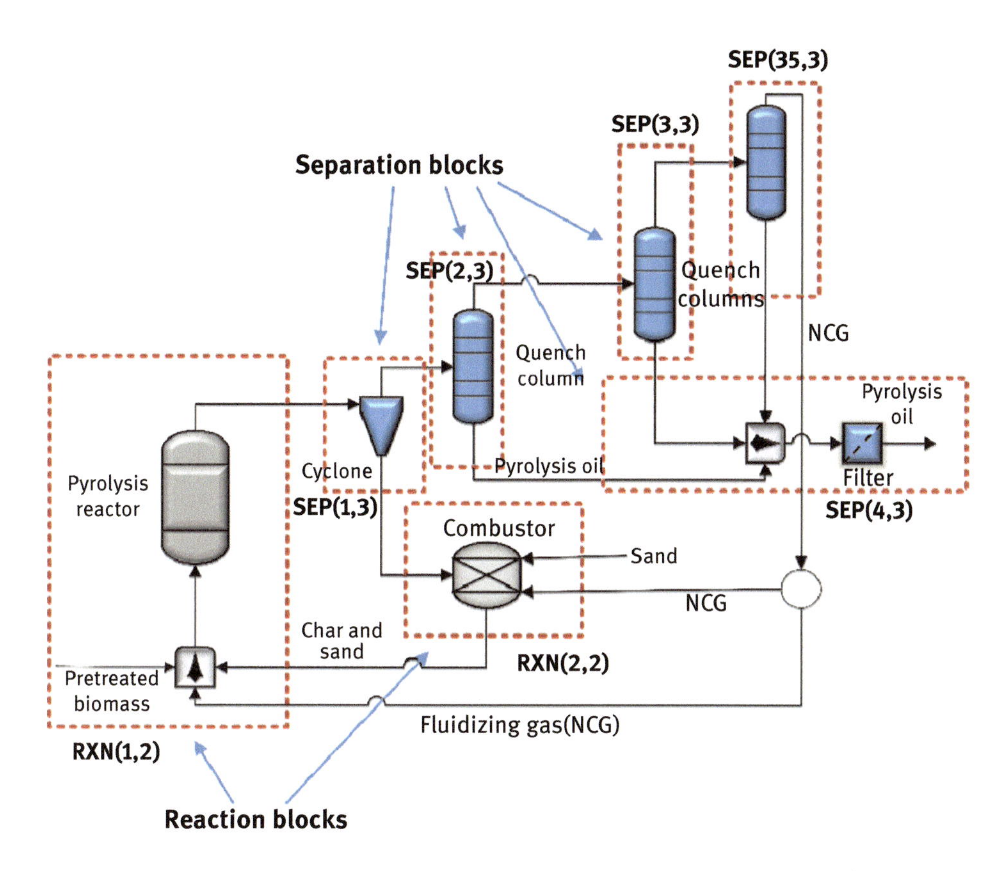

Figure 10.7: Formulation of reaction and separation blocks.

definition. This is done to keep the consistency between the unit operations and the processing blocks. Following these steps, the result is the formulation of processing blocks flowsheets, as shown in Figure 10.8 for the Pyrolysis-Hydroprocessing-Separation process route.

10.5.2 Detailed superstructure to simplified superstructure: Lumped process blocks definition

The next step in the superstructure formulation is the creation of a network diagram with the corresponding five processing blocks flowsheets. The network diagram or detailed superstructure (Figure 10.9), as will be called in this chapter, starts from the feedstock selection, continues with the selection of the thermochemical conversion technology, and the selection of the upgrading and separation technologies.

It can be already inferred from Figure 10.9 that different paths can be taken besides the proposed initial process routes. However, it is essential that the superstructure

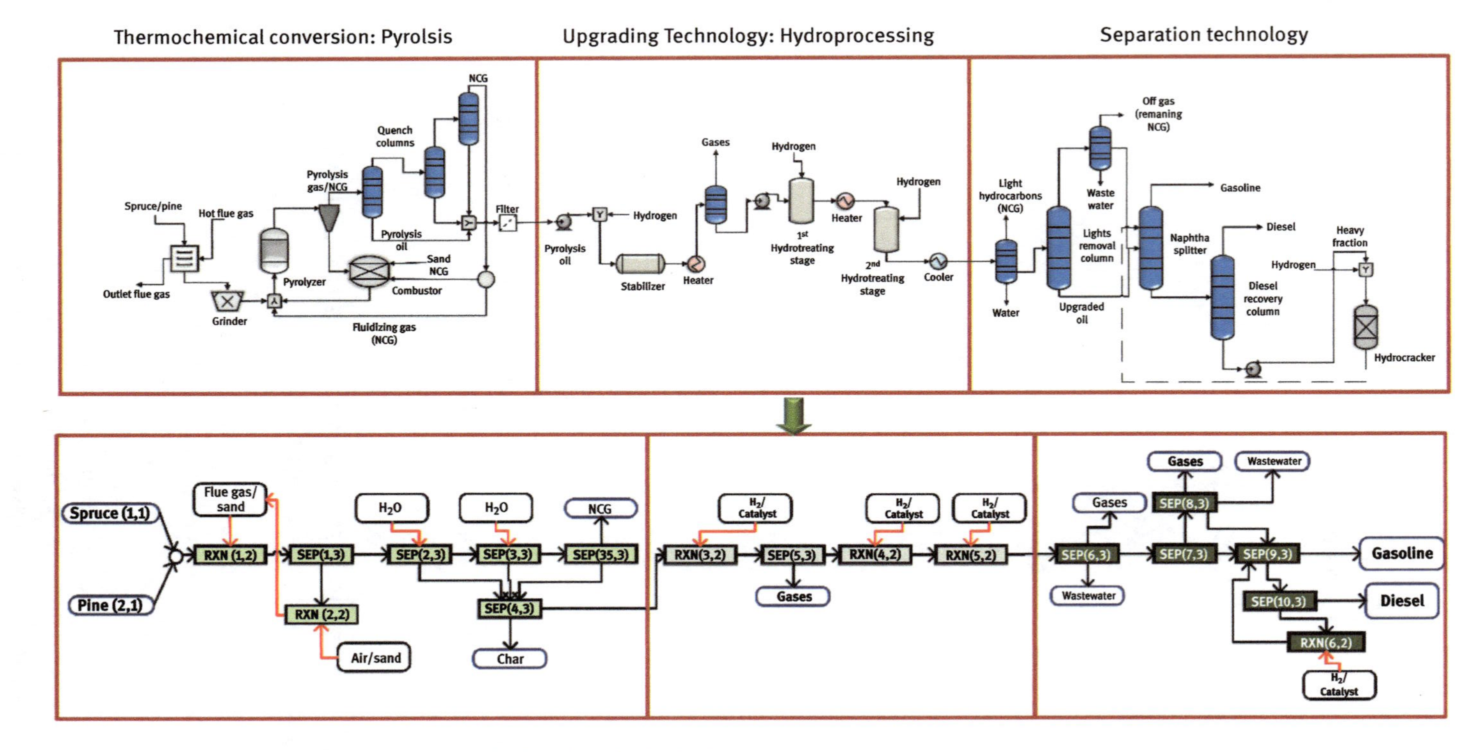

Figure 10.8: Formulation of the processing blocks flowsheet for PR-PHS.

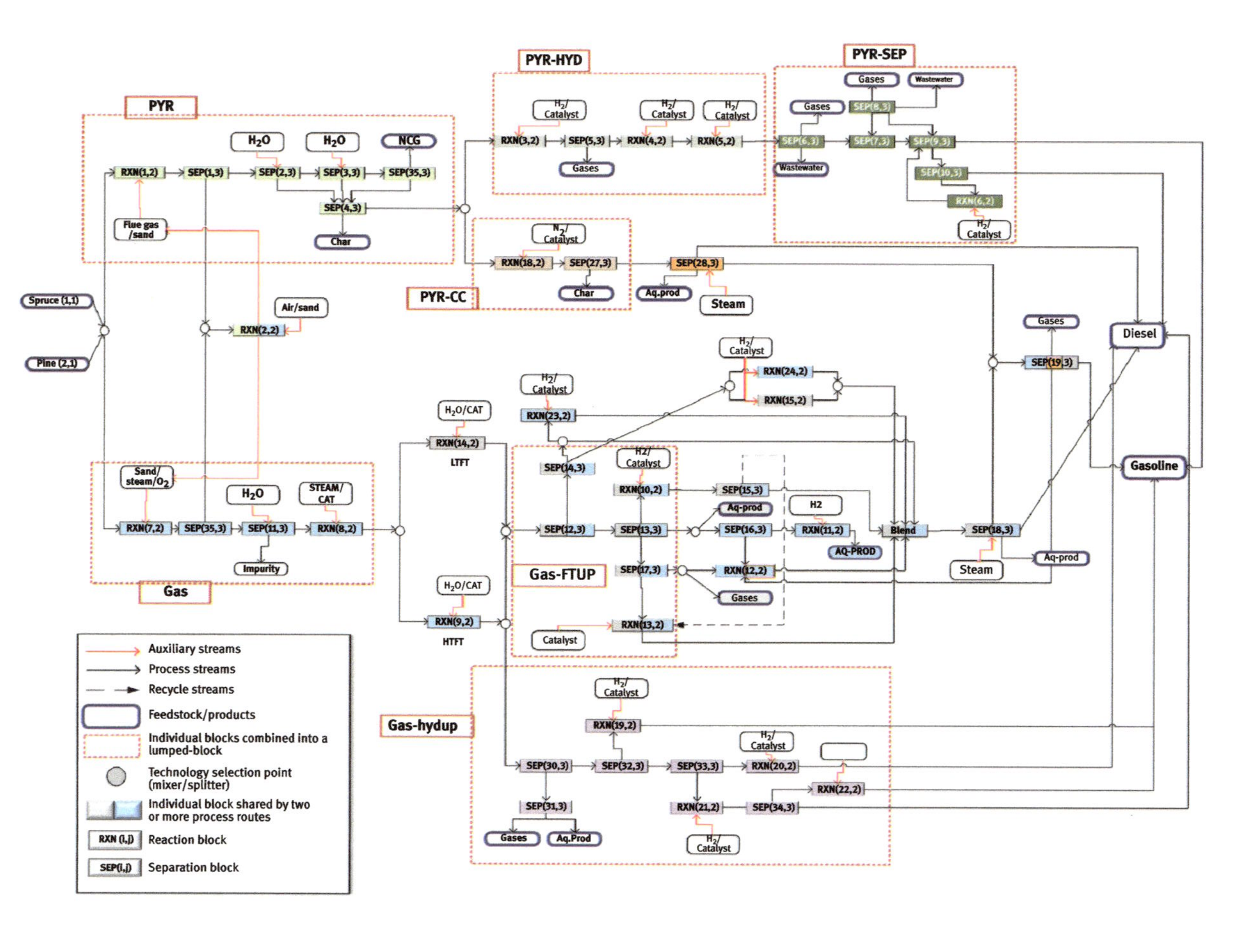

Figure 10.9: Softwood BtL detailed superstructure.

should remain uncluttered and be easy to follow, to avoid errors in interpretation and evaluation during superstructure modeling and programming. Therefore, the detailed superstructure must be simplified. The simplification can be accomplished with the support of the conceptual information and the simulation results. The possibilities of blocks combination between the process routes were analyzed by identifying blocks that perform similar tasks and present similar capital and energy costs.

It was found that most of the blocks are sequential, so that we proposed lumped-process blocks performing major tasks like biomass conversion to syngas via gasification (GAS) and bio-oil production via pyrolysis (PYR) and pyrolysis oil catalytic cracking (PYR-CC), pyrolysis oil hydroprocessing (PYR-HYD), upgraded product separation (PYR-SEP), and HTFT products hydroprocessing (GAS-HYDUP), as depicted in dotted-squared frames in Figure 10.9. In addition, it was found that in the HTFT and LTFT products upgrading sections the reaction blocks catalytic reforming (10,2) and alkylation (13,2), as well as the separation blocks (12,3), (13,3), (14,3), (15,3), (17,3), and (18,3) perform the same tasks and have similar capital and energy costs in both alternatives. Therefore, these process blocks were combined into one block (GAS-FTUP), leaving outside the reaction (oligomerization (12,2) and hydrogenation (11,2)) and separation blocks that are not in common between the process routes, as well as the ones that perform similar tasks but do not have similar costs. For representation purposes, in Figure 10.9, the combined blocks are colored in two or three colors depending on the number of routes they share. The simplified superstructure diagram considering the definition of the lumped-process blocks is presented in Figure 10.10.

As shown in Figure 10.10, when combining the processing blocks, the resulting lumped-process blocks were redefined by assigning them a new tag number. The final reaction and separation lumped blocks presented were tagged as follows; bio-oil production via pyrolysis RXN(1,2)-SEP(1,3); pyrolysis oil hydroprocessing RXN (3,2)-SEP(2,3); pyrolysis oil catalytic cracking RXN(5,2)-SEP(4,3); biomass conversion to syngas via gasification RXN(6,2)-SEP(7,3); upgraded product separation RXN(4,2)-SEP(3,3); HTFT hydroprocessing RXN(15,2)-SEP(10,3); HTFT and LTFT hydrocarbons upgrading section RXN(10,2)-SEP(8,3). Moreover, the reaction and separation blocks that were not combined were only renumbered. In Figure 10.11, the redefinition of the reaction and separation blocks conforming the lumped-process block PYR is presented.

The lumped-process blocks definition has the advantage to ease the evaluation through the superstructure. However, as mentioned, it is necessary to keep the consistency between the individual blocks' information and the final superstructure, which means that the performance and economic parameters should not be affected.

It implies that the mass balances, conversion factors, capital, and energy costs of the individual reaction blocks have to be considered in the definition of the lumped-reaction blocks. Similarly, the mass balances, separation factors, capital, and energy costs of the individual separation blocks must be considered for the lumped-separation blocks definition.

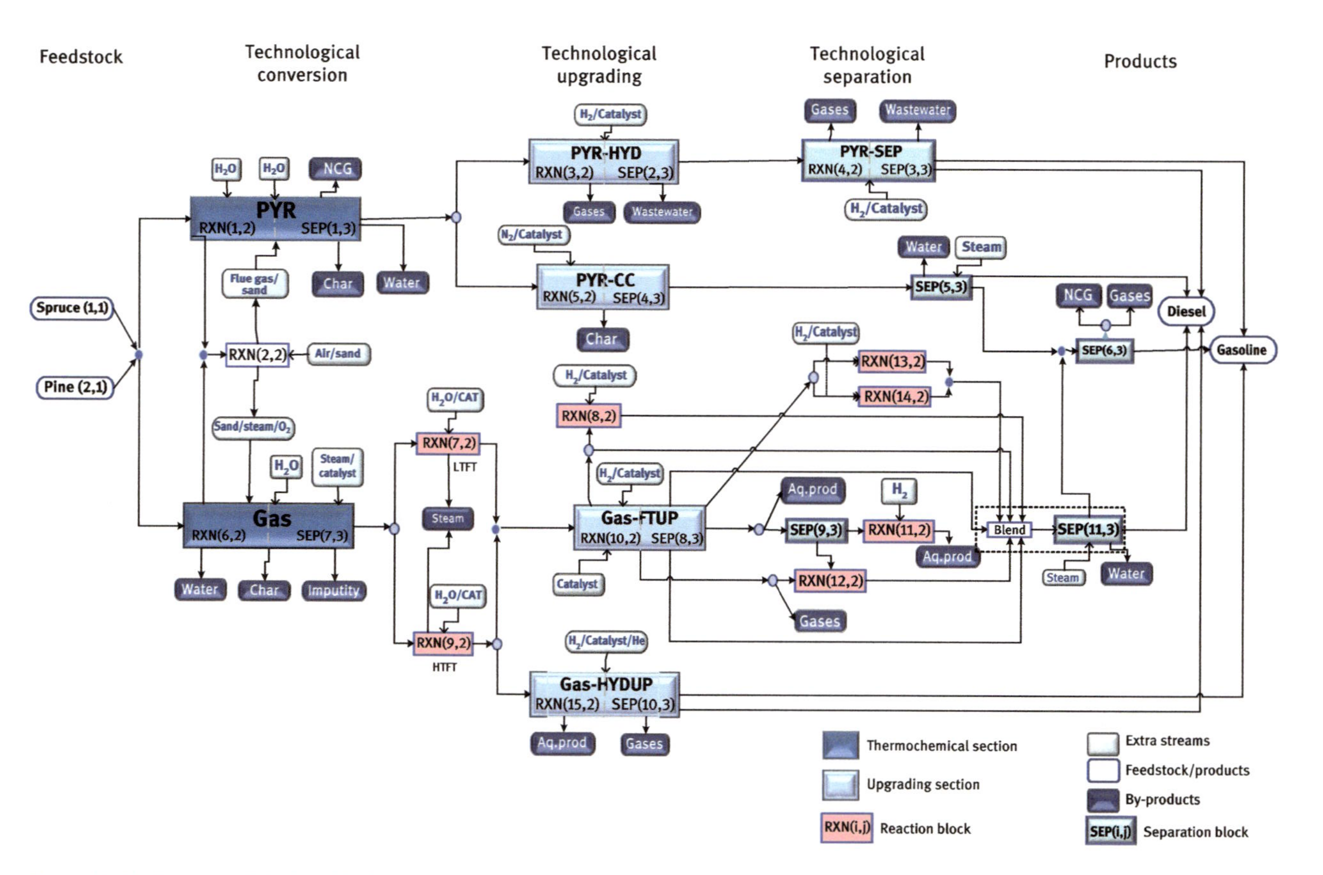

Figure 10.10: Softwood BtL simplified superstructure diagram.

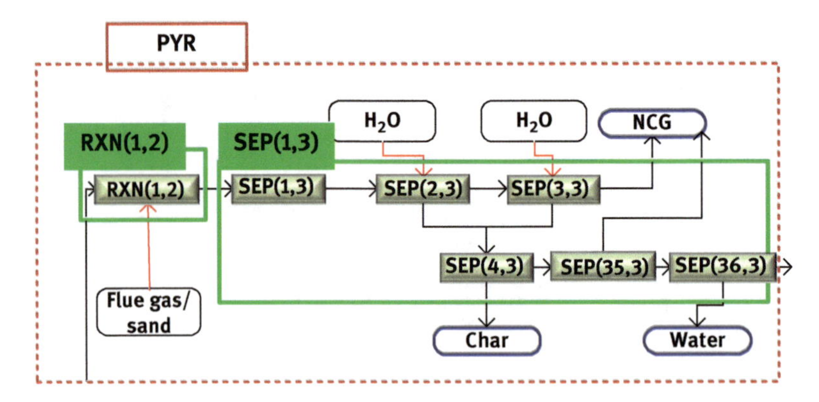

Figure 10.11: Reaction and separation lumped blocks definition for PYR.

10.5.3 Mass and energy integration

In addition to the processing blocks definition, mass and energy integration between the unit operations can be considered. The evaluation and implementation of different possibilities of integration in the proposed superstructure can increase the profitability of the BtL production processes.

From the rigorous simulations, liquid, gas, and solid emissions were quantified and it was found that an important amount of gases and water streams are produced through the processes. Therefore, process integration techniques can be used to reduce the liquid emissions and the overall energy consumption of the process routes. For instance, it might be possible to identify that a process can use the heat rejected by another unit or that a resource can be recovered from a waste stream and be later reused in the process. The integration possibilities are described in Sections 10.5.3.1 and 10.5.3.2.

10.5.3.1 Mass integration

Water integration: BtL processes consume and produce high amounts of water. For instance, during bio-oil and syngas production an excessive amount of water is required for product condensation and cleaning. In the case of study, water coming from the quench columns employed in the cooling and recovery of the pyrolysis oil could be processed and recycled back to the quench columns. Additionally, in the gasification-based routes, the wastewater stream coming from the scrubber can be purified (contains ammonia and CO_2) and sent back to the scrubber. The thermodynamic package Electrolyte NRTL model with Redlich–Kwong equation of state was selected for the simulation of the water treatment units.

Gases integration: During the pyrolysis reaction, NCGs are usually employed as fluidizing agents to increase the formation rates of the pyrolysis products. These gases (mainly CO, CO_2, CH_4, H_2, C_2H_4, C_2H_6) are also produced through the process and thus can be potentially recovered and recycled to the pyrolysis reactor. On the other hand, fuel gases such as methane, ethane, and propylene produced in both the pyrolysis-based and gasification-based process routes can be sent to a cogeneration plant for power generation.

10.5.3.2 Energy integration

Cogeneration through combined heat and power can simultaneously produce electricity with the recovery and utilization of heat. It is a highly efficient form of energy conversion and it can achieve primary energy savings compared to the separate purchase of electricity from the national electricity grid and a gas boiler for onsite heating. Therefore, to produce heat and electricity a cogeneration plant can be considered in the case of study. The cogeneration plant was simulated using the thermodynamic package Soave-Redlich-Kwong equation of state with Kabadi-Danner mixing rules and its application through the processes is described next.

Heating: The general objective is to reduce heating requirements in the process, which can be achieved by utilizing the waste process streams to provide energy for the system. In the cogeneration plant, pure water coming from the processes is sent to a steam generation section, where the water is sent to a series of pressure boilers (low, intermediate, and high-pressure boilers) and heat exchangers to produce steam. One part of the steam coming from the high-pressure boiler is sent to a steam turbine to produce electricity and the other part is exhausted to provide heat to the process routes. Simultaneously, the fuel gases (methane, ethane, and propylene) are compressed and burned in a combustion reactor to produce the heat required by the heat exchangers to increase the temperature of the pure water and produce the different levels of steam. Then, these gases are sent to the electricity generation section.

Electricity: In the electricity generation section, the electricity demand of each of the gasification-based and pyrolysis-based process routes is supplied if the fraction of steam produced in the high pressure boiler is sent to a steam turbine (as described before in heating section) to produce electricity and if the fuel gases (methane, ethane, and propylene) that were previously compressed and burned to provide the heat required to produce the steam are further sent to a gas turbine to generate more electricity.

10.6 Mathematical modeling: Superstructure setting and MINLP solution

From the superstructure, the optimal selection of technologies and integration scenarios can be determined with algorithms that are able to handle complicated nonlinear and continuous problems. The optimization of mass and energy transfer between processing blocks can be addressed by mixed integer nonlinear programming (MINLP) in which the total annual cost of the BtL fuels production processes is minimized and the biofuels production is maximized.

The MINLP model rigorously describes the input–output relationships of each processing block within the superstructure. Therefore, the MINLP problem consists of specifying the input (feedstocks, model compounds, auxiliary streams, flowrates, etc.) and output (desired products, minimum outlet flowrates, etc.) stream data collected from the rigorous simulations, and interconnecting the thermochemical conversion, upgrading, and separation technologies that convert the inputs into the desired outputs, while meeting the objective function.

Accordingly, the synthesis process can be decomposed into two levels. On the first level, the optimization problem selects the technological routes, which implies discrete decisions. On the second level, the decision on how these technologies should be interconnected is performed in a continuous space. The equations used to represent the BtL processing superstructure are mass balances, economics, capacity adjustments, etc., which are sets of linear and nonlinear equations. For instance, some of the nonlinearities are due to stream splittings and economic calculations. Considering this, it is clear that the optimal process route synthesis involves a discrete-continuous nonlinear problem.

10.6.1 Superstructure setting

As described, the BtL processing superstructure algorithm consists of the specification of input and output streams and processing blocks parameters, which are defined, based on data collected from the simulations, such as mass and energy balances, recycles, processing blocks costs, economic constraints, and product profile constraints.

Regarding the inputs, first, feedstock flowrates, stream compositions, and yields as well as market prices for raw materials, utilities, and intermediate chemicals were defined. Then, performance parameters such as conversion and separation factors were given together with auxiliary reagents ratios (fluidizing agents, catalysts, solvents, etc.). Concerning the economics, energy consumption, number of unit operations, and capital and energy costs of each technological processing block were specified. Likewise, cogeneration and wastewater treatment units cost and specifications were given.

After the definition of the overall process inputs and outputs, the equations to describe the superstructure were specified. These equations are the objective function, mass component balances, integrations (gases, water, and cogeneration), waste streams, and capital and energy costs calculations, which are given as follows.

Objective function:
The objective is to determine the optimal processing route that minimizes the total annual cost (TAC) of BtL fuels systems. The TAC values were evaluated considering 10 years as time for return of the investment. The objective function is depicted in Eq. 10.2:

$$\text{minimize TAC} = \sum\left[\left(\frac{\text{Capital Cost}}{\text{Time of investment}}\right) + \text{Energy costs}\right] \tag{10.2}$$

Mass balances:
The mass balances show the key inputs and outputs of the superstructure blocks. Each lumped-process block consists of an individual reaction block and a separation block and, therefore, individual mass balances for each block are carried out. The mass balance equations are given considering the respective block indexes (*i,j*) and a new index ***k*** is introduced, which stands for the mass component flow rate per hour. The mass balance equations for the reaction blocks in the conversion section (PYR and GAS) are depicted in Eqs. 10.3 and 10.4.

$$\mathrm{m_{in(RXN(i,j,k))} = m_{feedstock} + m_{new(RXN(i,j,k))}} \tag{10.3}$$

$$\mathrm{m_{out(RXN(i,j,k))} = m_{in(RXN(i,j,k))} + m_{in(RXN(i,j,k))} \cdot Conv_f} \tag{10.4}$$

Where $\mathbf{m_{in(RXN(i,j,k))}}$ is the reaction block total inlet flow rate in kg/h, $\mathbf{m_{feedstock}}$ is the mass flow rate of spruce and/or pine residues, $\mathbf{m_{new(RXN(i,j,k))}}$ is the auxiliary streams flow rate, such as, catalysts, hydrogen, steam, sand, fluidizing gas, and so on, and $\mathbf{m_{out(RXN(i,j,k))}}$ stands for the total outlet flow rate and $\mathbf{Conv_f}$ refers to the factor that represents the conversion of reactants to products.

For the subsequent reaction blocks belonging to an upgrading section, the mass balances are presented in Eqs. 10.5 and 10.6. Where the total inlet flow rate of the reaction block $\mathbf{m_{in(RXN(i,j,k))}}$ is equal to the outlet flow rate $\mathbf{m_{outPB}}$ coming from the previous block plus the auxiliary stream $\mathbf{m_{new(RXN(i,j,k))}}$, which refers to catalysts and other reagents. On the other hand, the outlet flow rate of a reaction block is equal to the inlet flow rate plus the inlet flow rate times the conversion factor, as presented in Eq. 10.6.

$$\mathrm{m_{in(RXN(i,j,k))} = m_{outPB} + m_{new(RXN(i,j,k))}} \tag{10.5}$$

$$\mathrm{m_{out(RXN(i,j,k))} = m_{in(RXN(i,j,k))} + m_{in(RXN(i,j,k))} \cdot Conv_f} \tag{10.6}$$

Concerning the separation blocks mass balances, the inlet flowrate $\mathbf{m}_{\mathbf{in(SEP(i,j,k))}}$ is equal to the outlet flowrate m_{outPB} from the previous block (the sum of all feed streams). The outlet flowrate $\mathbf{m}_{\mathbf{out(SEP(i,j,k))}}$ is equal to the sum of the separation streams $\mathbf{m}_{\mathbf{out_n}}$ ($\mathbf{n} = 1, 2, 3, \ldots$). The mass balances for the separation blocks are depicted in Eqs. 10.7 and 10.8.

$$m_{in(SEP(i,j,k))} = m_{outPB} \tag{10.7}$$

$$m_{out(SEP(i,j,k))} = m_{out_1(SEP(i,j,k))} + m_{out_2(SEP(i,j,k))} + \ldots + m_{out_n(SEP(i,j,k))} \tag{10.8}$$

In addition, from the separation units' rigorous simulations, separation or recovery factors $\mathbf{Sep_f}$ were calculated. These factors stand for the component mass flow rate that is being recovered in each outlet stream. These factors are implicit in the outlet stream $\mathbf{m}_{\mathbf{out_n(SEP(i,j,k))}}$ terms, as presented in Eq. 10.9.

$$m_{out_n(SEP(i,j,k))} = m_{in(SEP(i,j,k))} \cdot Sep_f \tag{10.9}$$

Integrations:
The different levels of integration considered are NCGs integration, water integration, and cogeneration. The general equation to represent the integrations is in the form of a mass balance where an extra term considering the integration is introduced, as depicted in Eq. 10.10.

$$m_{in(RXN(i,j,k))} = m_{outPB} + m_{new(RXN(i,j,k))} + m_{rec(SEP(i,j,k))} \cdot X_n \tag{10.10}$$

The term $\mathbf{m}_{\mathbf{rec(SEP(i,j,k))}}$ refers to the outlet stream coming from a separation block, that might be recycled, and the term $\mathbf{X_n}$ decides whether the stream is being recycled or not. $\mathbf{X_n}$ can take values of 1 if the stream is being recycled or 0 if it is considered as emission. This term is an integer variable because more than the total flowrate of the recycle stream is needed to fulfill the actual process requirements. For example, if terms like X_1 and X_2 are considered, and the term X_1 refers to a possible integration and X_2 considers it instead as emission. Then X_1 should be specified in the integration Eq. 10.10 and X_2 in the emissions Eq. 10.12, and thus an extra equation to relate them should be introduced, as depicted in Eq. 10.11.

$$X_1 + X_2 = 1 \tag{10.11}$$

Waste streams disposal:
The outlet streams coming from the separation blocks and that are not being recycled are considered emissions. The emissions or waste streams can be quantified with Eq. 10.12. In this equation, the term $\mathbf{X_n}$ takes values of 1 if is considered as emission or 0 if it is being recycled, these decisions are made for simplification of the model and for the same reasons as explained before in the integration part.

$$m_{in(WWD(i,j,k))} = \sum m_{out_n(SEP(i,j,k))} \cdot X_n \tag{10.12}$$

Capital cost:
The individual processing blocks' capital cost $\mathbf{CC_{base(i,j)}}$ given in the algorithm, was calculated in Aspen Process Economic Analyzer for a base-flow rate $\mathbf{m_{base}}$, and thus an equation to adjust the capital cost $\mathbf{CC_{adj(i,j)}}$ depending on the real mass input flow rate $\mathbf{m_{in(i,j,k)}}$ should be specified, as depicted in Eq. 10.13. The index k stands for the mass component flowrate per hour.

$$CC_{adj(i,j)} = CC_{base(i,j)} \left(\frac{m_{in(i,j,k)}}{m_{base}} \right)^{0.8} \tag{10.13}$$

The total capital cost $\mathbf{CC_{tot}}$ of a process route is equal to the sum of the individual capital cost of the processing blocks, as presented in Eq. 10.14.

$$CC_{tot} = \sum CC_{adj(i,j)} \tag{10.14}$$

Energy cost:
The energy cost $\mathbf{EC_{(i,j)}}$ described by Eq. 10.15 refers to the amount of electricity, steam, and thermal fluids required in each of the processing blocks. The electricity, heating, and cooling requirements were calculated with Aspen Economic Evaluation version 8.8 in the form of Eq. 10.15. For the plant operation, 8,000 hours were considered.

$$\begin{aligned} EC_{(i,j)} = {} & \text{Utility consumption}\left[\frac{\text{kg}}{\text{h}} \text{ or } \frac{\text{kW}}{\text{h}}\right] \cdot \text{Plant Operation}\left[\frac{\text{h}}{\text{year}}\right] \\ & \cdot \text{Utility cost}\left[\frac{\text{USD}}{\text{kg}} \text{ or } \frac{\text{USD}}{\text{kW}}\right] \end{aligned} \tag{10.15}$$

The total energy cost of a process route is equal to the sum of the individual energy cost of the processing blocks, as shown in Eq. 10.16.

$$EC_{tot} = \sum EC_{(i,j)} \tag{10.16}$$

Finally, the mathematical model evaluates the superstructure and selects the optimal technology within a set of technological selection constraints, mass balance constraints, and techno-economic evaluation constraints. The major decision variables are the following:
- Feedstock mass flowrate and composition
- Technological route selection
- NCGs integration
- Water integration
- Split of cogeneration into steam and/or electricity production
- Gasoline outlet flow rate

- Diesel outlet flow rate
- Product components flowrate

Some of these decision variables, but not all, are restricted to be integer. The branch-and-bound approach is recommended to solve problems in which some of the variables are constrained to be integral. Branch-and-bound is a strategy that considers the division of the feasible region into more manageable subdivisions. The subdivisions are generated solely by the integral variables. The general procedure is to subdivide based upon the variable with highest objective contribution [47]. In this case of study, it is subdivided in terms of the possible mass and energy integrations. The integration decision integer variables considered will be presented in Section 10.6.2.

On the other hand, to obtain the optimal technological route, the selection of processing blocks is performed by not specifying them as integer variables, and therefore the model can consider all the process routes simultaneously.

10.6.2 MINLP solution: Global optimal scenarios

The proposed MINLP model has the advantage of being able to find the optimal network flowsheet that fits a given set of constraints. Therefore, if the set of constraints is changed the problem solution might change as well. To prove the flexibility of the model, two different scenarios were evaluated.

The first scenario promotes the simultaneous production of gasoline and diesel and considers a set of initial values for the feedstock flowrate, feedstock composition, integrations, and biofuels production. The feedstock flowrate going to the pyrolysis section was initially set to 250 kg/h, likewise, the feedstock flowrate going to the gasification section was set to 250 kg/h. These initial values are given to guarantee that both processes are considered in the iterations. Concerning the feedstock composition, both feedstock streams contain 50 wt.% spruce and 50 wt.% pine. Additionally, all types of mass and energy integrations were considered, and boundaries concerning gasoline and diesel production were given based on product flowrates taken from the simulations. In this scenario, gasoline and diesel flowrates were set to be greater than or equal to 260 kg/h and 50 kg/h, respectively. On the other hand, for the second scenario, the production of gasoline is preferred and thus the diesel flowrate was set to be lower than or equal to 200 kg/h and the gasoline flowrate was set to be greater than or equal to 500 kg/h. However, the same initial values for the feedstock flowrate, feedstock composition, and integrations were given as in the first scenario. The initial values are the starting point that allow the algorithm to search the optimal solution within the limits.

The MINLP problem was solved on a HP EliteBook laptop with Intel(R) Core (TM) i5-6300U, 240 GHz CPU, 8 GB RAM, and Windows 10 64-bit. The model uses as solver the Branch-And-Reduce Optimization Navigator (BARON), which is recommended for the global solution of nonlinear programming (NLP) and MINLP. BARON implements deterministic global optimization algorithms of the branch-and-bound type that are guaranteed to provide global optima under fairly general assumptions [48].

The MINLP solution is determined by analyzing which processing blocks present flowrates and by confirming if the final product components are produced through those blocks. For the first scenario, the solution proposed a combined flowsheet of Gasification followed by simultaneous high and low temperature FT reactions and fractional upgrading units (distillate hydrotreating, wax hydrocracking, naphtha reforming, and tail gas alkylation) as the global optimal technological route, which is designated as CA1-GLTHT and is shown in Figure 10.12. The CA1-GLTHT could increase the liquids fuels production and reduce the TAC if mass and energy integration are considered. To analyze the given solution, the resulting decision variables' values are presented in Table 10.2.

As presented in Table 10.2, if the objective is to increase the production of biofuels and minimize the TAC, then gasification is the optimal thermochemical route and steam and electricity generation in the cogeneration plant should be considered.

On the other hand, for the second scenario, the MINLP solution proposed a combined flowsheet of gasification followed by simultaneous high and low temperature FT reactions and fractional upgrading units as the global optimal technological route, which is designated as CA2-GLTHT and is presented in Figure 10.13. However, notably, the upgrading units are different to the ones considered in CA1-GLTHT, since in this scenario the production of gasoline is preferred, and the diesel production is not a priority. In this scenario, tail gas alkylation and naphtha reforming, LPG oligomerization, and wax hydrocracking units are required, in which most of the components are upgraded to gasoline blends. Regarding the decision values of the integer variables, all values taken were the same as in the first scenario, except that in this scenario the electricity generated might be lower, which is understood given that most of the gases are upgraded to gasoline components and there is a low amount of waste gases available for power generation.

10.7 Optimal scenarios evaluation with process simulator

To benchmark the optimal solutions and prove that the objectives are being achieved, rigorous simulations of the proposed alternatives are carried out. From the simulations, product profiles, waste streams flowrates, energy costs, capital costs, and total annual costs are collected. The numerical evaluation is used to compare the new flowsheets with the initial five process routes proposed in Sections 10.4.1.1 to 10.4.1.5.

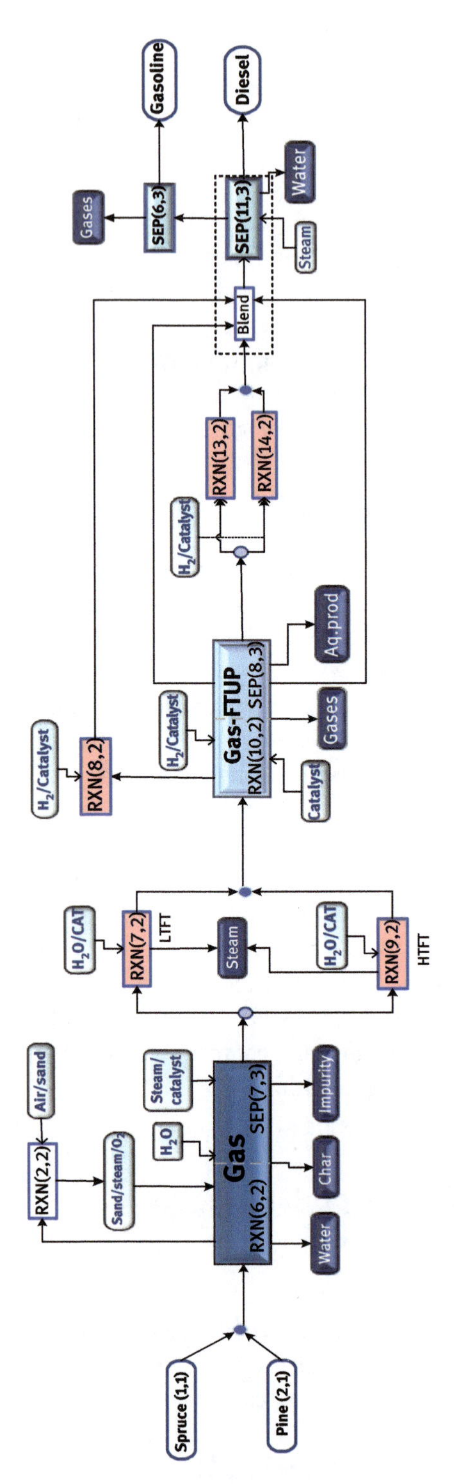

Figure 10.12: Scenario 1: Global optimal solution CA1-GLTHT.

Table 10.2: Decision values of integer variables.

Variable	Value	Description
X_1	0	Recycle of NCGs from SEP (1,3) to RXN(1,2)
X_2	1	No recycle of NCGs from SEP (1,3) to RXN(1,2)
X_3	0	Recycle of NCGs from SEP (2,3) to RXN(1,2)
X_4	1	No recycle of NCGs from SEP (2,3) to RXN(1,2)
X_5	0	Pure water integration between quench columns in SEP(1,3)
X_6	1	No pure water integration between quench columns in SEP(1,3)
X_7	0	Pure water recycled to the scrubber in SEP(7,3)
X_8	1	Pure water is not recycled to the scrubber in SEP(7,3)
X_9	0	Water integration between heat exchangers in SEP(7,3)
X_{10}	1	No water integration between heat exchangers in SEP(7,3)
X_{11}	1	Steam production in Cogeneration plant
X_{12}	1	Electricity production in Cogeneration plant
X_{13}	0	No production of steam in Cogeneration plant
X_{14}	0	No production of electricity in Cogeneration plant

The two technological routes flowsheets presented in Figures 10.14 and 10.15 were used in the evaluation of the CA1-GLTHT and CA2-GLTHT process routes. The rigorous simulations were performed in Aspen Plus V8.8 using the setups and designs of the gasification-based process routes as starting points. Then, because of changes in the composition and flowrates of the FT fractions, minor adjustments were done to the design parameters in order to achieve the desired conversions and recoveries. After achieving the objectives, Aspen Process Economic Analyzer V8.8 was used to calculate energy costs and equipment costs for both process routes (including integrations). The economic results obtained for both scenarios are presented in Table 10.3 and are compared with the initial five process routes. Total annual costs without and with heat and power integration are reported for all the cases.

The numerical results obtained for the CA1-GLTHT showed that the biofuels production could increase to 90% (wt) compared to the 87% (wt) reported for the PR-GLTUF. In addition, a product distribution of 69% (wt) gasoline and 21% (wt) diesel were achieved, and a TAC reduction of 26% if pure water and fuel gases integration are considered for the generation of heat and power. On the other hand, for CA2-GLTHT, the biofuels production was increased to 97% (wt) with a product

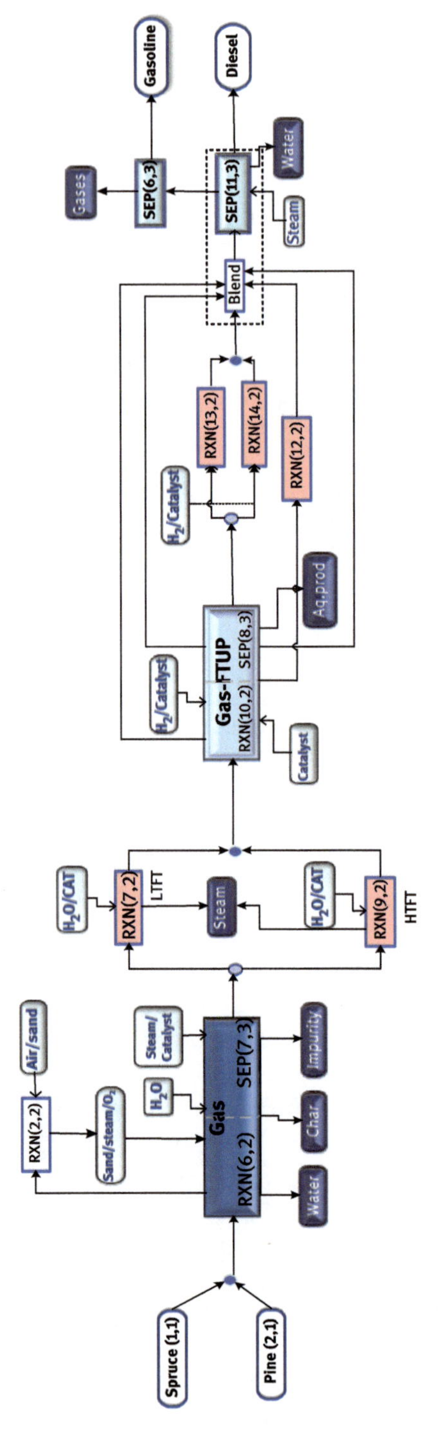

Figure 10.13: Scenario 2: Global optimal solution CA2-GLTHT.

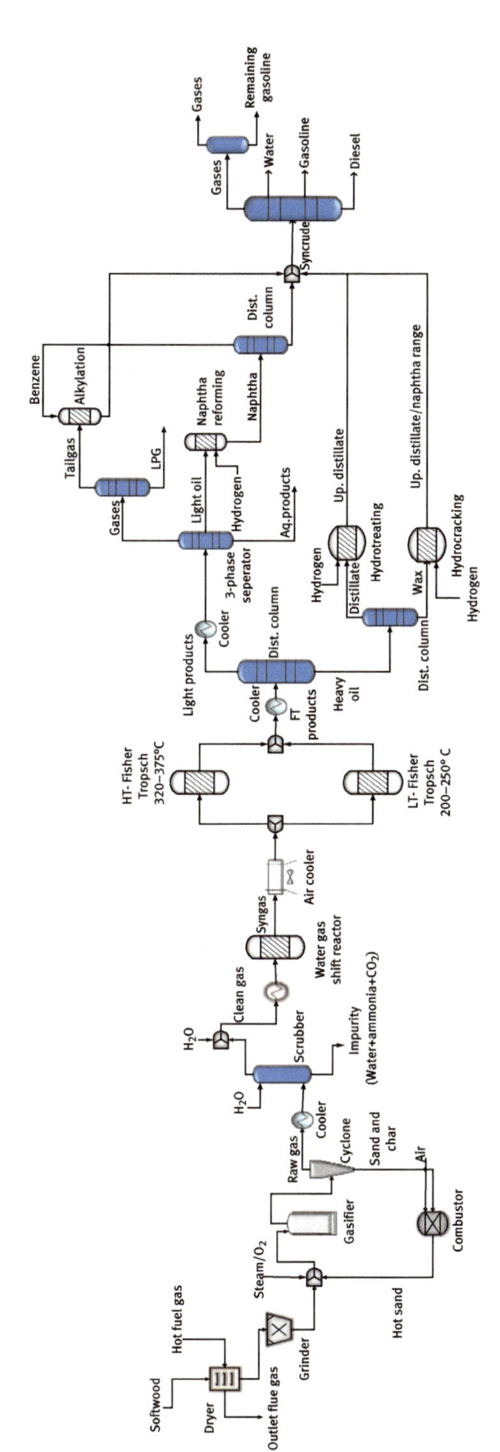

Figure 10.14: CA1-GLTHT process flowsheet.

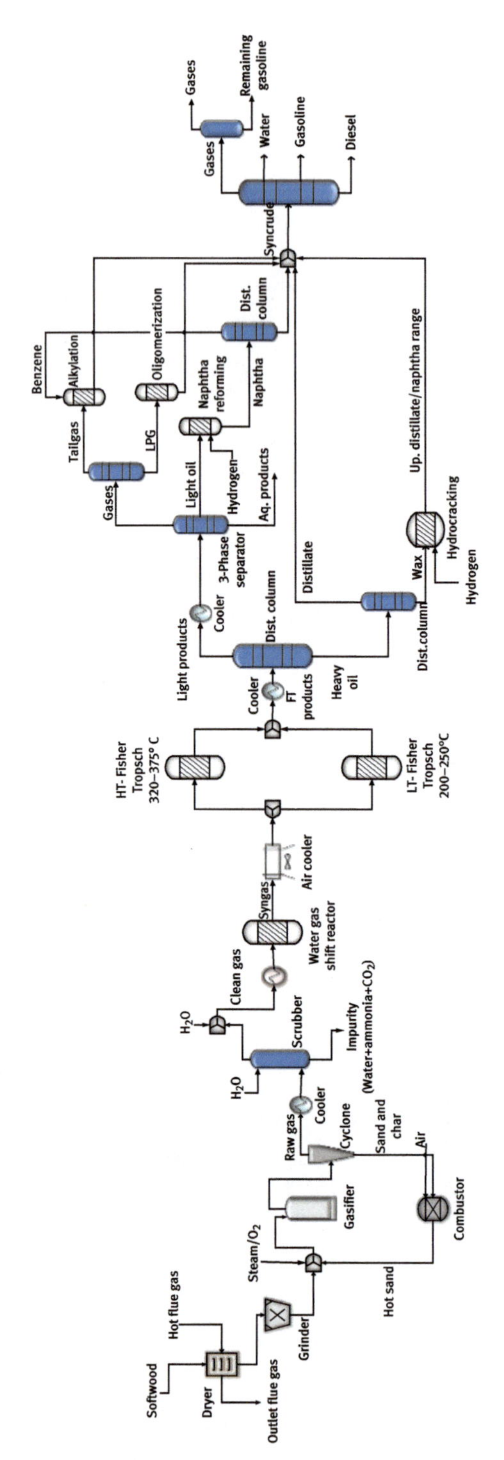

Figure 10.15: CA2-GLTHT process flowsheet.

Table 10.3: BtL technological routes analyzed during the optimization.

	PR-PHS	PR-PCCF	PR-GLTUF	PR-GHTUF	PR-GHTH	CA1-GLTHT	CA2-GLTHT
Capital Cost [$]							
Conversion section	861,557.6	861,557.6	221,642.4	203,561.3	203,561.3	230,638.2	230,638.2
Upgrading tech.	904,703.2	198,875.5	161,743.8	313,825.2	226,000.0	182,077.6	182,600.0
Separation tech.	842,606.9	73,400.0	73,000.0	74,400.0	250,575.6	74,400.0	75,200.0
Energy Cost [$/year]							
Conversion section	53,632.1	53,632.1	92,420.5	1,411,072	1,411,072	133,954.4	133,954.4
Upgrading tech.	141,082.3	50,748.6	38,078.4	39,598.8	39,103.4	39,598.8	40,105.63
Separation tech.	196,543.9	39,892.3	36,826.7	39,376.5	85,238.6	61,880.5	55,351.5
TAC [$/year]	652,145.1	257,656.3	212,964.2	1,549,225.8	1,603,427.7	284,145.3	278,255.4
TAC w/int. [$/year]	514,744.1	263,941.2	171,899.8	298,900.8	397,637.2	202,460.8	266,471.9
Upgraded prod. [kg/h]	942	902	700	818	985	1,022	1,069
Total fuel prod. % (wt)	78	60	87	70	89	90	97
Gasoline % (wt)	54	45	38	64	84	69	80
Diesel % (wt)	24	15	49	6	5	21	17

distribution of 80% (wt) gasoline and 17% (wt) diesel and a TAC reduction of 5% if heat and power generation in the cogeneration plant are included.

10.8 Upgrading section configuration alternatives and possible intensifications

After determining the global optimal technology scenarios for the simultaneous production of gasoline and diesel by the MINLP approach, the next step is to analyze in detail the unit operations involved in the optimal process routes. By analyzing the performance and economic evaluation of the process sections' units, it is possible to identify the intensification possibilities between the unit operations.

From the results reported in Table 10.3, it is observed that the sections accounting for the highest capital and energy costs are the conversion section and the upgrading section. However, as can be observed from the process flowsheets, there are limited possibilities of improvement or intensification in the conversion section unless further conversion kinetics with new reactor configurations are explored. On the other hand, for practical reasons, the upgrading section can be considered for further integration and intensification.

In the upgrading section, several separation units are required to recover the different fractions before being sent to their corresponding upgrading units. Moreover, as presented in Figures 10.14 and 10.15 the upgrading units required for each process route are different depending on the desired products. This is because in a FT refinery the composition of the products can be controlled according to the design of the plant. The production of a specific type of fuel can be maximized by manipulating in an efficient way the carbon number distribution of the syncrude, and therefore different plant designs can be explored. For instance, different separation configurations can be proposed to divide the FT product in several fractions depending on the carbon number and after the fractions recovery, different upgrading units can be employed depending on the desired product profile. For example, if jet fuel production is desired besides the main products considered in the case of study (gasoline and diesel), then a hydroisomerization unit to convert the C_{11}–C_{14} fraction to a jet fuel component can be considered (Figure 10.18). In Figures 10.16, 10.17, and 10.18 some of the proposed separation and upgrading configuration alternatives are presented.

In general, the different upgrading configurations might present high costs due to the several separation units, which include flashes, distillation columns, fractionation columns, and solvent extraction columns required for the recovery of the carbon fractions. The integration between unit operations, i.e., intensification by combining or merging the unit operations in multiple ways could improve the performance of the processes and/or reduce the costs. For this, the definition of column sections is important to allow changing the configuration of the separation trains

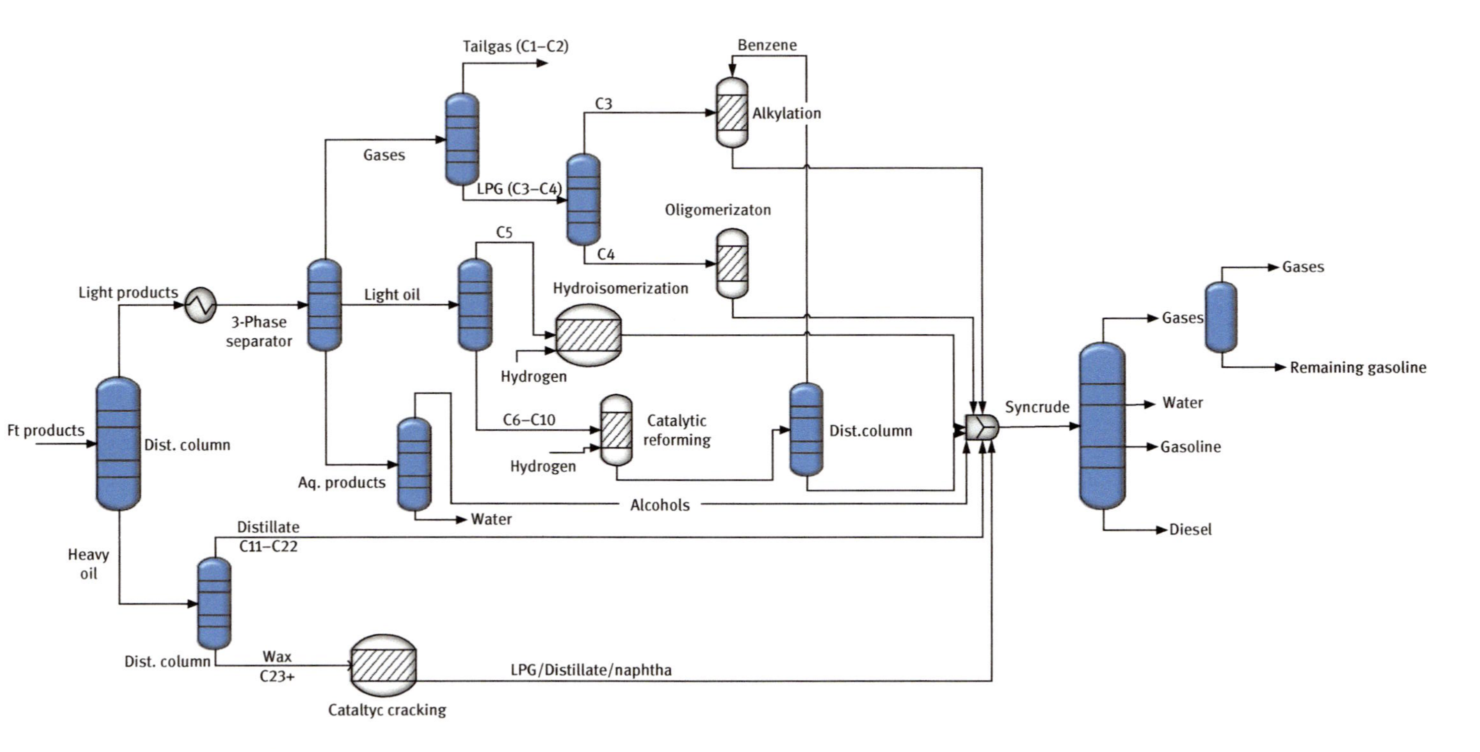

Figure 10.16: Upgrading configuration for the main production of gasoline.

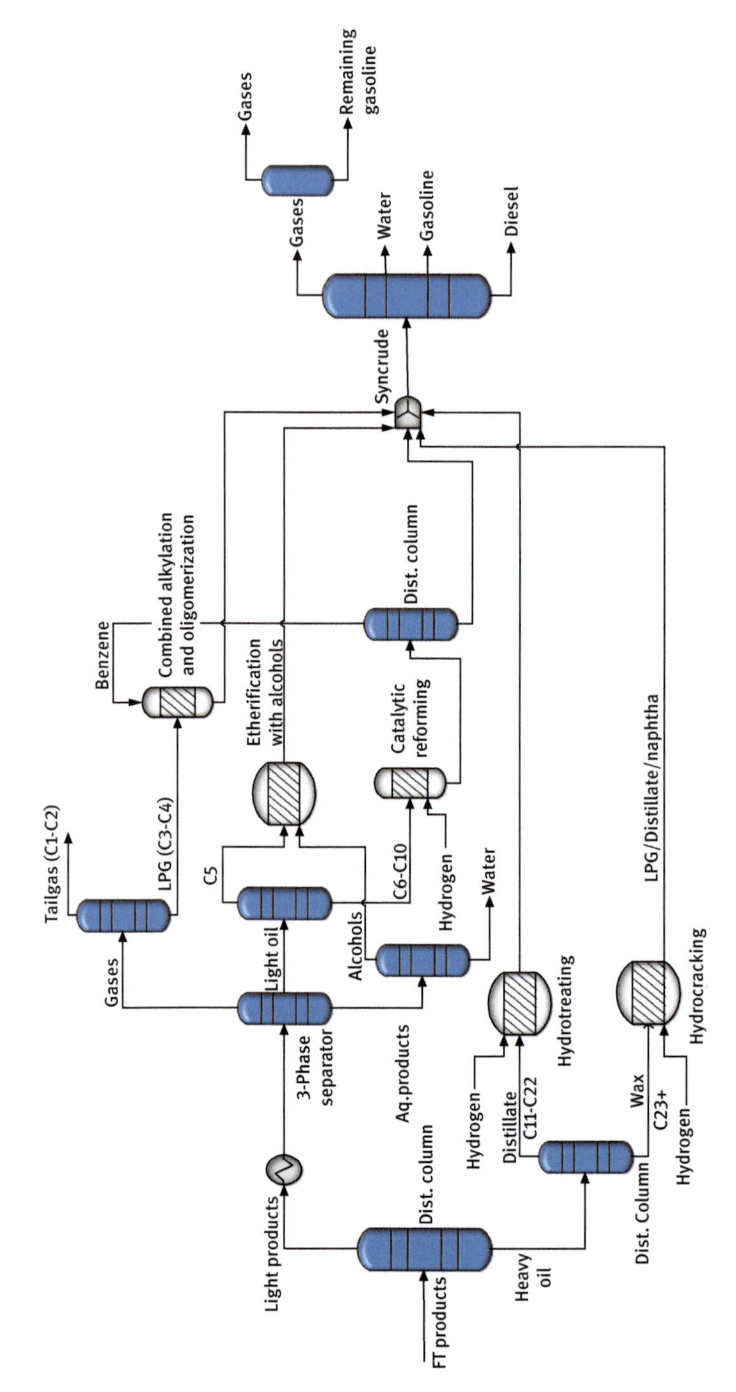

Figure 10.17: Upgrading configuration for the simultaneous production of gasoline and diesel.

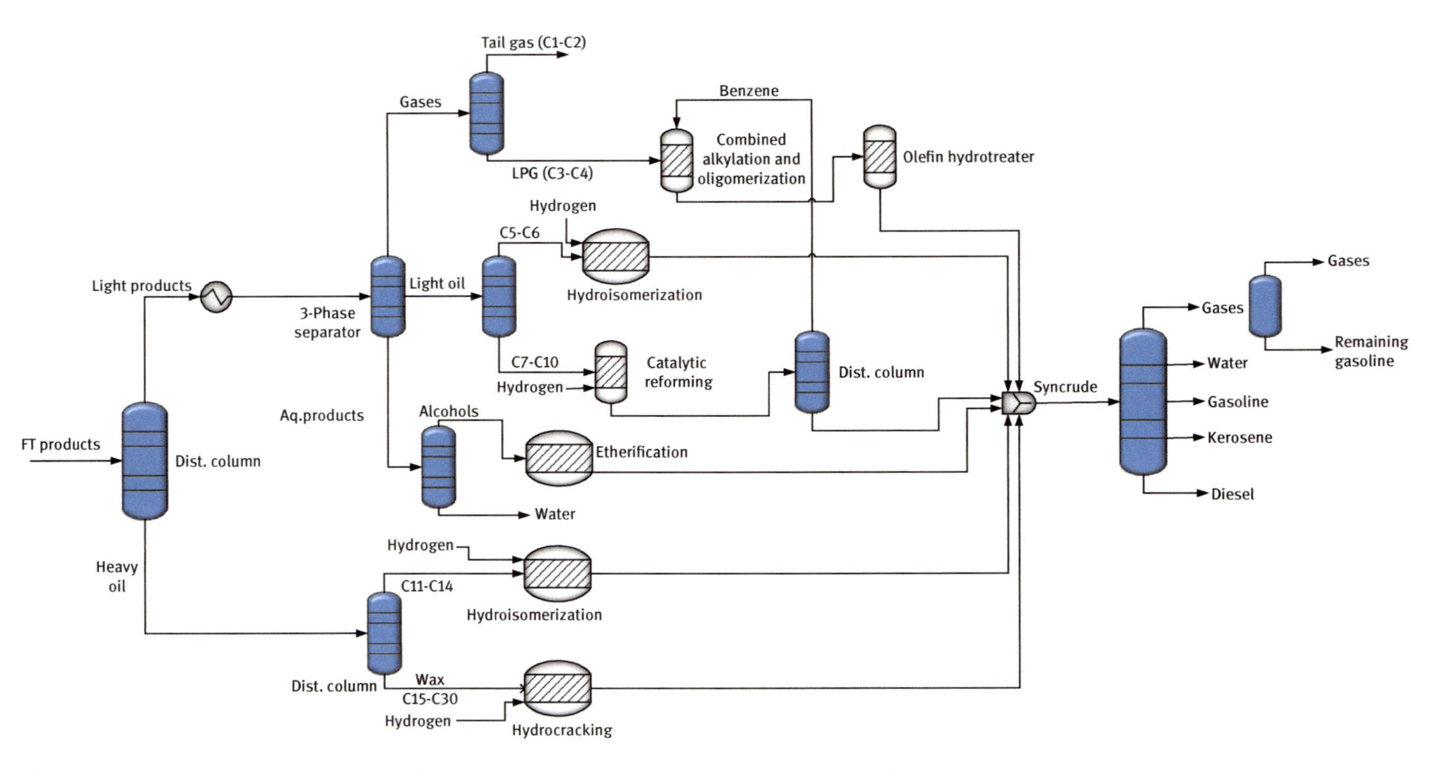

Figure 10.18: Upgrading configuration for the simultaneous production of gasoline, diesel, and jet fuel.

and the generation of subspaces with intensification potential. Besides the section's definition, the use of synthesis tools such as hybridizing unit operations, adding column sections to replace the flashes or reallocating column sections can be employed to reformulate the configurations. For instance, combining the three-phase separator with the gas recovery column and the light oil column or combining the fractionation column with the flash are intensification possibilities. However, separation process intensification is not the only possibility; reaction-separation can be also explored. The reactive-distillation process for the catalytic reforming of naphtha is one of the potential alternatives.

Further intensified methods and equipment are to be explored through rigorous simulations to examine the performance of the units and operating parameters.

10.9 Conclusions

Lignocellulosic biomass is the most abundant and biorenewable biomass on earth and has been presented as a promising source to produce transportation fuels and chemicals. It presents many advantages over fossil fuels and first-generation biofuels from food-derived sources. Lignocellulosic sources can decrease CO_2 emissions and atmospheric pollution, are a perfect equivalent to petroleum to produce fuels, their product properties can be adapted to the requirements of internal combustion engines, and they do not compete with the production of food. In general, the conversion of BtL transportation fuels can be achieved through different processes. Recent advances in BtL process synthesis attempt to design modern lignocellulosic biorefineries that parallel the work of a crude-oil refinery, where the abundant raw material is converted through different processes into a mixture of products, including biofuels, valuable chemicals, heat, and electricity.

However, the major obstacles in the production of biofuels by the BtL processes are the high investment costs and the lack of infrastructure. Therefore, the development of technology to produce biofuels will allow moving toward a sustainable and economically feasible production of biofuels.

In this chapter, a systematic methodology framework was proposed for the synthesis of BtL processes for the production of gasoline and diesel. The methodology was divided in three main stages. In the first stage, the synthesis of five technological routes based on experimental work was performed. Different thermochemical conversion, upgrading, and separation technologies were considered in the process routes definition. The process simulator Aspen Plus was used to predict mass and energy balances, thermodynamic properties, waste streams flowrates, equipment size, and costs. In the second stage, a BtL processing superstructure was defined considering the different reaction and separation units involved in the process routes. The superstructure was formulated by identifying the possible interconnections between the technological routes, as well as mass and energy integrations. In the last stage, the superstructure was defined as a Mixed Integer Non-Linear Programming (MINLP) problem, which sets the objective to minimize the TAC's of BtL fuels under different cases and integration scenarios. To achieve this, the superstructure was coded in GAMS 24.5.6 and the solver BARON 14.4 was employed to determine the process route that meets the objective function. From the superstructure optimization, optimal technological routes for the production of gasoline and diesel considering different integration possibilities were identified. These MINLP solutions were rigorously simulated to confirm design specifications and perform numerical comparison between systems. The optimal technology routes were then taken as starting points for the analysis of potential upgrading configurations and intensification alternatives.

By applying the methodology to the BtL process synthesis, different thermochemical conversion, upgrading, and separation technological routes that reduce the TAC of BtL transportation fuels and increase the fuels production can be proposed. Likewise, mass and energy integration scenarios to analyze their effect on the processes' TAC can be explored. Moreover, proposing alternative process configurations and intensifications is possible through this methodology.

The main findings from the systematic methodology framework application were that the CA1-GLTHT and CA2-GLTHT are the most favorable process routes under the desired product profiles even when they present higher costs compare to PR-GLTUF. The proposed alternatives produced more liquid fuels compared to the initial routes and the TAC can be reduced if mass and energy integration are considered. The results demonstrated that the proposed methodology applied to softwood BtL transportation fuels production can explore and propose optimal conversion pathways that reduce the costs of BtL-fuels and increase their viability.

10.10 Nomenclature

BARON	= Branch-And-Reduce Optimization Navigator
BtL	= Biomass to liquid
CA1-GLTHT	= Combined alternative one of Gasification followed by simultaneous high and low temperature FT reactions and fractional upgrading units
CA2-GLTHT	= Combined alternative two of Gasification followed by simultaneous high and low temperature FT reactions and fractional upgrading units
$CC_{adj(i,j)}$	= Adjusted capital cost
$CC_{base(i,j)}$	= Base capital cost
CC_{tot}	= Total capital cost
$Conv_f$	= Conversion factor
$cost_{adj}$	= Adjusted capital cost
$cost_{base}$	= Base capital cost
$EC_{(i,j)}$	= Energy cost
EC_{tot}	= Total energy cost
FT	= Fischer-Tropsch

GAS = Gasification
GAS-FTUP = FT products fractional upgrading
GAS-HYDUP = HTFT products hydroprocessing
GHG = Greenhouse gas
HHV = Higher heating value
HTFT = High temperature Fischer-Tropsch
LCVs = Commercial vehicles
LHV = Lower heating value
LPG = Liquefied petroleum gas
LTFT = Low temperature Fischer-Tropsch
MINLP = Mixed Integer Non-Linear Programming
m_{base} = Base inlet flowrate
$m_{feedstock}$ = Feedstock mass flowrate
$m_{in(i,j,k)}$ = Actual inlet flowrate
$m_{in(RXN(i,j))}$ = Reaction block inlet flowrate
$m_{in(SEP(i,j))}$ = Separation block inlet flowrate
$m_{in(WWD(i,j))}$ = Waste streams disposal inlet flowrate
$m_{new(RXN(i,j))}$ = Auxiliary streams flowrate
m_{outPB} = Previous block outlet flowrate
$m_{out(RXN(i,j))}$ = Reaction block total outlet flowrate
$m_{out(SEP(i,j))}$ = Separation block total outlet flowrate
$m_{out_n(SEP(i,j))}$ = Separation stream outlet flowrate
$m_{rec(SEP(i,j))}$ = Separation stream flowrate to be recycled
NCGs = Noncondensable gases
NLP = Nonlinear programming
NREL = National Renewable Energy Laboratory
PR = Process route
PR-GHTH = Gasification- HTFT Hydrocarbon Production-FT Hydrocarbon Hydroprocessing
PR-GHTUF = Gasification- HTFT Hydrocarbon Production-FT Hydrocarbon Upgrading
PR-GLTUF = Gasification- LTFT Hydrocarbon Production-FT Hydrocarbon Upgrading
PR-PCC = Pyrolysis-Catalytic Cracking
PR-PHS = Pyrolysis-Hydroprocessing
PYR = Pyrolysis
PYR-CC = Pyrolysis oil catalytic cracking
PYR-HYD = Pyrolysis oil hydroprocessing
PYR-SEP = Pyrolysis upgraded product separation
RD&D = Research, development, and demonstration
RXN = Reaction
SEP = Separation
Sep_f = Recovery factors
SUVs = Sport utility vehicles
TAC = Total annual cost
WGSR = Water gas shift reactor
X_n = Decision variables
τ_{base} = Base residence time
τ_{adj} = Adjusted residence time

References

[1] U. S. Energy Information Administration. Transportation sector energy consumption Transportation sector energy consumption by fuel. In: International Energy Outlook. 2016, pp. 127–137.
[2] Façanha C, Blumberg K, Miller J. Global Transportation Energy and Climate Roadmap: The impact of transportation policies and their potential to reduce oil consumption and greenhouse gas emissions. Washington DC: 2012.
[3] Transportation Sector Energy Consumption, Monthly Energy Review May 2018. 2018.
[4] U.S. Energy Information Administration. Biomass – Energy Explained, Your Guide To Understanding Energy [Internet]. 2017, ([Accessed Jun 11, 2018] at https://www.eia.gov/energyexplained/print.php?page=biomass_home)
[5] Greene DL, Wegener M. Sustainable transport. J Transp Geogr 1997, 5(3), 177–190.
[6] Brito Cruz CH, Mendes Souza G, Barbosa Cortez LA. Biofuels for Transport. In: Future Energy: Improved, Sustainable and Clean Options for our Planet. Elsevier; 2014, pp. 215–244.
[7] Lynd LR, Larson E, Greene N, et al. The role of biomass in America's energy future: framing the analysis. Biofuels, Bioprod Biorefining 2009, 3(2), 113–123.
[8] Youngs H, Somerville C. Biofuels for Europe: Advanced Biofuels – Biofuels for Europe [Internet]. 2017, ([Accessed Jul 16, 2018] at http://www.biofuelsforeurope.eu/advanced-biofuels/)
[9] Fiorese G, Catenacci M, Verdolini E, Bosetti V. Advanced biofuels: Future perspectives from an expert elicitation survey. Energy Policy 2013, 56, 293–311.
[10] Stöcker M. Biofuels and Biomass-To-Liquid Fuels in the Biorefinery: Catalytic Conversion of Lignocellulosic Biomass using Porous Materials. Angew Chemie Int Ed 2008, 47(48), 9200–9211.
[11] Jones S, Meyer P, Snowden-Swan L, et al. Process Design and Economics for the Conversion of Lignocellulosic Biomass to Hydrocarbon Fuels: Fast Pyrolysis and Hydrotreating Bio-oil Pathway, 2013.
[12] Wright MM, Daugaard DE, Satrio JA, Brown RC. Techno-economic analysis of biomass fast pyrolysis to transportation fuels. Fuel 2010, 89, S2–10.
[13] Betchel. Aspen Process Flowsheet Simulation Model of a Battelle Biomass-Based Gasification, Fischer–Tropsch Liquefaction and Combined-Cycle Power Plant. Pittsburgh, Pennsylvania: 1998.
[14] Martín M, Grossmann IE. Process Optimization of FT-Diesel Production from Lignocellulosic Switchgrass. Ind Eng Chem Res 2011, 50(23), 13485–13499.
[15] Baliban RC, Elia JA, Floudas CA. Biomass to liquid transportation fuels (BTL) systems: process synthesis and global optimization framework. Energy Environ Sci 2013, 6(1), 267–287.
[16] Baliban RC, Elia JA, Floudas CA, Gurau B, Weingarten MB, Klotz SD. Hardwood Biomass to Gasoline, Diesel, and Jet Fuel: 1. Process Synthesis and Global Optimization of a Thermochemical Refinery. Energy & Fuels 2013, 27(8), 4302–4324.
[17] Torres-Ortega CE, Gong J, You F, Rong B-G. Optimal synthesis of integrated process for co-production of biodiesel and hydrotreated vegetable oil (HVO) diesel from hybrid oil feedstocks. Comput Aided Chem Eng 2017, 40, 673–678.
[18] Torres-Ortega CE, Rong B-G. Integrated biofuels process synthesis: integration between bioethanol and biodiesel processes. In: Ben-Guang Rong, editor. Process Synthesis and Process Intensification: Methodological Approaches. De Gruyter Graduate; 2017, pp. 241–289.
[19] Carrasco JL, Gunukula S, Boateng AA, Mullen CA, DeSisto WJ, Wheeler MC. Pyrolysis of forest residues: An approach to techno-economics for bio-fuel production. Fuel 2017, 193, 477–484.

[20] Kumar AK, Sharma S. Recent updates on different methods of pretreatment of lignocellulosic feedstocks: a review. Bioresour Bioprocess 2017, 4(1), 7.
[21] Shinnar R, Citro F. A Road Map to U.S. Decarbonization. Science (80-) 2006, 313(5791), 1243–1244.
[22] Ullah K, Ahmad M, Sofia, et al. Assessing the potential of algal biomass opportunities for bioenergy industry: A review. Fuel 2015, 143, 414–423.
[23] Ullah K, Kumar Sharma V, Dhingra S, Braccio G, Ahmad M, Sofia S. Assessing the lignocellulosic biomass resources potential in developing countries: A critical review. Renew Sustain Energy Rev 2015, 51, 682–698.
[24] Ramirez J, Brown R, Rainey T. A Review of Hydrothermal Liquefaction Bio-Crude Properties and Prospects for Upgrading to Transportation Fuels. Energies 2015, 8(7), 6765–6794.
[25] Anwar Z, Gulfraz M, Irshad M. Agro-industrial lignocellulosic biomass a key to unlock the future bio-energy: A brief review. J Radiat Res Appl Sci 2014, 7(2), 163–173.
[26] Gollakota ARK, Kishore N, Gu S. A review on hydrothermal liquefaction of biomass. Renew Sustain Energy Rev 2018, 81, 1378–1392.
[27] Swain PK, Das LM, Naik SN. Biomass to liquid: A prospective challenge to research and development in 21st century. Renew Sustain Energy Rev 2011, 15(9), 4917–4933.
[28] Pandey A, Bhaskar T, Stöcker M, Sukumaran R. Recent Advances in Thermo-Chemical Conversion of Biomass. Elsevier, 2015.
[29] European Technology and Innovation Platform. FT-Liquids and Biomass to Liquids [Internet]. 2018, [(Accessed Jun 11, 2018] at http://www.biofuelstp.eu/btl.html)
[30] Gollakota ARK, Subramanyam MD, Nanda Kishore N, Sai Gu S. Upgradation of Bio-oil Derived from Lignocellulose Biomass—A Numerical Approach. In: Proceedings of the First International Conference on Recent Advances in Bioenergy Research. Springer, New Delhi; 2016, pp. 197–212.
[31] Mponzi P. Production of Biofuels by Fischer–Tropsch Synthesis, 2011.
[32] Ratnasamy C, Wagner JP. Water Gas Shift Catalysis. Catal Rev 2009, 51(3), 325–440.
[33] Maitlis PM, de Klerk A. Greener Fischer-Tropsch processes for fuels and feedstocks. First Edit. Wiley-VCH; 2013.
[34] Triantafyllidis KS, Lappas AA, Stöcker M. The role of catalysis for the sustainable production of bio-fuels and bio-chemicals. 1st Editio. Elsevier; 2013.
[35] Bridgwater AV, Peacocke GVC. Fast pyrolysis processes for biomass. Renew Sustain Energy Rev 2000, 4(1), 1–73.
[36] Lu Q, Li W-Z, Zhu X-F. Overview of fuel properties of biomass fast pyrolysis oils. Energy Convers Manag 2009, 50(5), 1376–1383.
[37] Xu X, Zhang C, Liu Y, Zhai Y, Zhang R. Two-step catalytic hydrodeoxygenation of fast pyrolysis oil to hydrocarbon liquid fuels. Chemosphere 2013, 93(4), 652–660.
[38] Elliott DC. Historical Developments in Hydroprocessing Bio-oils. Energy & Fuels 2007, 21(3), 1792–1815.
[39] Jones SB, Male JL. Production of Gasoline and Diesel from Biomass via Fast Pyrolysis, Hydrotreating and Hydrocracking: 2011 State of Technology and Projections to 2017. Richland, WA (United States): Pacific Northwest National Laboratory (U.S.), 2012.
[40] Liao HT, Ye XN, Lu Q, Dong CQ. Overview of Bio-Oil Upgrading via Catalytic Cracking. Adv Mater Res 2013, 827, 25–29.
[41] Hew KL, Tamidi AM, Yusup S, Lee KT, Ahmad MM. Catalytic cracking of bio-oil to organic liquid product (OLP). Bioresour Technol 2010, 101(22), 8855–8858.
[42] Paasikallio V, Lindfors C, Lehto J, Oasmaa A, Reinikainen M. Short Vapour Residence Time Catalytic Pyrolysis of Spruce Sawdust in a Bubbling Fluidized-Bed Reactor with HZSM-5 Catalysts. Top Catal 2013, 56(9–10), 800–812.

[43] ECN. Phyllis2, Spruce bark (#3124) [Internet]. 2007, [(Accessed Jul 21, 2018] at https://www.ecn.nl/phyllis2/Biomass/View/3124)

[44] ECN. Phyllis2, Forest residue chips, pine spruce (#3155) [Internet]. 1996, [(Accessed Jul 21, 2018] at https://www.ecn.nl/phyllis2/Biomass/View/3155)

[45] Ibarra-Gonzalez P, Rong B-G. Systematic Synthesis and Evaluation of Thermochemical Conversion Processes for Lignocellulosic Biofuels Production: Total Process Evaluation and Integration. Ind Eng Chem Res 2018,

[46] Ayoub N, Seki H, Naka Y. Superstructure-based design and operation for biomass utilization networks. Comput Chem Eng 2009, 33(10), 1770–1780.

[47] Bradley SP. Integer Programming. In: Addison Wesley, editor. Applied Mathematical Programming. 1977, pp. 272–319.

[48] Sahinidis N. Baron [Internet]. 2015, [(Accessed Jul 17, 2018] at https://www.gams.com/latest/docs/S_BARON.html)

Index

https://doi.org/10.1515/9783110596120-011